POLYMERS AND THEIR PROPERTIES
Volume 1: FUNDAMENTALS OF STRUCTURE AND MECHANICS

ELLIS HORWOOD SERIES IN CHEMICAL SCIENCE

KINETICS AND MECHANISMS OF POLYMERIZATION REACTIONS: Applications and Physicochemical Systematics
P. E. M. ALLEN, University of Adelaide, Australia and
C. R. PATRICK, University of Birmingham
ELECTRON SPIN RESONANCE
N. ATHERTON, University of Sheffield
METAL IONS IN SOLUTION
J. BURGESS, University of Leicester
ORGANOMETALLIC CHEMISTRY – A Guide to Structure and Reactivity
D. J. CARDIN and R. J. NORTON, Trinity College, University of Dublin
STRUCTURES AND APPROXIMATIONS FOR ELECTRONS IN MOLECULES
D. B. COOK, University of Sheffield
LIQUID CRYSTALS AND PLASTIC CRYSTALS
Volume I, Preparation, Constitution and Applications
Volume II, Physico-Chemical Properties and Methods of Investigation
Edited by W. GRAY, University of Hull and
A. WINSOR, Shell Research Ltd.
POLYMERS AND THEIR PROPERTIES
J. W. S. HEARLE, University of Manchester Institute of Science and Technology
BIOCHEMISTRY OF ALCOHOL AND ALCOHOLISM
L. J. KRICKA, University of Birmingham and
P. M. S. CLARK, Queen Elizabeth Medical Centre, Birmingham
MEDICAL CONSEQUENCES OF ALCOHOL ABUSE
P. M. S. CLARK, Queen Elizabeth Medical Centre, Birmingham
L. J. KRICKA, University of Birmingham
STRUCTURE AND BONDING OF SOLID STATE CHEMISTRY
M. F. C. LADD, University of Surrey
METAL AND METALLOID AMIDES: Synthesis, Structure, and Physical and Chemical Properties
M. F. LAPPERT, University of Sussex
A. R. SANGER, Alberta Research Council, Canada
R. C. SRIVASTAVA, University of Lucknow, India
P. P. POWER, Stanford University, California
BIOSYNTHESIS OF NATURAL PRODUCTS
P. MANITTO, University of Milan
ADSORPTION
J. OSCIK, Head of Institute of Chemistry, Marie Curie Sladowska, Poland
CHEMISTRY OF INTERFACES
M. J. JAYCOCK, Loughborough University of Technology and
G. D. PARFITT, Tioxide International Limited
METALS IN BIOLOGICAL SYSTEMS: Function and Mechanism
R. H. PRINCE, University Chemical Laboratories, Cambridge
APPLIED ELECTROCHEMISTRY: Electrolytic Production Processes
A. SCHMIDT
CHLOROFLUOROCARBONS IN THE ENVIRONMENT
Edited by T. M. SUGDEN, C.B.E., F.R.S., Master of Trinity Hall, Cambridge and
T. F. WEST, former Editor-in-Chief, Society of Chemical Industry
HANDBOOK OF ENZYME BIOTECHNOLOGY
Edited by A. WISEMAN, Department of Biochemistry, University of Surrey
INTRODUCTORY MEDICINAL CHEMISTRY
J. B. TAYLOR, Director of Pharmaceutical Research, Roussel UCLAF, Paris and
P. D. KENNEWELL, Deputy Head of Chemical Research and Head of Orientative Research, Roussel Laboratories Ltd., Swindon, Wiltshire

POLYMERS AND THEIR PROPERTIES

Volume 1: FUNDAMENTALS OF STRUCTURE AND MECHANICS

J. W. S. HEARLE, M.A., Sc.D., Ph.D., F.Inst.P., F.T.I.
Professor of Textile Technology
University of Manchester Institute of Science and Technology

ELLIS HORWOOD LIMITED
Publishers · Chichester

Halsted Press: a division of
JOHN WILEY & SONS
New York · Brisbane · Chichester · Toronto

First published in 1982 by

ELLIS HORWOOD LIMITED
Market Cross House, Cooper Street, Chichester, West Sussex, PO19 1EB, England

The publisher's colophon is reproduced from James Gillison's drawing of the ancient Market Cross, Chichester.

Distributors:

Australia, New Zealand, South-east Asia:
Jacaranda-Wiley Ltd., Jacaranda Press,
JOHN WILEY & SONS INC.,
G.P.O. Box 859, Brisbane, Queensland 40001, Australia

Canada:
JOHN WILEY & SONS CANADA LIMITED
22 Worcester Road, Rexdale, Ontario, Canada.

Europe, Africa:
JOHN WILEY & SONS LIMITED
Baffins Lane, Chichester, West Sussex, England.

North and South America and the rest of the world:
Halsted Press: a division of
JOHN WILEY & SONS
605 Third Avenue, New York, N.Y. 10016, U.S.A.

British Library Cataloguing in Publication Data
Hearle, J.W.S.
Polymers and their properties.
Vol. 1: Fundamentals of structure and mechanics
1. Polymers and polymerization
I. Title
547.7 QD380

Library of Congress Card No. 81-13316 AACR2

ISBN 0-85312-033-1 (Ellis Horwood Limited, Publishers)
ISBN 0-470-27302-X (Halsted Press)

Typeset in Press Roman by Ellis Horwood Ltd.
Printed in the USA by the Maple Vail Book Manufacturing Group, New York

There is this hope. I may communicate in part; and that surely is better than utter blind and dumb; and I may find something like a hat to wear of my own. Not that I aspire to complete coherence. Our mistake is to confuse our limitations with the bounds of possibility and clap the universe into a rationalist hat or some other. But I may find the indications of a pattern that will include me, even if the outer edges tail off into ignorance.

. . . from *Free Fall* by William Golding

Table of Contents

Author's Preface

When I was an undergraduate, polymers were not mentioned in the chemistry and physics courses, because the subject was too new. My professional career started on fibres at the Shirley Institute; where my initial indoctrination included a session on cellulose chemistry by G. F. Davidson; and my research led me into the physical properties of polymers, at first through electrical properties, then into mechanical properties and structure at UMIST in the 1950s.

Consequently, I came into the subject of polymer physics near the end of the first wave when the basic ideas of molecular structure were established and the chemistry and physics of rubbers, plastics and fibres were being studied. The next wave followed in the 1950s with the discoveries of details of polymer crystallisation; in the 1960s, the metallurgists started to become interested and brought in another wave of ideas from materials science. Although the subject now proliferates in many ways, it still remains important that workers should continue to retain a sound grasp of the principles of polymer physics.

The presentation of material in books – and lectures – can be made in two ways. For the expert, it is desirable to collect all the material relevant to a topic together in one compact sequence, and to separate the material strictly according to subject. But this is not the way in which we learn a new subject. One needs first to get an elementary understanding of the subject over a wide range, and then return – often again and again – to a refinement of understanding. The same pattern must be followed in a book which aims at starting at the beginning but going on to advanced topics by the end. Philip Morse makes the point well in his book *Thermal Physics* (concerning thermodynamics)

> '. . . it is not easy to understand one part until one understands the whole. In such a case it is better pedagogy to depart from strict logical presentation. Hence several derivations and definitions will be given in steps, first presented in simple form, and only after other concepts have been introduced, later re-enunciated in final, accurate form.'

In this book, the analagous intention was first to describe polymers in general, then to discusss the way in which a single molecule would behave, and to go on to the complexities of real assemblies of many molecules, before returning to the fundamentals of theory. In the event this first volume concentrates on the essential physics of the relation between structure in a variety of forms and mechanical and thermal responses. It is planned that another volume (hopefully not so long in the writing) will cover other physical properties, deal with some areas where knowledge is less and understanding is more speculative, and return to a discussion of fundamental aspects of the subject. My aim has been to give a straightforward, readable account of polymer physics, which is not obscured by too much detail of particular cases, but this remains for the reader to judge. For myself, there has been enjoyment writing a book like this through the development of a fresh coherence in my understanding of wide-ranging subject matter, and the opportunity to present in new ways such ideas as the hypothetical consideration of an isolated macromolecule in Chapter 2, followed by a switch from the conventional sequence of presenting the classical theory of rubber elasticity, and, in the last chapter, my own research ideas on a unified view of crystalline polymers through the definition of some useful general parameters.

Finally, I must express my gratitude to many students, colleagues, and others who have sharpened my comprehension; to the staff in my office for typing, copying and organising material; to my son, Adrian D. Hearle for converting obscure sketches into well-drawn diagrams; and to my wife for putting up with the piles of paper and my disappearance into my study over many years.

J. W. S. Hearle
Mellor, Cheshire
August 1981

ACKNOWLEDGEMENTS

Thanks are due to the following for permission to reproduce photographs (including, in many instances, the provision of prints), original versions of diagrams and other copyright material.

Dr. D. C. Bassett, University of Reading.
Prof. P. H. Geil, University of Illinois.
Dr. V. F. Holland, Monsanto (with Dr. P. H. Lindenmeyer and Dr. F. R. Anderson).
Prof. R. Hosemann, Berlin.
Dr. R. P. Kambour, GEC, Schenectady, New York.
Dr. B. C. Jariwala, Manchester.
Dr. D. C. Keith, Bell Laboratories, New Jersey.
Prof. A Keller, University of Bristol.
Dr. F. A. Khoury, National Bureau of Standards, Washington, DC.
Dr. R. J. Oxborough, Cambridge.

Dr. A. J. Pennings, Groningen.
Dr. A. Peterlin, Washington, DC.
Prof. R. D. Preston, Leeds.
American Institute of Physics (*J. Appl. Phys.*).
Marcel Dekker Inc. (*J. Macromol. Sci.*).
Faber and Faber (*Free Fall*, William Golding).
Macmillan Journals (*Nature*).
Royal Microscopical Society (*Proceedings*).
Pergamon Press (*Proc. Int. Conf. Crystal Growth*).
Taylor and Francis (*Phil. Mag.*).
Textile Institute (*Journal of Textile Progress*).
John Wiley and Sons Inc. (*J. Pol. Sci.* and *J. Appl. Pol. Sci.*).

Note that where original illustrations are used, the full reference is given below the caption. Where illustrations have been redrawn on the basis of published data, the authors' names are given, and the reference is included in Appendix C.4.

CHAPTER 1

Polymer Molecules and Materials

'. . . a few early investigators were led by their experiments to favour the view that cellulose, starch and rubber are composed of very large molecules. If the results of their molecular weight determinations had been accredited, the concept of giant molecular structures might have been established long before the 1930s . . . . The gap between molecules of ordinary size and polymers hundreds (actually thousands) of times as large was too great to be bridged in a single leap.'

Paul J. Flory, *Principles of Polymer Chemistry*

1.1 POLYMERS AND POLYMER PHYSICS

As a class, polymers are among the most important of all materials. They are a basis of life itself: among the many constituents of living organisms, proteins perform a great variety of functions in both animals and plants; more complicated polymers determine hereditary characteristics and control development; starch is a vital part of out diet; cellulose is the main structural material of plants; natural rubber and natural resins have their special functions; and so on. Furthermore, man uses many of these materials, not only as food, but also for clothing and construction. Other polymers occur naturally as minerals, such as diamond and graphite. Synthetic polymers, almost unknown fifty years ago, are now made and processed in the huge plastics, synthetic rubber and man-made fibre industries: the next fifty years will see their use expanding even more as they rival metals in versatility and usefulness.

The traditional chemical definition of a polymer was a material whose molecules were produced by the amalgamation of a number of small molecules of a simple substance. The number of molecules joining together could be few. But this terminology is no longer the general usage. The meaning of the word **polymer** (abbreviated from the older, more specific term, **high polymer**) is now more restricted, because it is commonly taken to imply only substances whose molecules contain a very large number of atoms linked by covalent bonds: but it is also less restricted in that it is no longer limited to molecules formed by the simple addition of smaller molecules. Polymers may be produced by more complicated reactions, any large molecule being included as a polymer. Another name for these giant molecules is **macromolecule**, so that a macromolecular substance is synonymous with a polymer. It may be noted that assemblies of atoms joined together by electrovalent bonds, for instance metals and ionic crystals, are not regarded as polymers.

While, in principle, a polymer molecule could be made up of a completely irregular collection of atoms, most actual polymers consist either wholly or mainly of identical, or sometimes merely similar, units joined together, or of simple mixtures of such units. The particular groups of atoms which form the repetitive pattern in the polymer molecule are known as **monomers, mers,** or **repeating units.**†

In the early days – and they are so recent that we have hardly grown out of them yet – the study of polymers was naturally dominated by chemistry. The subject did not exist until chemists had established that many natural materials were high polymers, truly giant molecules, and not, as was the common view before 1920, mere colloidal associations of small molecules. The chemical formulae had to be established. For example, it was only in 1926 that the formula of cellulose was settled, and even now the detailed formulae of most proteins are still unknown. Subsequently, chemists worked to discover new synthetic polymers. The chemical laboratory was the source of new materials.

At first, physics was of importance only to characterise the properties of the materials being produced by the chemists. But it is now realised that the physical structure of polymers, the way in which the molecules are packed together, is of as much importance as the chemical constitution in determining the properties of polymers. New materials can come from a control of physical structure of polymers, based on a study of polymer physics. For instance, cellulose fibres, all composed of the same chemical substance, exist in a wide variety of structural forms, natural and man-made, with differences in properties typified by the stress-strain curves in Fig. 1.1.

Besides its technological significance, polymer physics is an important academic study. Large molecules behave very differently to small ones. Polymers can indeed claim to be another state of matter, or rather to include several other states of matter, different from the simple gases, liquids, and solids which are traditionally studied by physicists. One of the aims of this book is to bring out the nature and characteristics of these states.

The subject of polymer physics includes the two branches of the study of structure and the study of properties. But there are interactions between the two since a knowledge of properties helps in elucidating structure, while a knowledge

† The strict chemical definition of a monomer is the chemical substance from which the polymer is made. If the polymerisation is by simple addition, this is identical with the structural repeating unit; but the two will be slightly different when polymerisation is by condensation with the elimination of a minor product of the reaction, such as water. Sometimes, also, the chemical precursor may contain more than one repeating unit. However, from a physical point of view, the precursor substances are irrelevant – especially as the same polymer may often be made from different starting substances – and it is the repeating unit, as defined above, which is important.

of structure helps to understand properties. Indeed the most rewarding aspects of the whole subject are the development of relations between properties and structure, and the description of how structure arises from the behaviour of individual molecules and assemblies of molecules: both of these aspects are susceptible to theoretical analysis. The theoretical physics of polymer systems is a developing subject.

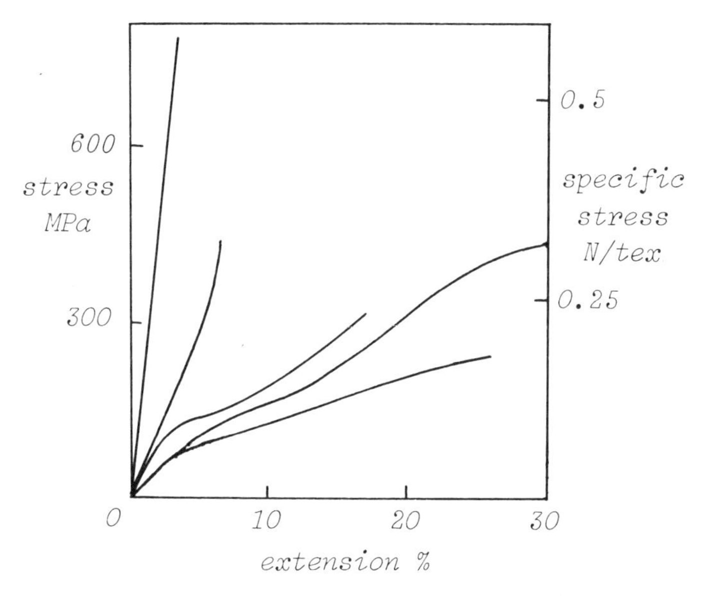

Fig. 1.1 – Stress-strain curves of different fibres, all composed of the same polymer cellulose. (For note on units of specific stress, see page 384.)

1.2 THE PHYSICAL TYPES OF POLYMER

1.2.1 Molecular networks

Polymer molecules are made up of thousands, or millions, of atoms joined together by strong covalent bonds. The biggest differences between different types of polymer are a consequence of differences in the type of network linking the atoms together; and the most important distinctions are firstly between three-dimensional, two-dimensional, and one-dimensional polymer molecules, and secondly between those which form a regular, crystalline lattice, and those which are irregular, amorphous structures.

Diamond is an example of a crystalline, three-dimensional polymer. Each carbon atom is linked to four neighbours at the corners of a tetrahedron, so that every individual diamond is a single giant molecule. Its structure is fully known as a result of orthodox crystallographic studies, and some of the interesting features are illustrated in Fig. 1.2.

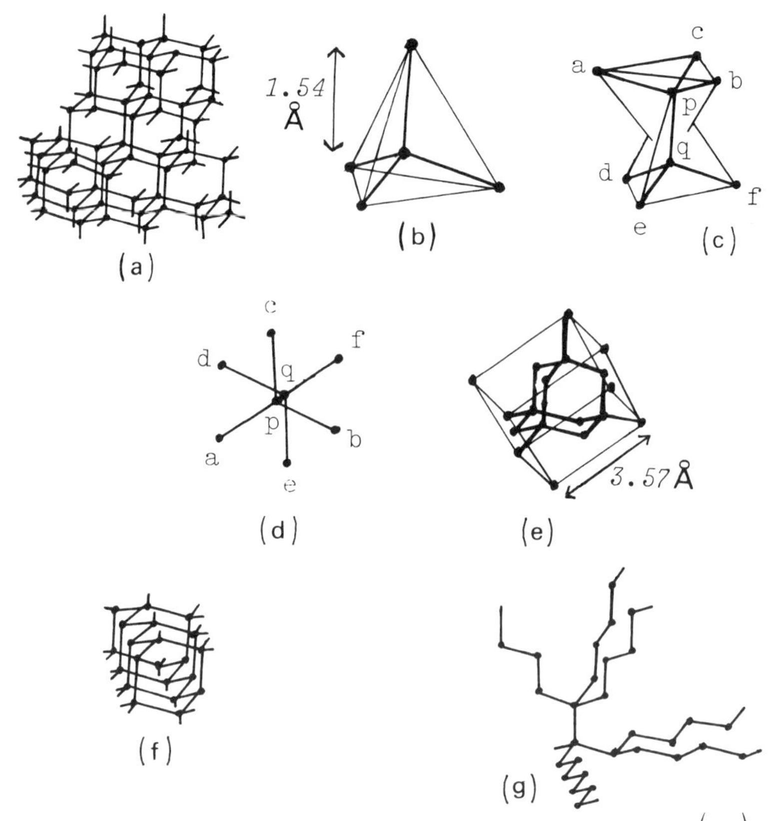

Fig. 1.2 – Views of the diamond crystal lattice. The chemical formula is $\left(-\text{C}-\right)_n$. All bonds are equivalent single covalent –C–C– bonds. (a) A portion of the three-dimensional single crystal molecule. (b) Each atom is linked to four neighbours at the corners of a regular tetrahedron. (c) Neighbouring tetrahedra are oriented in different directions, due to the staggering of the bonds on adjacent atoms. The bonds run in four directions because ap is parallel to qf, and so on. (d) The staggering is clearly seen in a view along a bond, pq. (e) The simplest crystallographic repeat is a cubic unit cell. The cell contains an effective total of eight atoms: four whole atoms; six halves at the centres of the cell faces; and eight eighths (portions at the corners). (f) An interesting feature of the structure is the presence of tunnels running through the lattice in six directions parallel to the diagonals of each face of the cubic cell. (g) Extended zig-zag chains run in six directions through the lattice.

Some synthetic polymers are made from basic units which possess three (or more) active groups capable of reacting together to form the polymer. Fig. 1.3 illustrates the way in which the branching in space results in a three-dimensional network being formed as these tri-functional components join together. The network will not be regular, nor as compact as that of diamond. These materials are the amorphous three-dimensional polymers. They are a difficult system for fundamental study, and little is known in detail about their structure.

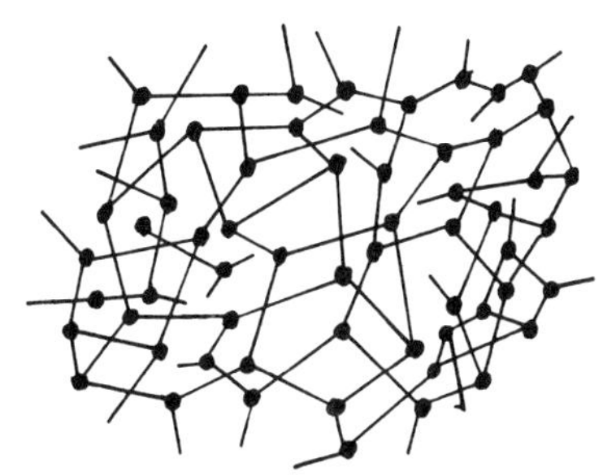

Fig. 1.3 - Portion of an irregular three-dimensional polymer network.

Two-dimensional polymers are rare. The condition for their formation is a basic unit with three or more active groups, *all directed in the same plane and capable of joining up to give a planar network.* This is a very restrictive condition, and such coplanar groups are unlikely to occur frequently. One example is graphite; the regular planar molecules pack in sheets to give the whole crystal lattice. The structure is shown in Fig. 1.4.

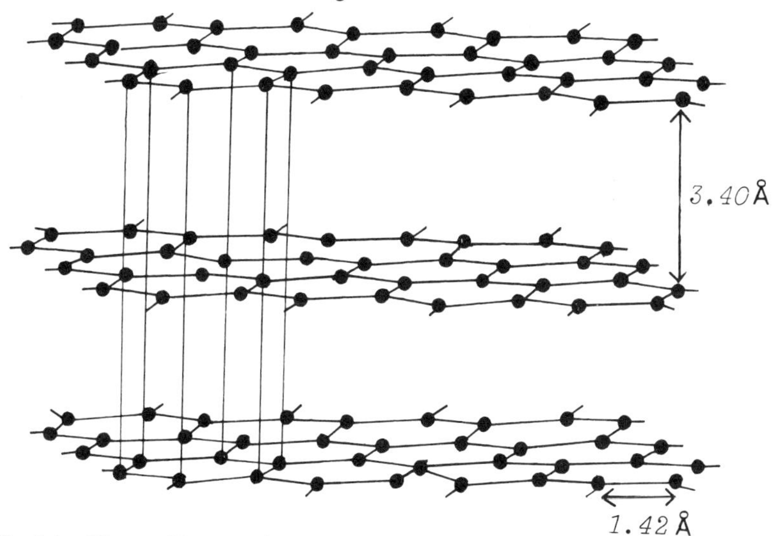

Fig. 1.4 – The graphite crystal lattice. The chemical formula is:

$$\left(\begin{array}{ccc} & C = = C & \\ = = C & & C = = \\ & C = = C & \end{array}\right)_n$$

In a more traditional representation, two-thirds of the bonds would be single C–C bonds and one-third would be double bonds C=C; but the double bond character is better regarded as uniformly distributed among all the bonds. The bond distance of 1.42 Å lies between the single-bond distance of 1.50 Å and the double-bond distance of 1.33 Å. Between planes, there are only van der Waals' forces and the spacing is much greater. The vertical lines show atoms which are in register. Note that alternate planes are displaced.

Taking a broader, topological, definition of two dimensions as meaning sheet systems, whether planar or otherwise, we see that it would also be possible to have a crystalline two-dimensional polymer in which each molecule was a regular curved sheet: corrugated iron would be a larger-scale analogy. An amorphous two-dimensional polymer would consist of an assembly of irregularly formed or distorted sheets: the large-scale analogy would be a lot of sheets of paper screwed up together. But while it is possible in principle to imagine structures such as these, it is difficult to suggest how they could be made or studied in practice, although some carbon fibres may contain molecules which are highly distorted graphitic sheets.

By contrast with the rarity of two-dimensional polymers, one-dimensional polymers occur in profusion. They will be formed whenever units with two reacting groups join up to make a chain. Here again, we are adopting a broad, topological definition, taking one-dimensional to mean any linear system, whether in straight lines or not. If the long-chain molecules are packed regularly side-by-side, as in Fig. 1.5, we have a crystalline polymer, while if they are irregularly tangled as in Fig. 1.6 the material is amorphous.

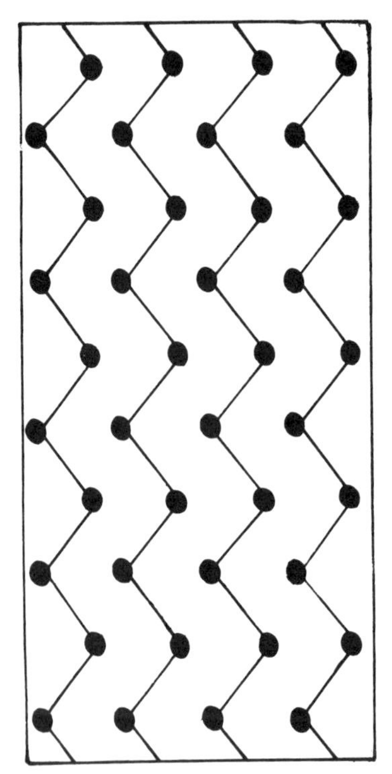

Fig. 1.5 – Schematic representation of the lattice of a one-dimensional crystalline polymer.

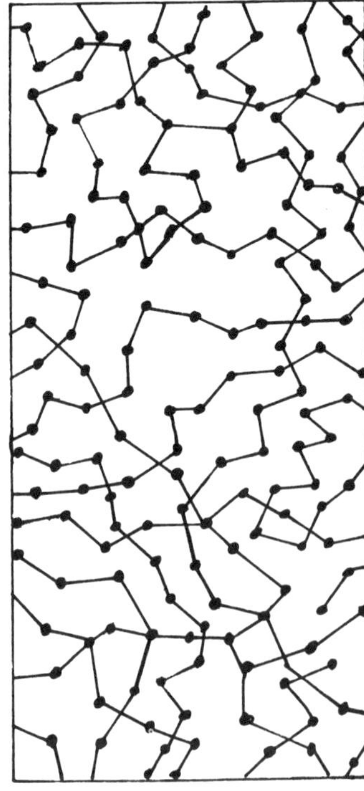

Fig. 1.6 – Schematic representation of the structure of a one-dimensional amorphous polymer.

1.2.2 A note on functionality

There is a basic topological reason why one-dimensional and three-dimensional polymers are the commonest. The functionality of a chemical group is the number of bonds by which it is joined to other groups. In the original molecules taking part in the synthesis, the sites for forming these bonds may be free radicals, double bonds which can open up, or groups which can react by a condensation reaction with the elimination of some other substance.

Table 1.1 shows the consequences of reaction involving units with different functionalities. Units of zero functionality cannot join up with one another; and two mono-functional units lead to another small molecule, reducing the functionality to zero. Bifunctional units are unique in that the products of reaction always remain bifunctional, and a long chain molecule develops. When the initial functionality is greater than two, there is a continual increase in functionality as the reaction proceeds. Unless the geometric form of the groups is very special, so that a planar network is formed, the continual branching must lead to a three-dimensional network. The network may tend to be tighter, the greater the functionality of the units; but otherwise, there is no difference in

Table 1.1 – Products of Successive Reactions.

Functionality of initial component		Products of reaction
zero 	⟶	no reaction 
one 	⟶	zero ⟶ no further reaction 
two 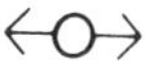	⟶	two ⟶ two ⟶ ------  ------ ⟶ long bifunctional chain 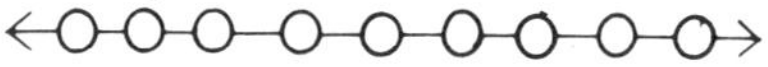
three 	⟶	four ⟶ five ⟶ six ⟶ ------ 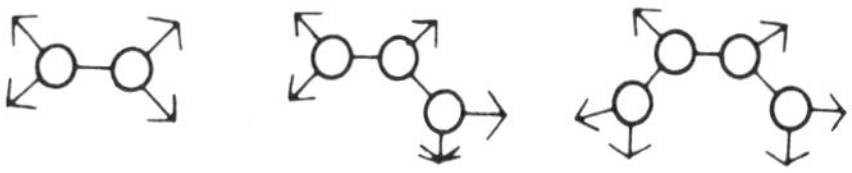 ------ ⟶ three-dimensional network
four 	⟶	six ⟶ eight ⟶ ------ 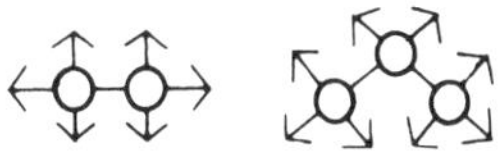 ------ ⟶ three-dimensional network

principle as the functionality rises above three, since a trifunctional unit leads to a sequence through all higher functionalities.

We can also see the effects of reactions involving units with a mixture of functionalities. Zero-functional and bifunctional units act only as external or internal dilutents of the network, since the former will not be incorporated and the latter will add on without changing the functionality of the units which they join. By contrast, mono-functional units will lead to a termination of growth, and trifunctional units (and units with higher functionality) will lead to branching into a three-dimensional network: the extent of either of these effects will depend on the proportions in which the mixed reaction occurs. Growth can also be terminated by the formation of closed rings with bifunctional groups.

The general formula defining the situation is:

$$p = m + n - 2 \tag{1.1}$$

where p is the functionality of the product of a reaction between units with functionalities, m and n.

By appropriate control of functionality in reactions, the chemist can determine the type of polymer network which is formed.

1.2.3 Major differences in properties

The fundamental differences in form are reflected in major differences in properties. The perfection of crystallinity in a diamond causes its valuable optical properties; and the compact three-dimensional network of strongly bonded atoms gives a substance which is immensely hard, infusible, chemically inert, and generally intractable. To a lesser extent this is also true of other three-dimensional polymers. The irregular, highly cross-linked three-dimensional networks are **thermo-setting resins**. Once their polymerisation is completed and the network established, the polymer cannot be softened, melted or moulded. It is hard, infusible, and insoluble: it cannot be changed without destroying the molecule. Consequently thermo-setting plastics must be shaped into their final form at the same time as they are being cured by the application of heat, or while a catalyst, which has been mixed in, is still promoting the reaction.

Two-dimensional polymers, on the other hand, have characteristic properties which derive from the way in which molecular sheets can split apart, slide over one another, or roll up: hence the lubricating properties of graphite.

Linear polymers are **thermoplastics**: at elevated temperatures, the molecules will slide over one another and flow as a liquid. They can be moulded, and then be remoulded in new forms; they can be melted and extruded as fibres or films; they can be dissolved in suitable solvents and then precipitated in desired forms. In practice, not all linear polymers display their thermoplastic character – some, like cellulose, decompose chemically before they reach what would be their melting point – but it is always inherently there, and solution is usually possible even if melting is not.

Looking at polymers which chemistry shows to be linear, we find five or more different forms of solid material. In two classes, there is no evidence of crystallinity. Firstly, there are rubbers – soft, elastic, and highly extensible – in which the chains of an amorphous network must be flexible in order to give easy deformation. Secondly, there are rather hard, rigid, glassy plastics like polystyrene: here the chains of the amorphous network must be rigid. The X-ray diffraction pictures of these polymers show a diffuse halo, which is characteristic of a disordered structure.

Then there are linear polymers which crystallise. Only when prepared under very special conditions, have these been found in the third class with the regular shapes, angles and faces of single crystals. More commonly they are found in a fourth class as tough plastics like polythene, softer and less brittle than the glassy thermoplastics, but much stiffer and less extensible than rubbers. X-ray diffraction studies of these materials give sharp rings; and the sharpness of the diffraction is clear evidence that the molecules are, at least in part, packed in crystalline (or near-crystalline) order, while the circular symmetry is evidence that there is no preferred orientation. The diagram is similar to the powder diagram obtained from a mass of small crystals, except that the presence of some diffuse scattering indicates the occurrence of some appreciable disorder as well. The crystallisation can only be partially complete. Furthermore, the milky appearance of polythene is evidence of units within the structure large enough and different enough to scatter light, just as the particles in milk do.

The fifth form of linear polymers is as fibres and films – very fine units of matter, perhaps no more than a few wavelengths of light in thickness, showing great strength along their length. These also show a sharp X-ray diffraction picture, but now the circles have been reduced to spots, to what is indeed known to X-ray crystallographers as a fibre diagram, characteristic of crystals lined up with a preferred orientation. There may also be other patterns of order and disorder, which would give different physical characteristics.

Although the five forms of linear polymer are physically distinct, the same polymer can occur in more than one form, and there are some intermediate states. On cooling, a rubber transforms to a glassy polymer; while, on heating, a material like polystyrene becomes rubbery. On rapid quenching, a polymer may remain amorphous whereas under other conditions it may crystallise. Stretching may make crystallisation easier, giving an oriented crystalline polymer. Similarly stretching – drawing – an extruded, unoriented, partially crystalline monofilament by 400 or 500% converts it to an oriented fibre. The various transformations between the different forms will be important parts of our study.

Natural rubber is a good example. Ordinarily it is a rubber. But if cooled rapidly below −70°C it forms a brittle glassy material capable of being shattered with a hammer. By contrast, if it is held for a few hours at around −20°C, it will crystallise to give a tough plastic. At 0°C, the crystallisation is very slow, but it can be speeded up by stretching the rubber giving an oriented, fibre-like crystalline polymer. No doubt, good single crystals could be made by crystallisation from dilute solution at a low temperature. Finally rubber can also be transformed into a three-dimensional polymer by vulcanisation, which, if carried far enough, converts rubber into ebonite – a hard, infusible, set material; or, alternatively, rubber may be melted into a liquid form.

1.2.4 A summary of polymer classification

The classification of polymers is summarised in Table 1.2. By far the most important are the linear polymers, which are the most amenable both to study and to manipulation, and which make up the larger part of the naturally occuring polymers and of the synthetic polymers used in technology. Consequently, this book will mostly be concerned with linear polymers. The crystalline materials like diamond and graphite will hardly be discussed further, as they have a special character of their own; and the thermo-set resins will receive only brief mention, since there is relatively little fundamental knowledge of their behaviour.

Herman Mark has made the useful diagrammatic classification of polymers shown in Fig. 1.7. Taking the flexible chains of rubber as the parent form, this shows that the material as a whole can be stiffened and modified in three ways:

Table 1.2 – Classification of polymers by structure and properties.

Form of network		Form of molecule		
		one-dimensional linear	two-dimensional sheet	three-dimensional bulk
		thermoplastics		
amorphous irregular	**flexible chain**	*rubbers* natural, synthetic		
	rigid chain	*glassy plastics* e.g. polystyrene, Perspex		*thermoset plastics* e.g. polyester and epoxy resins, Bakelite, ebonite
crystalline regular	**partial unoriented**	*tough plastics* e.g. polythene, nylon		
	partial oriented	*fibres and films* natural, man-made	e.g. carbon fibre	
	perfect	*single crystals* e.g. polyethylene	e.g. graphite	e.g. diamond

by stiffening individual chains, by cross-linking, and by crystallisation. In Chapter 2, we shall follow the fundamentals of this by studying the behaviour of ideal single ploymer molecules. In later chapters, we shall follow the same sequence in the study of real materials.

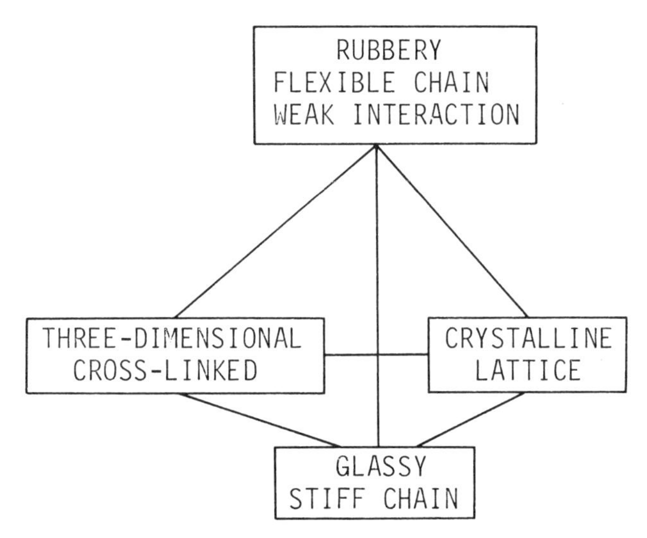

Fig. 1.7 – Diagrammatic classification of ways of stiffening polymer materials. The rigidity of the bulk polymer increases with the distance from the rubber, in any of the three directions or in combinations of them.

1.2.5 Variants of linear polymers

The simplest form of a linear polymer is shown in Fig. 1.8(a). But there are many variants of this basic form. The monomer units themselves will invariably have side-groups of one sort or another, as in Fig. 1.8(b), and may contain rings of atoms, as in Fig. 1.8(c): these features have an important influence on behaviour. If the rings are all linked up, we get the ladder structure shown in Fig. 1.8(d). We might even have a rigid girder structure as in Fig. 1.8(e): this would be a very stiff molecule.

In some circumstances, longer branches may be joined on to the chains as in Fig. 1.8(f). If these branches, or simple side-groups, join up with other chains then three-dimensional networks are formed as in Fig. 1.8(g) and Fig. 1.8(h). However, provided the branching and cross-linking are not too extensive, it is easier to think of these forms as modified linear polymers than as true three-dimensional polymers. A small amount of cross-linking will result only in some chain molecules being linked together in small groups; but a higher degree of cross-linking will give infinite networks.

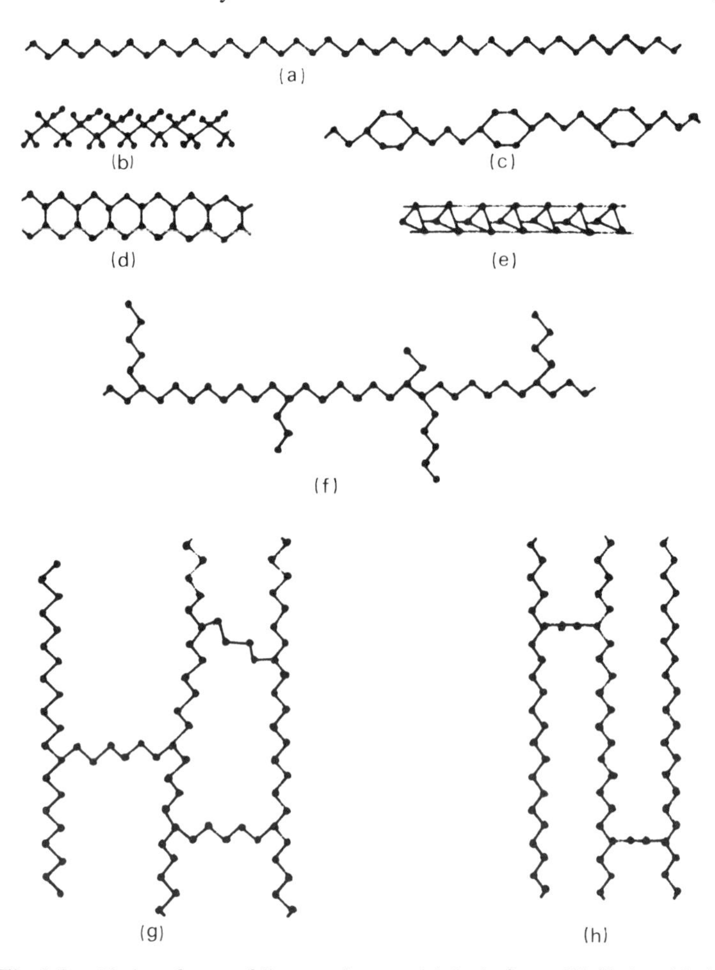

Fig. 1.8 – Variant forms of linear polymers. (a) Basic form. (b) Chain with side-groups. (c) Chain with rings. (d) Ladder molecule. (e) Girder molecule. (f) Chain with side branches. (g) Chains linked by side branches. (h) Chains linked by specific side-groups.

1.3 THE ARCHITECTURE OF THE SYSTEM

1.3.1 Some important molecular units

Although two themes of this book are the importance of physical form in determining polymer properties and the feasibility of discussing polymer physics in terms of purely physical definition of the polymer molecules, without becoming involved in their chemistry, it is nevertheless useful to have some knowledge of chemical constitution. In this section, we shall describe some of the materials which will be used as examples later in the book. A listing of the chemical constitution of some important polymers is given in Appendix C.

The vast majority of polymers contain carbon as an element in the chain. Its valency of four leads easily to the growth of large molecules, as we have already noted in diamond and graphite. The simplest linear polymer consists of a chain of carbon atoms, with hydrogen atoms attached to the other two bonds on each atom:

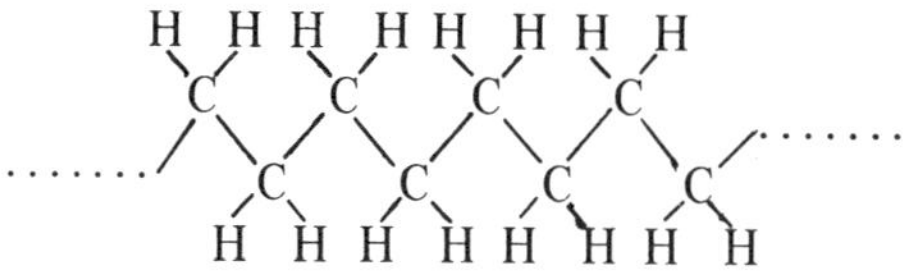

```
           H  H H  H H  H H  H
            \ /   \ /   \ /   \ /
             C     C     C     C
            / \   / \   / \   / \
.......  /     \ /   \ /   \ /   \ /.......
               C     C     C     C
              / \   / \   / \   / \
             H   H H   H H   H H   H
```

This substance is variously known as polymethylene (regarding $-CH_2-$ as its basic repeating unit), polyethylene (since it can be made by the addition polymerisation of ethylene molecules, $CH_2{=}CH_2$), and by trade names such as polythene and Alkathene. The first commercial process, developed in the 1930s, yielded a product with appreciable chain branching: this is known as branched, high-pressure or low density polyethylene – the first description coming from its structure, the second from its mode of formation, and the third from its properties. The discovery of new catalysts by Ziegler, led, in the 1950s to another process producing almost completely linear chains with no branching and giving a product known as linear, low-pressure or high-density polyethylene.

Derived from polyethylene we have a very large family of polymers, in which some of the hydrogen atoms are replaced by other groups. In the vinyl polymers one out of every set of four hydrogen atoms is replaced. Thus polyvinyl chloride is made by the polymerisation of vinyl chloride, $CH_2{=}CHCl$. Since the repeating unit is asymmetric, various regular and irregular forms are possible, as illustrated geometrically in Fig. 1.9. Complete head-to-tail addition will yield a regular molecule with the simplest possible repeat unit, namely $-CH_2CHCl-$ in polyvinyl chloride. A random choice between head-to-tail, head-to-head, and tail-to-tail addition will give an irregular molecule, while head-to-head and tail-to-tail, without any head-to-tail, will give a regular molecule in which the repeat unit contains two basic units. A small amount of head-to-head or tail-to-tail addition in an otherwise head-to-tail polymerisation (or vice versa) will cause an occasional irregularity in the chain. These different forms, which can occur with any asymmetric monomer units, are really different chemical substances: they will have different properties; and, by varying the conditions of preparation, it may be possible to influence the forms which result. Perhaps surprisingly to the physicist, regular head-to-tail addition usually predominates. There are other differences related to the geometry of the addition: this stereoisomerism is discussed in section 1.3.3.

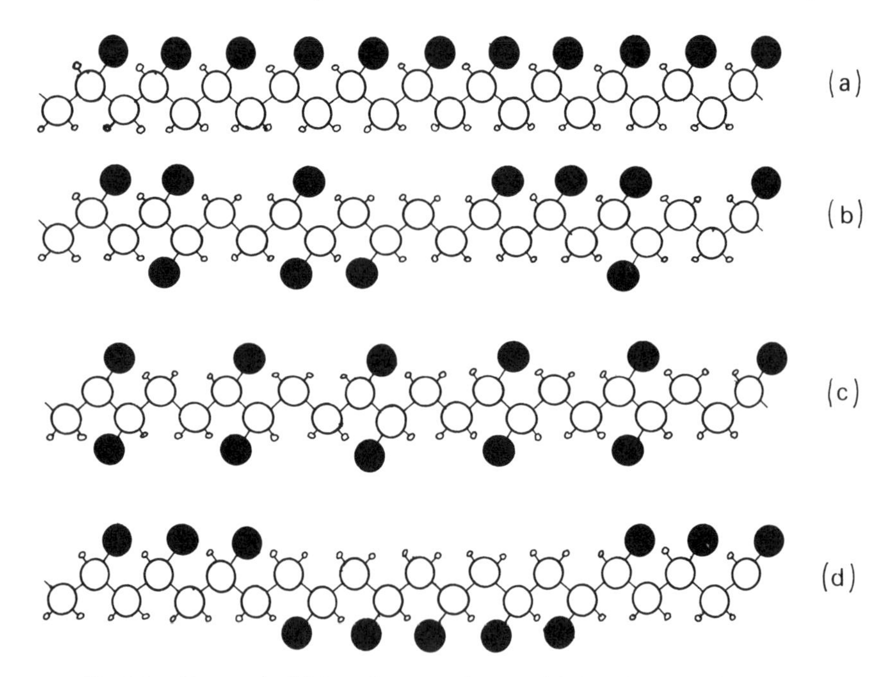

Fig. 1.9 – Forms of addition of symmetric units. (a) Completely head-to-tail. (b) Random. (c) Completely head-to-head and tail-to-tail. (d) Mainly head-to-tail, occasional head-to-head or tail-to-tail.

Where the starting substance contains two double bonds, such as butadiene $CH_2{=}CH{-}CH{=}CH_2$, other variations are possible. The addition may take place at only one of the double bonds in each molecule: this gives a vinyl polymer, with $-CH{=}CH_2$ as the side-group. Alternatively, the double bond may change position by resonance and yield a polymer with a double bond in the main chain, namely $(-CH_2{-}CH{=}CH{-}CH_2-)_n$. Once again this gives geometric differences, discussed in section 1.3.3. Finally it is possible for both double bonds to open, which would lead to a three-dimensional network since the repeat unit would have a functionality of four:

$$\leftarrow CH_2 - \underset{\downarrow}{CH} - \underset{\downarrow}{CH} - CH_2 \rightarrow$$

The polybutadienes with a double bond in the main chain belong to an important group of polymers. Natural rubber, gutta-percha, and some synthetic rubbers are other members of the family.

There are other synthetic polymers, usually made by condensation reactions in which some substance such as water is also formed, which have much longer repeating units, and contain different groups along the main chain. Thus the simplest polyamides (nylons) contain sequences of $-CH_2-$ groups joined by peptide links:

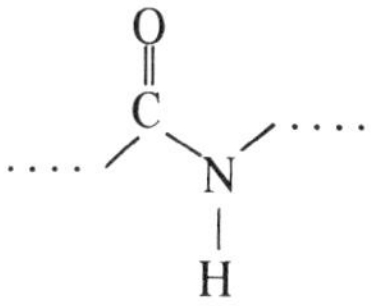

The polyesters contain the ester linkage:

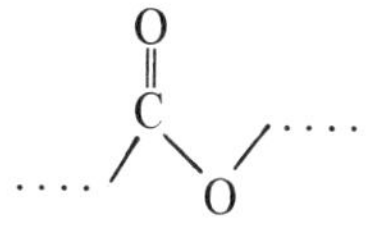

Commercially, the most important of the linear polyesters is polyethylene terephthalate (Terylene, Dacron, Mylar, etc.), which is notable in having a benzene ring in the main chain.

Most natural polymers have even more complicated structures. The chemical formula of cellulose is shown in Fig. 1.10, together with a diagram of its essential physical features. The chain is directional, and ribbon-like in shape. The rings lie in the same somewhat crumpled plane, giving a stiff resistance to bending in one direction, although the links between the rings allow the ribbon itself to

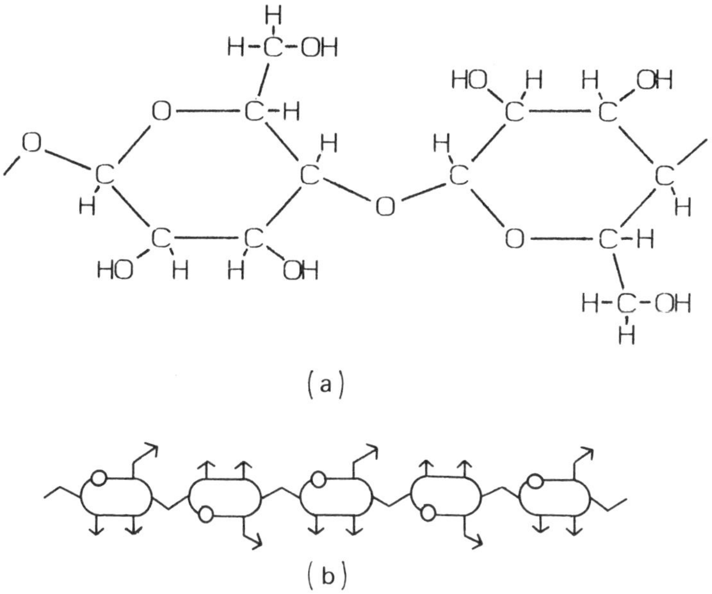

(a)

(b)

Fig. 1.10 – (a) Chemical formula of repeating units in cellulose molecule. (b) Essential physical features of the chain. The arrows represent hydroxyl groups which can form hydrogen bonds.

bend over out of the plane or to twist quite easily. There is thus a measure of flexibility. Sticking out as side-groups are many hydroxyl groups: in cellulose derivatives these are replaced by other groups.

The **proteins** have a rather simpler basic building unit:

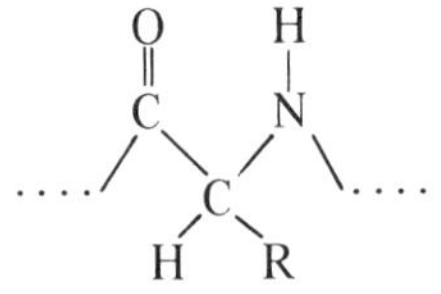

```
        O    H
        ‖    |
        C    N
     /    \ /  \
....       C      ....
          / \
         H   R
```

But complexity arises because the side-group R may be any of about twenty different types. This forms a code – or alphabet – and the particular properties of any protein depend on the particular sequence of side-groups which are present. The simplest side-group is just a hydrogen atom -H; some contain carbon and hydrogen, for example $-CH_3$; some contain hydroxyl groups, for example $-CH_2OH$; some are acidic, for example $-CH_2.COOH$, or in the ionised form $-CH_2.COO^-$; some are basic for example $-CH_2.CH_2.CH_2.CH_2.NH_2$ or $-CH_2.CH_2.CH_2.CH_2.NH_3^+$; and some contain sulphur, for example $-CH_2.CH_2.S.CH_3$. There are two odd ones, which are different from the basic form above. Cystine contains a double group, linking two chains together: $-CH_2.S.S.CH_2-$, and proline has a ring, joining up to the -N-, replacing the -H, and distorting the simple form of the chain:

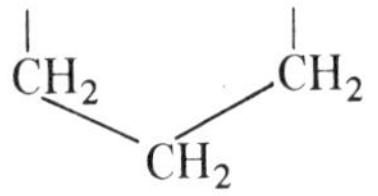

```
 |              |
CH2            CH2
    \         /
       CH2
```

Some other natural polymers, like DNA and RNA, which are important in control of the development of living organisms, have formulae rather similar to proteins, though they are not of exactly the same form and contain other types of atom.

There are a number of other linear polymers which are of special interest because of their commercial importance or because they have been used for important experiments in polymer physics.

Three-dimensional polymers used as thermo-setting resins contain tri-functional groups. Many of the forms are very complex, but two common examples are shown in Fig. 1.11.

We have already noted examples of polymers where oxygen or nitrogen appear in the main chain as well as carbon. There is a growing interest in inorganic polymers, without carbon in the main chain. These may be either linear or three-dimensional. The most important commercially are the silicones. One of the reasons for interest in inorganic polymers is that even the most stable bonds in organic compounds break up at about 500°C: some inorganic polymers may be stable at higher temperatures. However, one of the most familiar inorganic polymers is rather unstable: this is the allotropic form of sulphur known as **plastic sulphur**, obtained when molten sulphur is rapidly cooled. It is soft, elastic and translucent, composed of long chain molecules without side-groups, -S-S-S-S-S-, and is quite different from the usual forms of crystalline sulphur, composed of small molecules such as S_2.

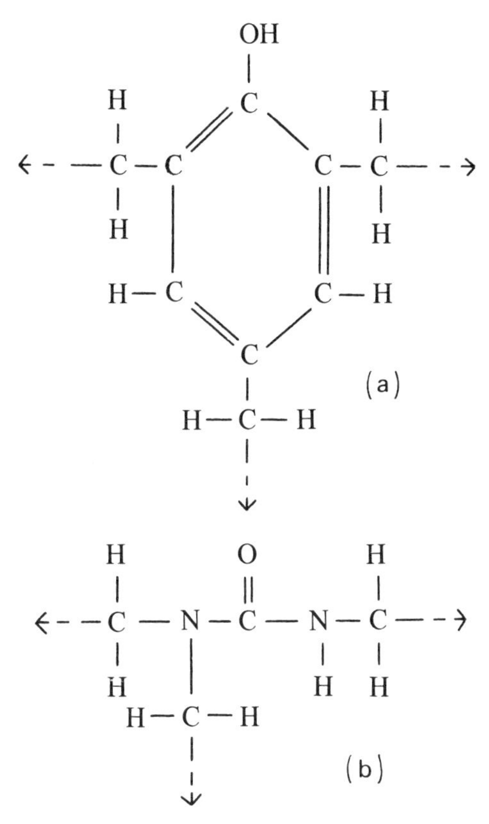

Fig. 1.11 – Trifunctional units appearing in two three-dimensional polymers. (a) Phenol-formaldehyde. (b) Urea-formaldehyde.

1.3.2 Copolymers

Tacitly, we have assumed in most of the last section that the chemical units making up a polymer molecule are identical. This is not necessary. **Copolymers** can be formed from mixed units. For simplicity, we consider copolymers containtaining two monomer units A and B: for example, -CH_2-CHCl- and -CH_2-CH(CN)-.

The various forms of copolymers are shown in Fig. 1.12. A regular alteration does not really give a copolymer at all, since this can be regarded as a polymer with a regular repeat of a single monomer unit *AB*. Random mixing of *A* and *B* gives a very irregular chain. However, if some of the qualities of a regular sequence are required, together with the different characters of different chemical forms, then block or graft copolymers can be made: in the latter case a trifunctional unit *C* has to be present at the junction. Finally, for comparison, Fig. 1.12 also shows a simple mixture of polymers.

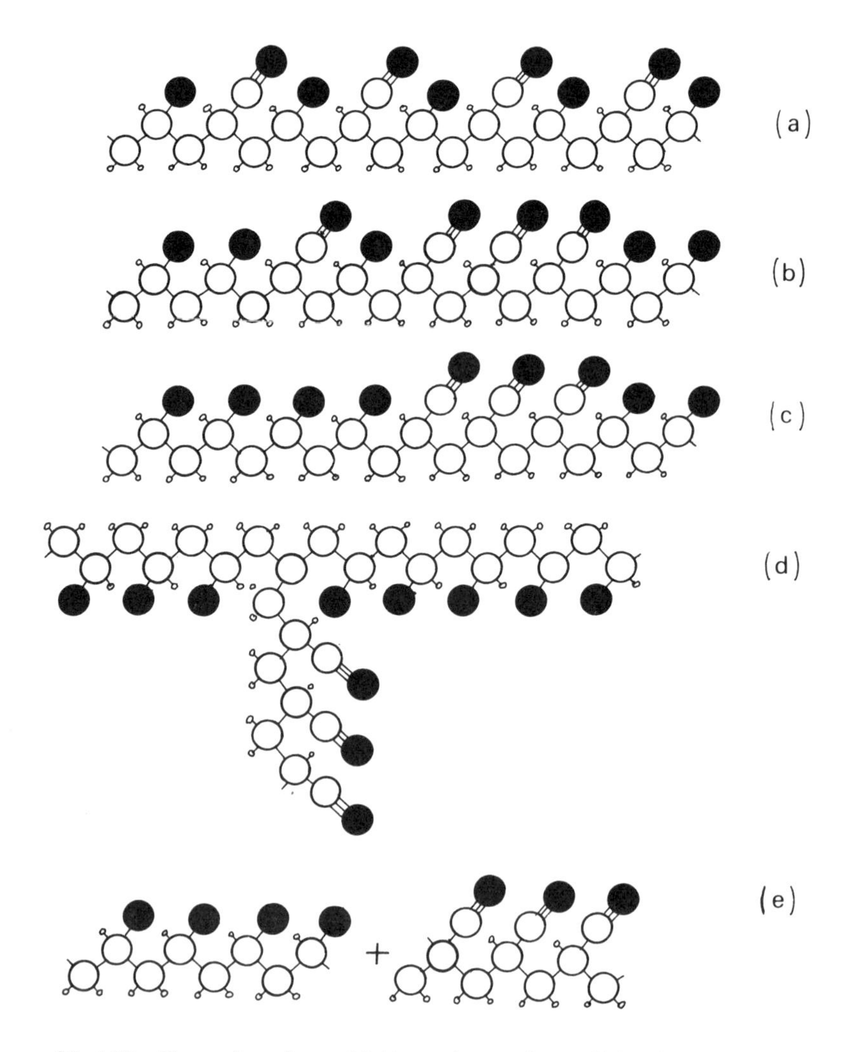

Fig. 1.12 – Types of copolymer. (a) Alternating copolymer. (b) Random copolymer. (c) Block copolymer. (d) Graft copolymer. (e) Mixture of homopolymers.

1.3.3 Molecular shape and geometric isomerism

The specific chemical interactions between atoms and groups in polymer molecules do, of course, play an important part in determining properties. But it is interesting to the physicist to note how large a part is played simply by the shape of the molecules.

Some molecules like polytetrafluorethylene are extremely compact, with fluorine atoms packed around the carbon chain. Where the attached atoms are much smaller, as in polyethylene, the molecules will be less impeded in movement and so will be more flexible. Where there are bulky side-groups as in polystyrene, these force the chains into particular conformations and give an irregular surface. In ordinary physical language it would be preferable to use the word **configuration**, but since polymer chemists have given this term another specific meaning described on page 38, we have to adopt the term **conformation** to describe the particular geometric arrangement taken up by a given polymer molecule.

Some aspects of the shape of polymer molecules are inherent in their make-up. However, the usual easy rotation about single bonds enables them to take up many different forms. For example, a polyethylene chain can take up a great variety of irregular forms.

Although many forms may be possible, it often happens that a particular conformation (or sometimes different particular conformations at different temperatures) is favoured, either for the isolated chain, or, more commonly, when the chain is associated with others in a crystal. Thus polyethylene molecules assume a regular zig-zag. Many other polymer chains take up a helical form such as the very long helix of polytetrafluorethyene, the shorter one of isotactic polystyrene, or the famous α-helix, proposed by Pauling for synthetic polypeptides and occurring in a slightly distorted form in many proteins.

Some proteins and other biological polymer molecules assume very detailed specific forms. For example, Fig. 1.13 shows the form taken up by the molecules of the globular protein myoglobin. Working out the detailed pattern of packing of the atoms in this polymer chain was a triumph of crystallographic study.

There are restrictions on the forms, regular or irregular, which can occur. Except for small displacements under external or internal strain the sizes of atoms, the bond-lengths between atoms within a molecule, and the bond angles remain constant. The easiest mode of deformation of a molecule is by rotation about bonds: this gives rise to the different conformations.

Asymmetry in a monomer unit, such as that of polystyrene, leads to a variety of geometrical forms. The asymmetry arises whenever we have two different side-groups on a carbon atom in the chain. Coming along the chain in one direction, we then find three different groups attached to the carbon atoms – the continuation of the main chain (M), a particular side-group (S), and another side-group such as hydrogen (H). These may be arranged in either of the two ways shown in Fig. 1.14. Arbitrarily defining them as clock-wise (c)

Fig. 1.13 – Conformation of the protein myoglobin which adopts a specific but complicated three-dimensional shape.

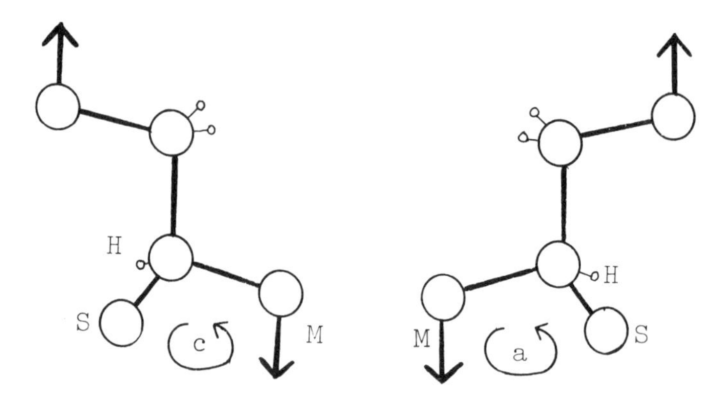

Fig. 1.14 – Asymmetric units, namely H (for example, hydrogen), S (for example, a bulky side-group) and M (main chain), attached to a carbon atom in a polymer chain; arbitrarily termed clockwise (c) and anticlockwise (a).

and anti-clockwise (a), we see that it is impossible to superpose the two, no matter how we rotate them about the main-chain bonds – we are not allowed to turn over the H and S positions since this would imply breaking the main-chain bonds. Clearly there are two distinct geometric forms, which are just as distinct as different chemical forms in the chain. The particular sequence of forms along a chain is termed the **configuration** of the chain.

A schematic way of representing the differences in a chemical formula is shown below:

```
---C     C---            ---C     C---
    \   /                    \   /
      C                        C
    /   \                    /   \
   S     H                  H     S
```

The lower left and right pointing groups are assumed to lie respectively above and below the plane.

From monomer units of the same type, all joined head-to-tail, we can thus get the different types of molecular configuration shown in Table 1.3. The isotactic form of polystyrene can take up a regular shape going into a helix in order to fit in the bulky side-groups; but no matter how we twist the model of the atactic form, we can never make it assume a regular shape – there is a fundamental asymmetry in the way in which the groups are joined together.

Another example of purely geometric difference occurs in polymers with a double bond in the main chain. The four groups attached to a C=C double bond usually lie in a plane, as indicated in Fig. 1.15(a). In a chain molecule, we therefore have the two choices for the main chain, namely **cis** and **trans** illustrated in Fig. 1.15(b) and (c). Since rotation about double bonds cannot occur, or at least is extremely difficult, these forms remain fixed as different geometrically isomeric forms in the chain. The chain as a whole may take up a variety of paths, but the double-bond links must remain in one shape or the other. For example, natural rubber is *cis*-polyisoprene and gutta-percha is *trans*-polyisoprene.

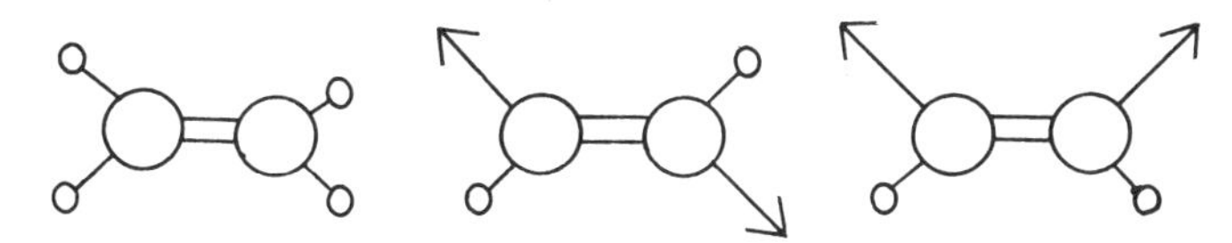

Fig. 1.15 – (a) Planar arrangement of groups attached to a C=C bond, for example in ethylene. The double bond itself is better regarded as lying in the plane perpendicular to the paper. (b) *trans* form for a polymer chain. (c) *cis* form for a polymer chain.

Geometric isomerism is a consequence of asymmetry in the chain, and the two examples quoted are merely two of the simplest forms of asymmetry. With more complicated chemical constitutions of repeat units, a greater variety of geometric forms will be possible. Freedom of rotation about single bonds, and bond straining, allows polymer molecules to take up many geometric **conformations**, but an isomeric difference in **configuration** is found when there is no way in which the forms can become identical without breaking the chain and re-forming it in a new configuration or forcing unallowable changes such as rotation about a double bond.

Table 1.3 – Different tactic forms.

Characteristic	Type	Sequence and representation
regular	isotactic	c c c c c c c c c c c c c c c c c c c c c c c c c c c c c c
		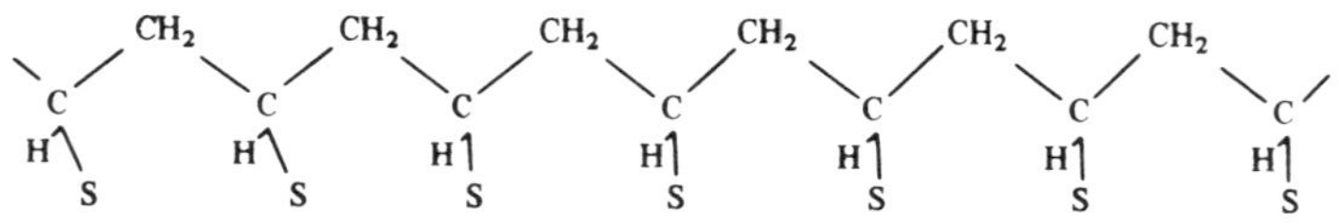
		a a a a a a a a a a a a a a a a a a a a a a a a a a a a a a
		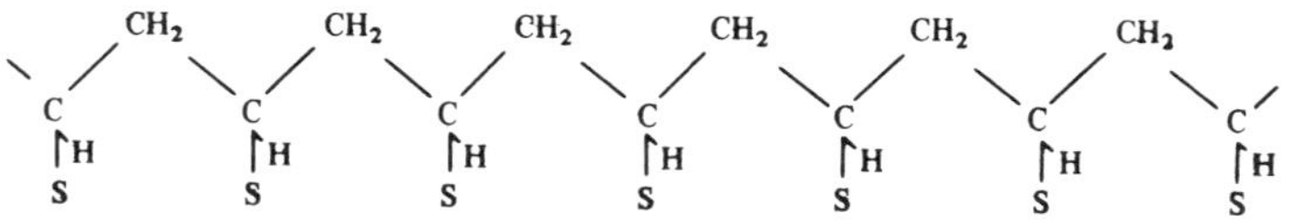
	syndiotactic	a c a c a c a c a c a c a c a c a c a c a c a c a c a c a c
		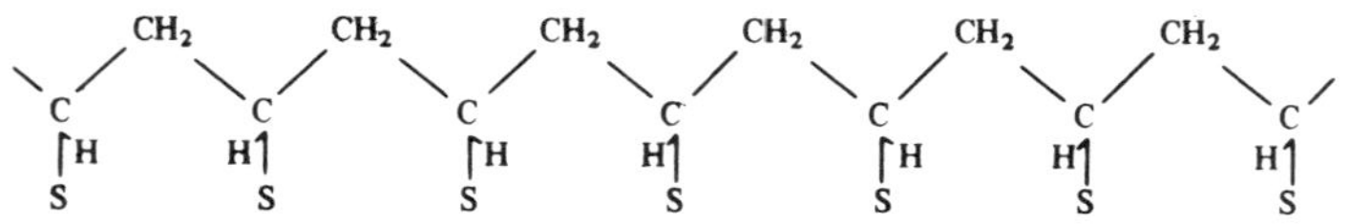
irregular	atactic	a a c a a a c c a c a a c a c c c a c c a c a a a c c a a
		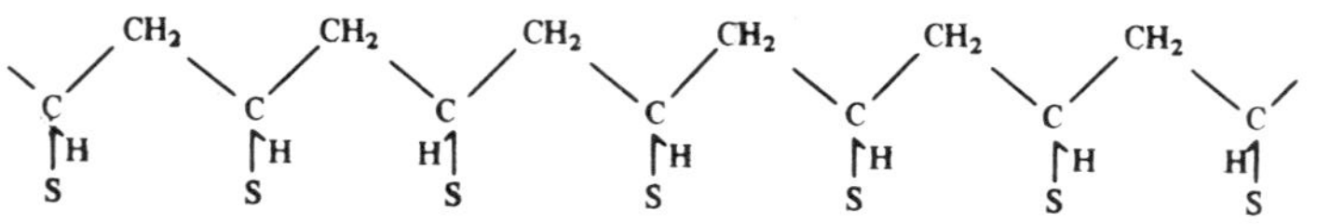
	block	a a a a a a a c c c c c c c c c c c a a a a a a a a

1.3.4 Flexibility, conformation, and folding of molecules

One of the most important characteristics of a linear polymer is the flexibility of its chain. What freedom of movement do the units in a polymer molecule possess? The covalent bonds fix the distances between atoms, save only for very small elastic strains, and the angles between bonds are also rather firmly fixed: this leaves rotation about bonds as the only easy way to get major changes in the conformation of chains. Another way of understanding this sequence of mobility is to note, as illustrated in Fig. 1.16, that bond lengths depend on interactions between two nearest neighbours, but bond angles require three atoms to be located, and bond rotations need four atoms.

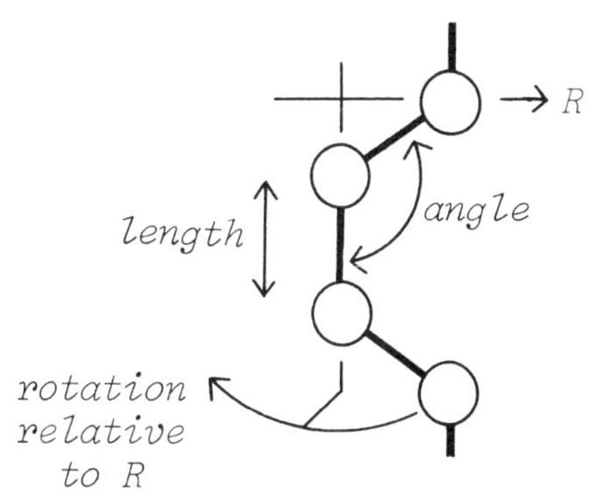

Fig. 1.16 – Bond length defined by two neighbouring atoms; bond angle by three atoms; and bond rotation by four.

Some groups in a chain are inherently rigid: this is true of any ring structure, which will be very difficult to distort, except perhaps by bending out of the plane of the ring as in Fig. 1.17(a), or conceivably by flipping over between different favoured forms such as those shown in Fig. 1.17(b). If the rings are

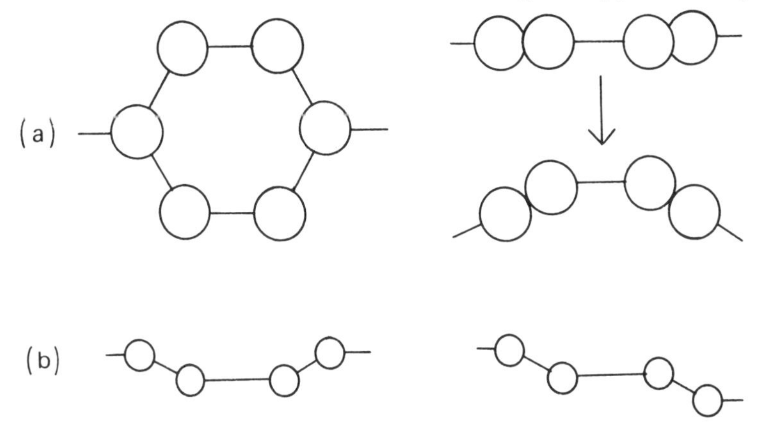

Fig. 1.17 – (a) A planar ring is stiff, except perhaps to a limited extent by bending of the plane. (b) A non-planar ring may have alternative conformations.

linked together, as in Fig. 1.8(d), the whole chain will be stiff; and, even the limited flexibility in one plane would be absent in the girder structure of Fig. 1.8(e). Multiple bonds can be regarded as a special sort of ring, and are stiff, resisting rotation strongly: indeed in classical stereochemistry, rotation about a multiple bond was regarded as completely impossible, but this extreme view is not now held.

Rotation about separate bonds is the chief source of chain flexibility, although, even here, rotation is by no means free and unimpeded. It is simplest to consider first a small molecule: ethane, illustrated in Fig. 1.18. During rotation around the C-C bond, the hydrogen atoms are alternately in staggered and opposed positions. This results in a variation of potential energy as shown in Fig. 1.18(c). Crudely this can be thought of as interference between the hydrogen atoms when they come into opposed positions – this would obviously be so if the hydrogen atoms were electrically charged, or if we had a model made of rubber balls so large that mechanical distortion occurred in the opposed positions. For a precise interpretation it would be necessary to do a quantum-mechanical analysis of the various conformations of the ethane molecule.

As a result of the symmetry of the ethane molecule, there are three equal minima and three equal maxima, with a difference in height of 12 kJ/mol.

At low temperatures, when there is little energy in the system, the molecules will all be in the staggered conformation corresponding to minimum potential energy. But as the temperature rises, more molecules acquire sufficient energy (at particular random times) to flip over the energy barrier; and a new degree of freedom appears in the system, with a consequent change in specific heat. At low temperatures, another important parameter is the resistance to a small degree of bond rotation close to the position of minimum energy: this rotation would be an elastic, recoverable deformation, giving a small distortion of the molecules, and the resistance to it is analogous to a modulus of elasticity or a spring constant.

When we turn to the simplest polymer molecule, polyethylene, the situation is already more complicated, as illustrated in Fig. 1.19. One of the hydrogen atoms on each carbon atom in ethane can be regarded as replaced by another carbon atom leading to the rest of the chain. This causes asymmetry in the variation of potential energy with rotation about the bond, as shown in Fig. 1.19(e). In comparison with ethane the positions of the minima are unchanged, but one is depressed to a value lower than the other two, which are equal; similarly, one of the maxima is raised. The preferred conformation, Fig. 1.19(a), is one in which the main chain leaves adjacent carbon atoms in opposite directions: this gives the chain a planar zig-zag form, which can also be regarded as a two-fold helix with half a turn at each bond. At angles of $\pm 2\pi/3$ there are two slightly less favourable positions, Fig. 1.19(b). The least favourable position is when the main chain atoms are opposed at the angle π Fig. 1.20(c) and other energy maxima occur at $\pm\pi/3$, Fig. 1.19(d).

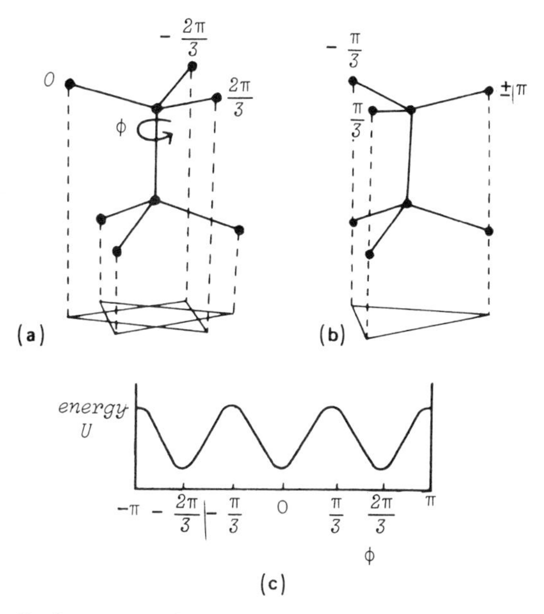

Fig. 1.18 – Conformations of an ethane molecule, CH_3–CH_3. (a) Hydrogens staggered. (b) Hydrogens opposed. (c) Variations in potential energy U with rotation.

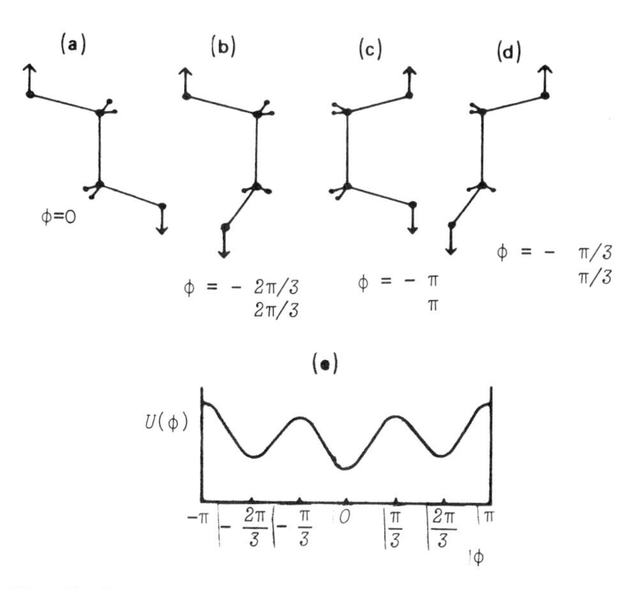

Fig. 1.19 – Conformations of units in polyethylene chain. (a) Most favoured, staggered, *trans*-conformation. (b) Less favoured, staggered, *gauche*-conformation. (c) Most unfavourable opposed conformation. (d) Conformation at other opposed energy maxima. (e) Variation in potential energy.

A comment on terminology of bond conformations should be made. The two planar conformations, shown in Fig. 1.20(a) and (b) are *trans* and *cis*. It is a sequence of *trans* conformations that gives the preferred planar zig-zag of polyethylene. In this simple chain, the *cis* conformations, with the neighbouring carbon atoms eclipsed, would be unfavourable, but in other types of chain (for example, in chains containing double bonds) it does occur as a stable state. The angular direction of a bond is conventionally measured from the *trans* direction, so that the *cis* direction is $\pm 180°$. The other named direction corresponds to the two other higher minima in polyethylene, is called the **gauche** conformation, and sticks out of the plane with angles of $+120°$ or $-120°$ as shown in Fig. 1.20(c). It will be noted that a complete rotation is given by two *trans* bonds or by three *gauche* positions (provided the three are all positive or all negative). Fig. 1.20(d) shows these conformations as viewed along the chain, and makes it clear that a continuous range of other bond conformations is possible, apart from those specifically named.

With other side-groups present, the potential energy diagrams will be still further distorted: the heights of maxima and minima will be altered and some may even be lost, being smeared out into points of inflexion as in Fig. 1.21.

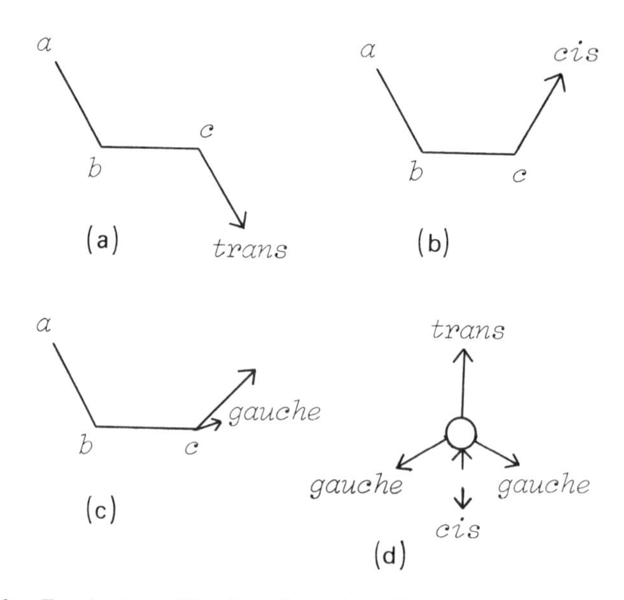

Fig. 1.20 – Terminology of bond conformations. The atoms abc, with the tetrahedral bond angle, define a plane. (a) The *trans* bond, parallel to ab, gives the planar zig-zag. (b) At 180°, the *cis* bond is an unfavourable position in polyethylene. (c) At 120°, the *gauche* conformations are minimum energy states, less favourable than *trans* in polyethylene, but more favourable in some other polymers. (d) Viewed along bc, with a direction defined by ab, the three conformations are identified.

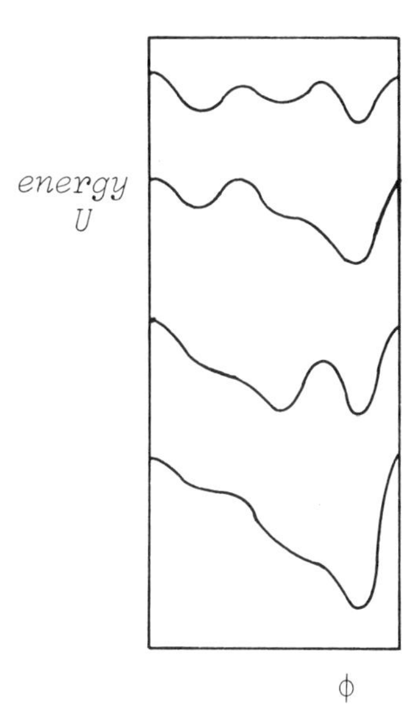

Fig. 1.21 – Variation of energy with bond rotation, with increased distortion from the symmetrical diagram for ethane shown in Fig. 1.18(c).

The relative positions may also be changed. Detailed discussion of the calculation of these energy diagrams lies in the field of theoretical chemistry: but some general points will be brought out here.

If the three symmetrical positions of the ethane diagram are maintained, but one of the *gauche*-configurations, Fig. 1.19(c) has the lowest potential energy,† then the chain will tend to take up the form of the three-fold helix, since there is a rotation of 120° in the direction of the main chain for each *gauche* bond. Most commonly this occurs in polymers with the formula $(\text{-CH}_2\text{-CHX-})_n$ when there is a *trans* bond at $\text{-CH}_2\text{-}$ and *gauche* at -CHX-, since the bulky X side-groups need to be staggered rather than sticking out in the same direction as they would after two *trans* bonds of a planar zig-zag. This alternating sequence of *trans* and *gauche* gives a three-fold helix with one complete turn for every three repeat units, each containing two main chain atoms. This happens in isotactic polystryrene with either right- or left-handed helices being possible.

If the angle of rotation, governed by the position of the minimum is not 120° then other helices will be formed: for example in isotactic vinyl polymers, we find $3\frac{1}{2}$-fold (two turns in seven repeating units) and 4-fold helices while the slight distortion in polytetrafluorethylene twists the planar zig-zag into a very long helix. In general, of course, there is no reason to expect the position of the

† In polyethylene, the two *gauche*-positions are similar, but this will not be so when there are other side-groups present and the main-chain carbon atoms are asymmetric.

minima to be such as would make the rotation for the addition of one repeating unit correspond to any simple fraction of a full turn: helices with a simple concurrence between the geometrical repeat and the chemical repeat are the exception, resulting from some symmetry in the molecular constitution, and not the general case, although we shall see later that the association of many chains in a crystal often favours the formation of a regular helix.

Returning to a discussion of chain flexibility, we are interested in the heights of the barriers between different conformations. If these are low, little energy will be needed to overcome them and the chain is likely to be flexible: if they are high, the chain is likely to be rigid, at least unless the temperature is particularly high and more energy is available.

In general, the presence of bulky side-groups, such as $-CH_3$, or -Cl, makes rotation more difficult, since the opposed conformation will be very unfavourable and the maxima will thus give high barriers. Sometimes, this trend is not observed, and an apparently anomalous easy rotation is found with a very bulky side-group whose shape and size is such that an easy fit is not possible even in the staggered position: the minima of the potential energy are thus raised, and the height of the barrier is reduced.

Any specific dipolar interaction between side-groups will also tend to favour certain positions, and raise barriers; though, here again, we must remember that it is the *difference* in interaction between different positions which is of real importance.

Adjacent to a multiple bond, rotation is usually easy. The form of the molecule $H_3C\text{-}C{\equiv}C\text{-}CH_3$ shown in Fig. 1.22, makes it very clear why the barriers

Fig. 1.22 – Illustration of ease of rotation around bonds which are adjacent to a triple bond.

to rotation about a bond adjacent to a triple bond should be so low: there are no side-groups on the next two carbon atoms. Even with double bonds, barrier heights are often halved. Chains containing double bonds can thus be expected to be very flexible.

Rotation about the multiple bonds themselves is very difficult; but the values for the potential barriers, high though they are, show that it is not completely impossible.

We have seen that various conformations of a polymer chain are possible. One consequence of this, important in polymer crystallisation, is that most polymer chains can easily fold back on themselves. Taking polyethylene as an example, the occurrence of a few appropriately chosen *gauche* conformations (not the regular sequence which gives a spiral) will lead to a reversal of chain direction. The potential energy of this form will be only slightly higher than the usual planar zig-zag form of *trans* conformations. Even some distortion from the exact position of the minima may not involve too much raising of potential energy, if folding is favoured for other reasons. Fig. 1.23 shows a chain fold in a model of polyethylene: there is obviously little awkward distortion.

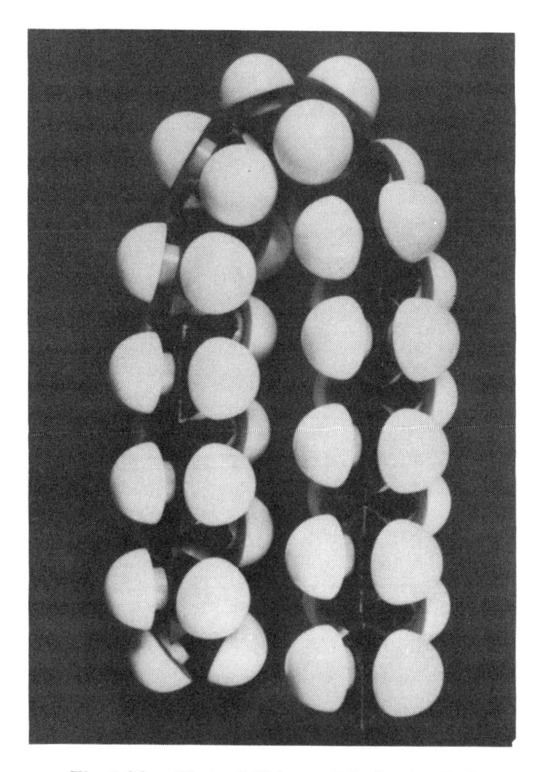

Fig. 1.23 – Chain fold in model of polyethylene.

One rather simple way of thinking about the different possible chain conformations is to consider the diamond lattice, shown in Fig. 1.2. Any sequence of atoms in the diamond lattice is a possible conformation for a linear chain of carbon atoms, provided the same atom is not used twice and provided room is left for the side-groups. In actual chains, there will be important differences in detail due to the presence of other groups. Nevertheless visualisation of paths in a diamond lattice does show up qualitatively the wide-range of possibilities: the millions of different irregular paths; the simplest regular form, namely the planar zig-zag; other regular forms, with helices or more complicated reversals; the possibility of folding back; and so on. It is also possible to visualise chain branching and cross-linking.

The conformational situation corresponds to the different polymer forms already discussed. In a rubber, there will be a continual and easy switching between varying irregular conformations; in a glassy plastic, the irregular conformations will be rigid; in a crystalline polymer, there will be regular conformations regularly packed; in a thermoset plastic there will be a three-dimensional conformation.

1.3.5 Interactions and cross-linking

Cross-linking is important, because it turns an assembly of separate chains into a network structure with a resultant change of properties. As we have indicated, this can be achieved by introducing a tri-functional component during polymerisation: this was done, for example, with a particular variety of acrylic fibre in order to give it greater stability. Alternatively, in hair, crosslinking occurs when the cysteine units in the protein chain open up and join together as cystine. Artificially, the same thing is done during vulcanisation of rubber, when bridges containing sulphur atoms are formed between the chains – a reaction which can occur through opening of double bonds in the chain (see Fig. 1.15). A similar reaction occurs during resin treatments of celluose fibres.

Apart from these covalent linkages, and other strong linkages such as electrovalent salt linkages between acidic and basic groups on protein chains, weaker intermolecular forces also play an important part in determining polymer properties. These weak forces, van der Waals' forces, are due to electromagnetic interactions between atoms in each molecule. A schematic indication of the interactions is given in Fig. 1.24. Where the packing of negative electrons around the positive nuclei is highly symmetrical and stable, as in polytetrafluorethylene, the van der Waals' forces will be very weak. They will be slightly stronger in polyethylene due to the asymmetry of C–H bonds, and much stronger where very asymmetric double and triple bonds, such as $>C{=}O$ in polyester and $-C{\equiv}N$ in polyacrylonitrile are present.

There is one bond of intermediate strength which is important in many polymer systems: the hydrogen bond. This is not the place to discuss this bond in detail; but Fig. 1.25 gives a suggestive picture showing why it may be strong.

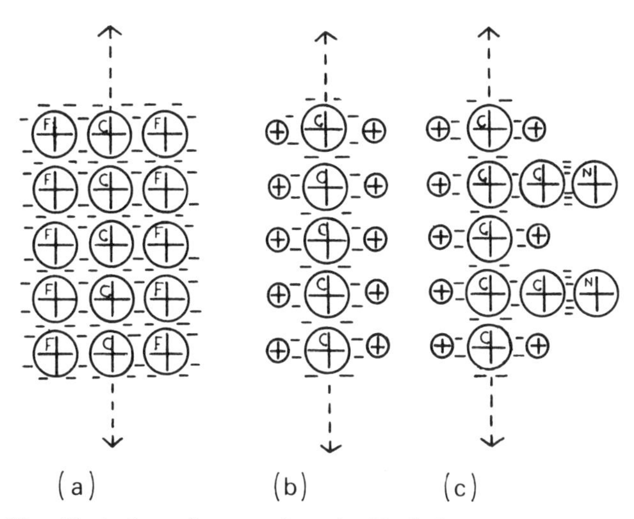

Fig. 1.24 – Illustrations of causes of van der Waal's forces. (a) In polytetrafluorethylene, weak electric dipole forces due to even distribution of negative and positive charge. (b) In polyethylene, medium electric dipoles. (c) In polyacrylonitrile, strong electric dipole forces due to strong concentrations of negative and positive charge in the -C≡N group.

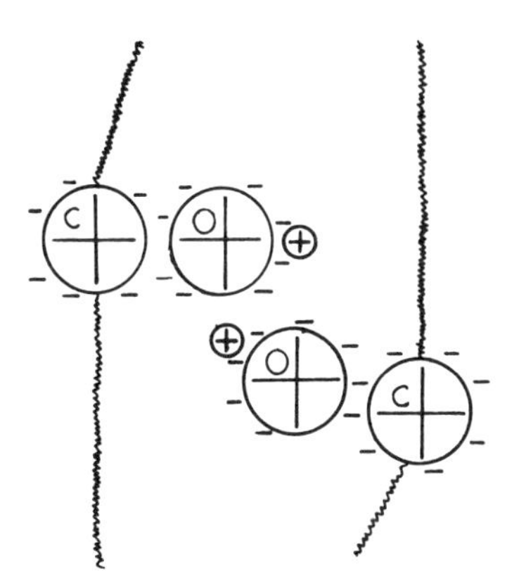

Fig. 1.25 – Schematic view of hydrogen bond.

Associating two electrons with a hydrogen atom, we can regard each hydrogen atom as belonging to either chain: there is consequently an effective sharing, and a moderately strong bond results. Hydrogen bonds can form between hydroxyl (-OH) groups on side chains or peptide (-CO.NH-) links in main chains.

1.3.6 Chain length

Two other features which influence the behaviour of polymer systems are the lengths of the chains and, in some cases, the extent to which they are branched.

A polymer chain length may be expressed either as an absolute value of the length of the extended chain, L; or as a degree of polymerisation, D, giving the number of repeating units joined together; or as a molecular weight, M. Obviously, we shall have:

$$D = L/a = M/M_0$$

where a is the length and M_0 is the molecular weight of the monomer unit.

In any actual polymer material, there will be a range of lengths present. This may be quite wide in a sample as prepared, and may be increased by degradation, or reduced by fractionation. In order to characterise the system completely, it is necessary to plot a distribution of chain lengths. One important branch of polymer physical chemistry is concerned with methods of determining such distributions, and of predicting what distribution will be obtained in particular circumstances. A few typical distributions are shown in Fig. 1.26, illustrating the great variety found in practice.

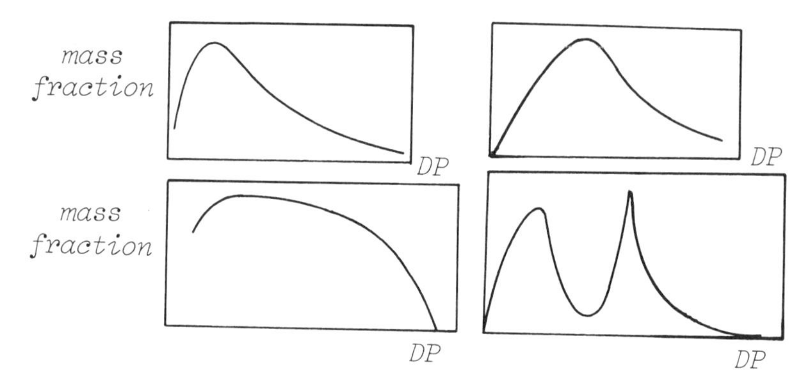

Fig. 1.26 – Four different distributions, by mass fraction of degrees of polymerisation. (a) In a condensation polymer, polyhexamethylene adipamide, nylon 66. (b) In an addition polymer, polymethyl methacrylate. (c) In thermally polymerised polystyrene. (d) In a polyvinyl acetate sample.

In assessing the contribution of chains to a frequency distribution of chain lengths, we may either count each chain as a unit contributing to the number of molecules, or we may count its contribution to the mass of material. In order to get the latter distribution, which is more realistic for many purposes since the long molecules contribute more to the total mass and bulk of the material, it is necessary to multiply each length group by its molecular length or molecular weight. Fig. 1.27 shows a comparison between distributions based on relative number and relative mass: obviously the shorter molecules will appear more predominant in the former.

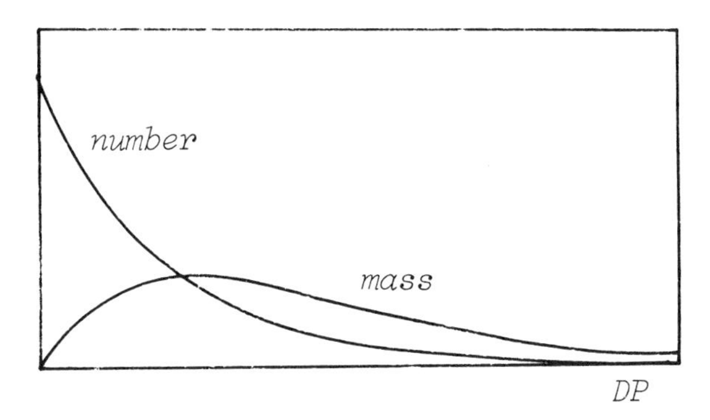

Fig. 1.27 – Comparison of typical number and mass distributions, with frequency plotted against degree of polymerisation (*DP*).

Corresponding to these two distributions, we can get the usual statistical measures such as the number-average M_n and the mass-average M_w molecular weights.

Values of M_n can be found by chemical end-group analyis, or from solution properties such as osmotic pressure. Values of M_w are given by light scattering or sedimentation methods applied to polymer solutions. Another measure of average molecular weight may be obtained from measurement of the viscosity of solutions, but since this depends in a more complicated way on the distribution, it is a different average value, M_v, the viscosity-average molecular weight. Viscosity methods are however very convenient when comparative values need to be obtained quickly and easily, as an estimate perhaps of degree of degradation of polymer chains.

Complete molecular weight distribution curves may be obtained by fractionation into a series of specimens, each having only a narrow range of molecular weights, followed by the use of one of the above methods. Distributions may also be obtained from sedimentation experiments using an ultracentrifuge, since the position of each individual molecule in the force field depends on its molecular weight.

Various chemical methods are available for the estimation of chain branching. The extent of branching depends greatly on the nature of the polymer and on the conditions of polymerisation. Thus one set of tests showed scarcely detectable branching in polystyrene, one branch per 10,000 monomer units in polymethylmethacrylate, and one branch per 1000 monomer units in polyvinyl acetate. The different extents of branching in polyethylene have already been mentioned in section 1.3.1. Branching can have a considerable influence on physical properties.

1.3.7 Absorption and other complexities

In addition to the molecular units which are incorporated into the polymer network, other substances may be present and may influence physical properties.

If there are groups such as -OH or -CO.NH- in the polymer molecule, they will attract water molecules and absorb them by forming hydrogen bonds, often with breaking of hydrogen bonds previously present between chains. The amount of water absorbed depends on the relative humidity of the atmosphere, but can be up to about 15% at 65% relative humidity – or much more at saturation, when the chain molecules can almost be regarded as swimming about in a sea of water. Absorption, with limited swelling, is only possible when cross-links or unattached crystalline regions maintain the coherence of the network, and prevent solution.

Water is a natural plasticiser; but other plasticisers may be deliberately added to polymer materials. For instance, di(isooctyl) phthalate, trixylyl phosphate, dibutyl sebacate and other non-volatile solvents are used to plasticise polyvinyl chloride (PVC). These molecules tend to reduce the attraction between chains, and thus act as internal lubricants. There are also many additives used to modify various properties of polymer materials: these range from dyestuffs to anti-static agents.

Other polymer systems of technical importance consist of a dispersion of one material in another: either or both may be polymers. Textile fibres reinforce rubber in tyres; glass fibres reinforce cross-linked resins; dispersed rubber toughens polystyrene; paper and cotton reinforce phenolic resins in materials like Formica; carbon black is added to rubber; and so on. All of these materials raise interesting problems of the physics of composite materials.

1.4 WHY ARE POLYMERS DIFFERENT?

1.4.1 Effects within the molecules

It may seem a rather worthless truism to say that polymers are different because they have large molecules. But this is the essential point behind all the distinguishing features of polymers. The same thought is expressed in the more sophisticated language of theoretical physics in the quotation at the head of Chapter 2.

In dealing with simple substances we can usually regard the molecule as a well-defined entity, with only one possible state, or perhaps a very few states; and we can make a clear distinction between effects which are due to external reactions of the molecules and effects within the molecules such as atomic vibrations. The different physical states – gases, liquids and crystalline solids – depend only on interactions and packing between molecules. But in polymers, we have to consider both the internal state and the external state. Even an isolated polymer molecule can exist in a variety of states depending on temperature and other conditions. Consequently, in any study of polymer physics, we must pay great attention to effects within a molecule in a way which has no analogy in the simplest molecules, though the beginnings of the problem are seen in the change of specific heat as molecules acquire additional degrees of freedom of movement. What is a minor detail there becomes a dominant feature in the study of polymers, and is the first reason why polymers are different.

1.4.2 Thermodynamics

A second reason why polymers are different is that they are not easily defined thermodynamically. Most polymer systems consist of a mixture of molecules, all of which are different, in length if not otherwise. Instead of the usual single parameter defining constitution (or the few parameters needed for simple mixtures) we have a host of parameters. This can perhaps be seen most dramatically by considering the phase rule:

$$F = C - P + 2 \tag{1.3}$$

where F is the number of degrees of freedom; C is the number of components; and P is the number of phases in equilibrium with one another.

But if each molecule is different, C must equal the number of molecules – and either P or F must also be very large! Intuitively, we feel that the situation cannot be quite as bad as that; but clearly the way in which the fundamental laws of thermodynamics should be applied to polymers needs examination; and the elaborate detailed relations worked out as valid for single or just a few component systems should not be used uncritically.

We must also consider what is meant by equilibrium. In fact, polymer systems are often a very long way from the ultimate equilibrium which occurs at the position of lowest free energy, and are merely in a state of metastable equilibrium. And it is very easy to shift from one metastable state to another. We have already indicated that a great many parameters would be needed to define the state of a polymer system. All these parameters can be regarded as giving co-ordinates in a multi-dimensional space. If the vertical co-ordinate represents free energy, there will be a great many minima in the multi-dimensional surface defining the free energy of the polymer system – considering only two other variables, it would be like a landscape with many hollows. All of these minima will represent metastable equilibrium states for the system, with varying degrees of stability depending on their depths. Mechanical action, change of temperature and other influences can easily shift the system from one hollow to another. In polymers, we cannot ignore this problem as we can often do, with varying degrees of justification, when dealing with simple gases, liquids and crystalline solids.

1.4.3 Molecular and macroscopic

The third aspect in which polymers are different is that effects normally regarded as molecular and effects regarded as macroscopic merge together. This is already true of the above discussion on metastable states.

For metastable states are really commonplace: any constructed article is in a metastable state; indeed anything which can be regarded as large on an atomic scale of sizes, such as the separate domains in a polycrystalline material, is in a metastable state. But we can usually ignore this, or perhaps deal with it by introducing some special mechanism such as diffusion or collapse or erosion or annealing. Any system which is internally in molecular equilibrium is still only in metastable equilibrium with its environment; but provided the system is large enough, the stability is so effective that the system comes to be regarded by the quite different concepts of large-scale macroscopic phenomena. As far as molecular phenomena in systems of small molecules are concerned, we can ignore metastable states except for some special cases like supercooled liquids. But in polymers, many detailed local conformations and packings of molecules define important metastable states.

There are other more practical examples of the merging of molecular and macroscopic phenomena. In dealing with the viscosity of simple liquids, we can consider either the molecular effects, governing events over distances up to a few atomic diameters, (say 20 Å), or we can consider the macroscopic phenomena, covering effects over distances (greater than say 100 Å) in which the behaviour of the individual molecules can be regarded as averaged out. There is a somewhat awkward intermediate region (say 20–100 Å), but the two ways of looking at the problem do not significantly overlap. But in polymers, we can get a direct force transmitted along a single molecule over a length of 10,000 Å or more: the molecular effects have penetrated well into the region where macroscopic phenomena are also important.

Similarly, in dealing with solids, it is possible with simple systems to separate the detailed mechanical behaviour of a group of molecules in equilibrium from the large-scale effects, which can be handled by elasticity theory and which, especially in composite systems, give important stress and strain concentrations at discontinuities in the material. But, in polymers, the important discontinuities and differences in material may go right down to units only a few molecules across: once again the molecular and macroscopic effects merge and present a challenge to polymer physicists.

1.4.4 The chemical diversity

Polymers differ from ordinary substances composed of small molecules for the reasons just described. But they also differ among themselves due to the variety of chemical groups which occur in the main chains and the side-groups of polymer molecules. The features which are important in determining the character of a polymer depend on geometry and on energies of interaction, and are:

1. the size and shape of the chemical groups;
2. the ease of extension, bending, and rotation of inter-atomic bonds within molecules;
3. the possibility of bond breakage;
4. the strength and directionality of interactions between molecules.

However, despite the importance of these features in polymer science and technology, they are really a part of general chemistry, and not solely of polymer chemistry, since particular groups will behave similarly whether they are in small molecules or large ones. The chemical composition of polymers mentioned in this book are included in Appendix C, which also lists some comparative data on atomic dimensions and energies of interaction.

CHAPTER 2

Basic Concepts: Ideal Polymeric States

'. . . a polymeric molecule can be regarded as a macroscopic system: it is possible to talk about its **macro state**, which is characterized, for example, by the end-to-end distance of the molecule . . . and its **micro state**, characterized by the mutual positions of all the bonds within the molecule . . .'

T. M. Birshtein and O. B. Ptitsyn, *Conformations of Macromolecules*

2.1 THE IDEA OF AN ISOLATED POLYMER MOLECULE

The real polymer systems, on which experiments are done, are either condensed states of the polymer molecules (solids and melts) or solutions. But a much simpler system for theoretical study would be an isolated polymer molecule – or a collection of them, which would be a polymer gas. Unfortunately, except at excessively low pressures and high temperatures, such molecules would always condense on the walls of any container. Nevertheless, it is possible to imagine an isolated polymer molecule in space, having just enough interaction with its surroundings to take up a well-defined temperature. It is even possible to imagine how such a molecule might get there – by an astronaut spraying out a fine dispersion of a dilute polymer solution. How will such a molecule behave? What will be the effect of changes in temperature? These are questions which will be discussed in this chapter. The answers will help us to understand the behaviour of the real condensed systems more clearly.

With a little more ingenuity, one can imagine the ends of such an isolated molecule being held, so that its load-elongation properties can be measured. This may seem a wild fancy from a practical point of view, but a polymer molecule contains so many atoms that it involves no theoretical or conceptual difficulty. A suitable system for experimental study could consist of a long polymer chain (containing 10,000 or more atoms in the chain, and at so at least 1 μm long) with metal balls (each made up of say 10,000 atoms) attached to each end. Forces could be exerted on the balls by electric and magnetic fields: so the molecule could be held in the centre of a container at low pressure and high temperature, and the mechanical and thermal response and other properties of the single molecule could be determined. From a statistical view-point, this would be a large enough system. Certainly it would be so massive that there would be no appreciable errors due to the uncertainty principle. Practically, it may be noted that the balls would be just about resolved in an optical microscope, and various other techniques could also be used to follow the system. If the resources devoted to some other scientific experiments were available, it is probable that an experiment on the above lines could really be performed. For the present, however, we must merely perform it in our imagination. For a simple quasi-static tensile test, it would perhaps be more realistic to think of the combined effects of a number of polymer chains held between the 'jaws' of a load-elongation tester, but separated so that there was no mutual interference.

2.2 THE BEHAVIOUR OF FLEXIBLE CHAIN MOLECULES

2.2.1 The application of kinetic theory to a polymer chain.

The prediction of the behaviour of an ideal gas is one of the classical studies of theoretical physics. The methods (described in Appendix A3) may be modified to apply to an ideal linear polymer molecule by supposing that the individual elastic point particles are not free as in a gas, but are joined up into a chain by weightless, inextensible links, each having complete freedom of rotation around a sphere. This gives an ideal flexible chain molecule, as indicated schematically in Fig. 2.1: such a molecule is alternatively termed a freely orienting molecule or a random chain. It is assumed that there are no other forces between the units of the chain. Because of their energies the point particles will still try and shoot about in all directions; but they will be constrained by their attachments to the other particles. In real polymer molecules there will, of course, be various other restrictions on chain flexibility.

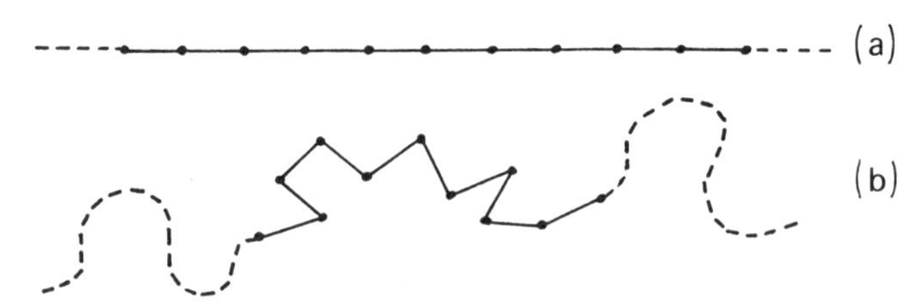

Fig. 2.1 – Two forms of a short portion of an ideal freely orienting chain molecule. (a) Fully extended. (b) A more-or-less random form.

If we could observe in detail the Brownian movement of such a random chain, we should see it apparently wriggling and writhing about and taking up all manner of different patterns in rapid succession. A series of instantaneous pictures of a single molecule, or a single picture of a large assembly of molecules in a polymer gas, would show a distribution of molecular conformations. Because we assume that there are no internal energy differences, all particular conformations are equally probable – just as all particular hands in a deal of cards are equally probable. But if we are more general in description, then we note that some types of conformation are more probable than others. A fully extended chain as in Fig. 2.1(a) will be very rare (literally unique), just as there is only one way of dealing a hand of thirteen spades in order; but conformations like the one shown in Fig. 2.1(b) will seem common because there are many other possible conformations of the same general character, just as there are many possible hands of cards containing a mixture of cards of different suits. We shall return to a quantitative discussion of the statistics of chain conformations later: at present, we wish to understand more about the physical properties of the ideal flexible polymer chain.

Suppose that, instead of allowing the chain to be free to assume any conformation and position, both ends are held as in Fig. 2.2. The particles in the chain will be vibrating about in all directions, and so a succession of varying impulses will be exerted on the two supports at the ends of the chain, in much the same way that the molecules of a gas exert impulses on the walls of a container. From the number and mean value of the impulses we should, in principle, be able to calculate the mean force on the support, just as we can calculate the mean pressure of the gas: in practice the details of the analysis are much more complicated. Qualitatively we can expect that the magnitude of the force will vary with the distance between the supports, since, when we extend the chain, we change the nature of the movements which take place.

When the chain ends are close together, there are many possible conformations; the system is disordered; and entropy is high – this is a condition for equilibrium under zero tension. By contrast, in the fully extended state, there is only one possible conformation; the system is highly ordered; entropy is low; the free energy, given by $-TS$, is high; and work must have been done in order to reach the extended state. Applying (A.3) and remembering our postulate that there are no internal energy changes, we get:

$$\text{tension in chain} = F = -T(\partial S/\partial l)_T \tag{2.1}$$

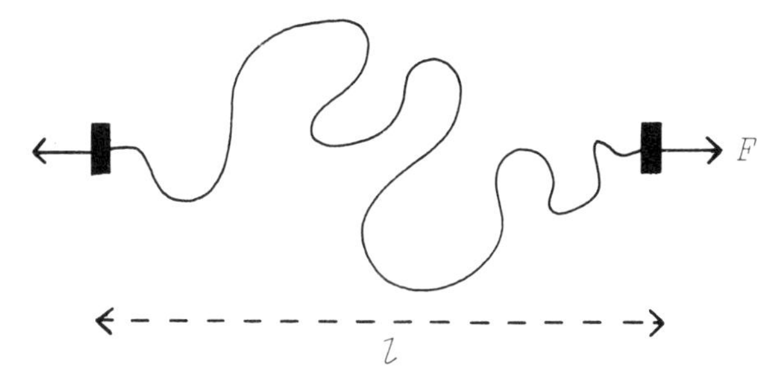

Fig. 2.2 – Simplified form of an ideal freely orienting molecule, represented, because of its length, as a line instead of the sequence of point particles shown in Fig. 2.1. The chain is held at a length l under a force F, which results from thermal vibration impulses on the ends.

where S is the entropy of a chain with a length l between its end, and T is the absolute temperature.

Since the entropy decreases as the length increases, the extended chain must be under a positive tension. We shall now try to calculate an expression for this variation of tension with extension of the chain, adopting first a kinetic argument and then using statistical arguments.

2.2.2 A simple mechanical approach

In order to bring out the essential physical principles of the kinetic theory of the behaviour of flexible polymer chains, and to show the close analogy with the kinetic theory of gases, without getting involved in technical mathematical difficulties, we shall first consider a drastically simplified and somewhat incorrect model.

We consider a chain of mass per unit length m having a length L when fully extended, as shown in Fig. 2.3(a). We assume that when the ends of the chain are separated by a distance l the chain oscillates from side to side along a one-dimensional path, as indicated in Fig. 2.3(b) with the individual units moving with a velocity u. Physically, it is important to note that the chain is regarded as elastic, with a very high modulus, as completely flexible at the links between each of the large number of units, and as having no intermolecular forces between the units. The supports at the chain ends are rigid.

When the chain has moved to its extreme position on one side, as in Fig. 2.3(c), it will pull tight, and the force exerted by the supports will cause it to reverse direction. Because the chain is elastic and the supports are rigid, there will be no loss of energy and the velocity will be altered in sign but not in

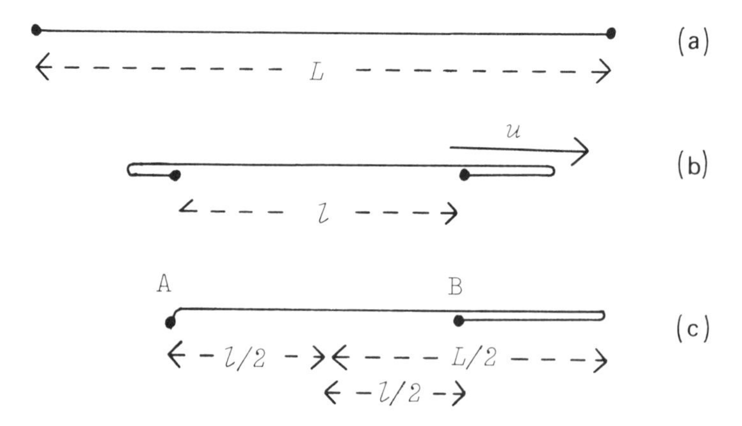

Fig. 2.3 – Idealised one-dimensional chain motion. (a) Fully extended chain, length L. (b) Chain with ends held at distance l, and moving to right. (c) Extreme position for reversal of direction of motion.

magnitude. The longer length $(L + l)/2$ will exert an inward force on the distant support, and the shorter length $(L - l)/2$ an outward force on the near support. Taking forces to the right as positive, we can summarise the changes in momentum in one complete cycle of two reversals as follows:

change of momentum acting on	A	B
reversal at right:	$2\left[\frac{(L+l)}{2}mu\right]$	$2\left[\frac{(L-l)}{2}mu\right]$
reversal at left:	$-2\left[\frac{(L-l)}{2}mu\right]$	$-2\left[\frac{(L+l)}{2}mu\right]$

The time for one cycle $= 2(L - l)/u$.
Therefore, force on $\mathbf{A} = F =$ rate of change of momentum

$$= \frac{2[(L+l)m/2]u - 2[(L-l)m/2]u}{2(L-l)/u}$$

$$= mu^2[l/(L-l)] \tag{2.2}$$

specific stress in chain $= f = F/m$

$$= u^2[l/(L-l)] \tag{2.3}$$

If there are N units along the chain, and we ascribe an energy $\frac{1}{2}kT$, to the one degree of freedom available to each, we get:

$$\text{kinetic energy of whole chain} = \tfrac{1}{2}Lmu^2$$
$$= \tfrac{1}{2}NkT \tag{2.4}$$

Hence:

$$u^2 = (N/Lm)kT$$
$$= kT/m_0 \tag{2.5}$$

where $m_0 = Lm/N =$ mass of one unit of chain.

But we also have:

$$m_0 = M_0/N_0, \quad \text{giving} \quad m = M_0N/N_0L \tag{2.6}$$

where $M_0 =$ molecular weight of a chain unit, and $N_0 =$ Avogadro's number.

Thus, by substitution in (2.3):

$$f = \frac{N_0kT}{M_0}\left[\frac{l}{(L-l)}\right] = \frac{N_0}{M_0}kT\left[\frac{l/L}{1-l/L}\right] \tag{2.7}$$

This equation has some obvious similarities with (A.15) for an ideal gas. The differences are that specific stress f on the chain replaces pressure, and that a function of the length of the chain replaces the volume.

Equation (2.7) gives the variation of stress with length shown in Fig. 2.4 varying from $f = 0$ at $l = 0$ to $f = \infty$ at $l = L$.

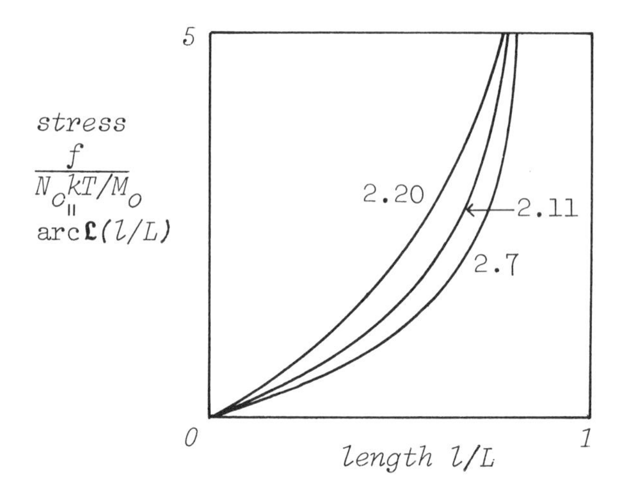

Fig. 2.4 – Force-elongation relations for two approximate models of a polymer chain, namely (2.7), one dimensional kinetic model, (2.11), statistical volume model; and for the more rigorous inverse Langevin function form, (2.20).

Summarising we can say that:

(a) the equilibrium state under zero stress occurs when the ends are brought together, at $l = 0$:

(b) for small separation of the ends, $l \ll L$,
we have a linear relation, $f = (N_0kT/M_0)\,(l/L)$

(c) as the separation increases, the stress rises more rapidly.

The kinetic argument given here is adequate to sustain the qualitative prediction that the mean stress on an isolated polymer chain will be zero when the ends are coincident, and will rise as the chain is pulled out towards a fully extended conformation. However, the model is clearly unsatisfactory for quantitative prediction: no real polymer chain would be constrained to move along one dimension alone. It is interesting to note that the system is so unreal that it is not possible to develop a corresponding argument from statistical thermodynamics without getting rid of the limitation to one dimension.† Any attempt to work out a detailed kinetic analysis, allowing for movement in three dimensions would clearly be very difficult.

† The discerning reader may also discover that, even in one dimension, the motion could not be quite as simple as is described here.

2.2.3 An approximate statistical derivation

We next consider a model in which the chain is allowed to vibrate so that it moves through all the available space in three dimensions. As indicated in Fig. 2.5, the volume which it sweeps out will be an ellipsoid with a semi-major axis of $L/2$, and semi-minor axes of $(L^2/4 - l^2/4)^{\frac{1}{2}} = \frac{1}{2}(L^2 - l^2)^{\frac{1}{2}}$. This ellipsoid will have a volume of $\pi L(L^2 - l^2)/6$.

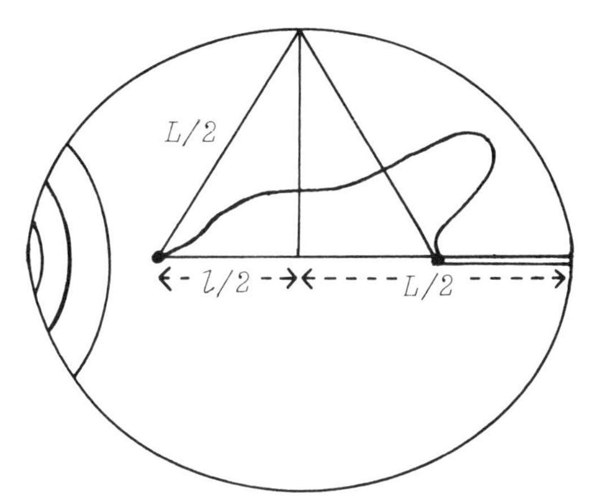

Fig. 2.5 – Ellipsoidal volume which could be swept out by a chain of length L when its ends are a distance l apart.

We can now follow (with an error which will be pointed out later) the same argument as was used in deriving the gas laws. We imagine the whole ellipsoid to be divided into elements of volume, v_0. The number of elements will be $\pi L(L^2 - l^2)/6v_0$, and if we assume that each of the N units in the chain can be placed in any of the volume elements, we find:

$$\text{number of possible arrangements} = W = [\pi L(L^2 - l^2)/6v_0]^N \qquad (2.8)$$

The Boltzmann relation states that entropy $S = k \log W$, but, since we can arbitrarily select the reference state from which entropy values are measured, we can drop the constant term $\log(\pi L/6v_0)^N$ and put:

$$\begin{aligned} S &= k \log_e [(L^2 - l^2)]^N \\ &= Nk [\log_e (L^2 - l^2)] \end{aligned} \qquad (2.9)$$

Applying the thermodynamic relation, (A.3), we get an expression for the tension in the chain:

$$\begin{aligned} F &= -T(\partial S/\partial l)_T \\ &= -NkT(-2l)/(L^2 - l^2) \\ &= 2NkT\, l/(L^2 - l^2) \end{aligned} \qquad (2.10)$$

Reducing this as before, to an expression for specific stress, we have:

$$f = \frac{2NkT}{m}\left[\frac{l}{(L^2 - l^2)}\right]$$
$$= 2\left(\frac{N_0 kT}{M_0}\right)\left[\frac{l/L}{[1-(l/L)^2]}\right] \quad (2.11)$$

Apart from a difference in the form of the function of the length of the chain, this result is similar to that obtained by the kinetic argument of the previous section, the form of the relation is also shown in Fig. 2.4. As before we have $f = 0$ at $l = 0$; $f = \infty$ at $l = L$; and for $l \ll L$ differing only in a numerical factor, $f = 2(N_0 kT/M_0)\,(l/L)$. This confirms that the resistance to extension of a polymer chain is a kinetic, entropy-dependent effect very similar to the resistance to volume reduction in an ideal gas. The molecule, in fact, acts like an entropy spring – on removal of tension, the chain returns to its disordered conformations, so the spring is elastic.

However, once again, there is a serious flaw in the detailed argument. We have allowed the N units of the chain to take up any distribution among the volume elements of the whole ellipsoid – but the vast majority of these distributions will fail to satisfy the requirement that the polymer chain must be continuous. In reality, only a limited number of arrangements are possible, and we must look for a way of computing the number which retain the identity of the chain molecule.

2.2.4 The form of the relation between chain stress and chain extension

Both the kinetic argument in section 2.2.2 and the statistical argument in section 2.2.3 yield similar relations between specific stress f and length l of a chain with a fully extended length L. Both expression are of form:

$$f = (N_0 kT/M_0)f(l, L) \quad (2.12)$$

The factor $(N_0 kT/M_0)$ arises from the fundamental basis of the kinetic and statistical thermodynamic theories: it will clearly appear in any formula. The problem is to derive the correct form of the function $f(l, L)$. Since $f(l, L)$ must be dimensionless it follows that it must be of the form $f(l/L)$.

Arguments based on symmetry enable some other features of the function to be rigorously predicted. Firstly, since the whole set of possible chain conformations will be symmetrically distributed about the line joining the chain ends, the mean force on the chain must act along this line. Secondly, an interchange of position between the two ends will not alter the physical reality; nor will a translation or rotation of the molecule as a whole: consequently the magnitude of the mean tension must be a function only of the distance between the end-points. Furthermore if we regard the end B as fixed, and then allow position of the end A to vary along a given line, the tension must merely change sign for corresponding distances on either side of B — in other words, if (f,l) is a solution then so is $(-f,-l)$. Thus, thirdly, $f(l,L) = -f(-l,L)$; and, if there is to be continuity through the origin, we must have $f(l,L) = 0$ at $l = 0$. Consequently the mean tension must be zero when the chain ends are coincident. These three conditions will be satisfied by any function, $f(l/L)$ which can be expressed as a polynomial containing only odd powers of (l/L):

$$f(l/L) = c_1(l/L) + c_3(l/L)^3 + c_5(l/L)^5 + \ldots \tag{2.13}$$

If c_1 is finite, it follows that for small enough extensions, when $l \ll L$:

$$f(l/L) \simeq c_1 l/L \tag{2.14}$$

and we have a linear relation:

$$f \simeq (c_1 N_0 kT/M_0)(l/L) \tag{2.15}$$

We note that the factor (N_0kT/M_0) or (RT/M_0) shows that the specific stress on a chain increases proportionately with absolute temperature and inversely with the molecular weight of a chain unit.

2.2.5 A more exact derivation

The problem of making a reasonably correct derivation of the relation between stress and strain in a flexible polymer chain has exercised many physicists. One of the best equations was derived independently by different methods by Kuhn and Grun and by James and Guth: the present account follows a form of the analysis given by Flory.

We consider one link in a freely joined chain under a tension F, as indicated in Fig. 2.6(a). If the length of the link is a, and it makes an angle ψ with the direction of F, we have:

$$\text{component of displacement of the link in direction of } F = x = a\cos\psi \tag{2.16}$$

A change in ψ and hence in x will raise or lower the load F and we can associate the value of x with a potential energy $-Fx$: the negative sign is put in because an increase in x causes a fall in the weight, and thus a lowering of potential energy. The geometry of Fig. 2.6(b), as set out in the caption shows that, in the absence of the energy differences, the *a priori* probability of x having a value between x and $(x + \mathrm{d}x)$ would be proportional to $\mathrm{d}x$ but independent of x. Consequently it follows from the Maxwell-Boltzmann law that under a load F, the probability that x has a value between x and $(x + \mathrm{d}x)$ is proportional to $\exp[-(-Fx/kT)]\mathrm{d}x$. Since x can only vary between $-a$ and a we have:

$$\text{mean value of } x = \bar{x} = \int_{-a}^{a} x\exp(Fx/kT)\mathrm{d}x \Big/ \int_{-a}^{a} \exp(Fx/kT)\mathrm{d}x$$

$$= a[\coth(Fa/kT) - (Fa/kT)^{-1}]$$

$$= a\mathcal{L}(Fa/kT) \tag{2.17}$$

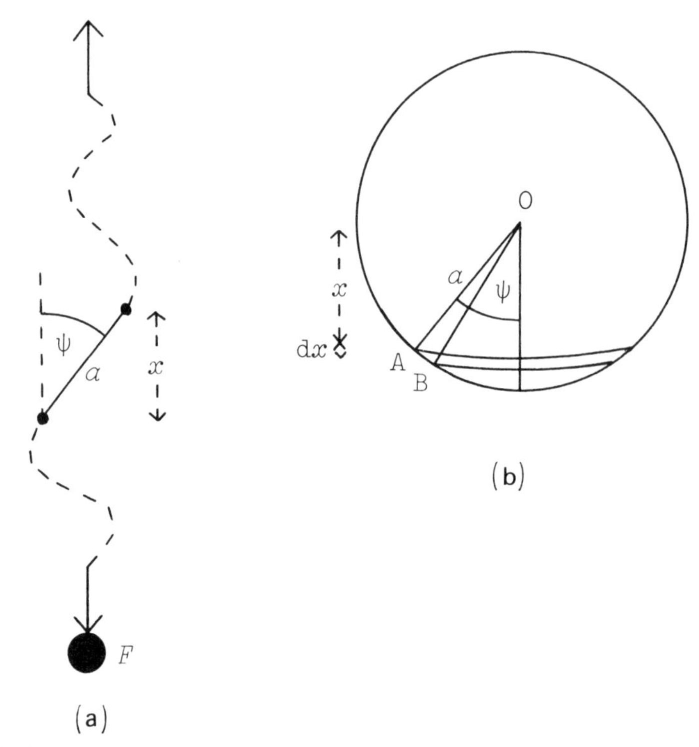

Fig. 2.6 – (a) Typical link in a freely jointed chain under a load F. (b) Sphere defined by position of one end of a link relative to a fixed position O for the other end. The portion of the sphere between x and $x + dx$ has: circumference $= 2\pi x \tan\psi$; width AB $= dx \operatorname{cosec}\psi$; area $= 2\pi x \tan\psi \, dx \operatorname{cosec}\psi = 2\pi a \cos\psi \tan\psi \operatorname{cosec}\psi \, dx = 2\pi a dx$. In the absence of energy differences, the free end would take all positions on the sphere equally, and hence, probability that x will have a value between x and $x + dx = 2\pi a dx/4\pi a^2 = dx/2a$.

The integral occurring in the above expression had previously been found by Langevin to appear in the theory of the alignment of electric and magnetic dipoles. This led to its evaluation, and the form shown above was given the name **Langevin function** represented by $\mathcal{L}$.

The mean total length l of the chain will be N times the mean projected lengths of each of the N units in the whole chain. Thus:

$$\begin{aligned} l = N\bar{x} &= Na\mathcal{L}(Fa/kT) \\ &= L\mathcal{L}(Fa/kT) \end{aligned} \tag{2.18}$$

since the extended chain length $= L = Na$.

Turning this expression round, we get:

$$F = (kT/a)\,[\text{arc}\,\mathcal{L}(l/L)] \tag{2.19}$$

where arc $\mathcal{L}$ is the **inverse Langevin function** (analagous to arc sin or $\sin^{-1}$).

We should note that the last step is not rigorously correct: the mean length at a fixed force does not necessarily invert exactly to the mean force at a fixed length. However, the error is likely to be small.

Dividing by the mass per unit length and substituting from (2.6) as before, we get an expression for specific stress:

$$f = (N_0 kT/M_0)[\text{arc}\,\mathcal{L}(l/L)] \tag{2.20}$$

This expression is of the general form needed to satisfy (2.12) and is shown in Fig. 2.7. For small values of l/L, expansion gives a polynomial of the right form:

$$\text{arc}\,\mathcal{L}(l/L) = [3(l/L) + 9/5(l/L)^3 + (297/175)(l/L)^5 + \ldots] \tag{2.21}$$

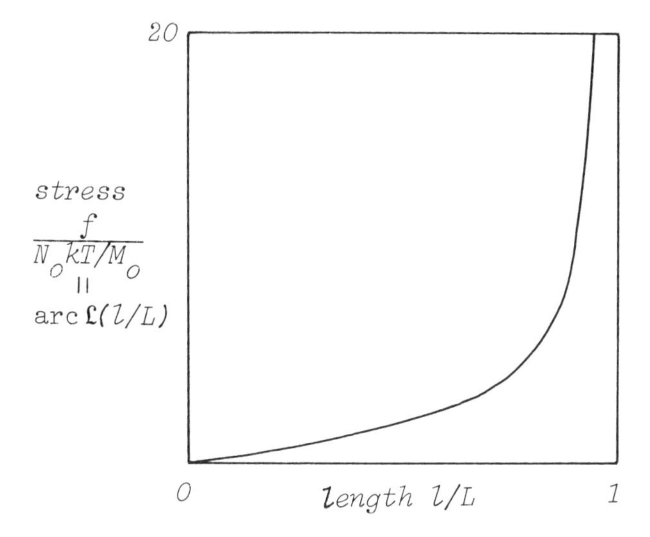

Fig, 2.7 – The inverse Langevin function form for the stress-elongation relation of an ideal polymer chain.

The first linear term gives a resonable approximation up to about $l/L = 0.4$.

Within the limitation of our definition of an idealised polymer molecule, the above derivation is sound if it can be assumed that the fluctuations in the value of x for each link are independent of one another. There is no geometrical inconsistency in this, as there was in the two previous derivations. Starting at the top end of the chain in Fig. 2.6(a) it is possible to give the first link any value x_1 within the range $-a$ to a; and then to follow this with any value x_2; and so on through the N links. Consequently all the conformations included implicitly in the expressions given in (2.16) and (2.17) are geometrically permissible and will occur; and there are no other permissible forms. But, when we consider the dynamics of the situation we see that the movements of the links must be coupled, since a change in position of a particle at the end of the link must also alter the conformation of the next link. A complete solution of the problem would demand an analysis of the wave mechanics of the whole molecule – a formidable task! Insofar as we can ignore the effects of the coupling as in many other physical problems, (2.20) will be a reasonable representation of the behaviour of an idealised flexible chain molecule. Intuitively, looking at Fig. 2.1 one can expect the coupling to have a progressively greater influence as the chain is pulled closer to its fully extended state.

2.2.6 Chain conformations

The theory of the behaviour of freely-orienting molecules may be developed to give expressions for the conformations which a free chain will take up. Because the free energy of the chain increases as the ends are pulled apart, it follows that if one end of a chain is fixed, the most probable location for the other end is at the same point. Paradoxically, this does not mean that the most probable end-to-end distance in free chains is zero. If the one chain end is at A in Fig. 2.8 then the end-to-end distance will lie between l and $(l + \mathrm{d}l)$ if the other end B lies anywhere within a spherical shell of volume $4\pi l^2\,\mathrm{d}l$. The increase in this *a priori* probability more than offsets the effect of the increase in free energy with increase of l.

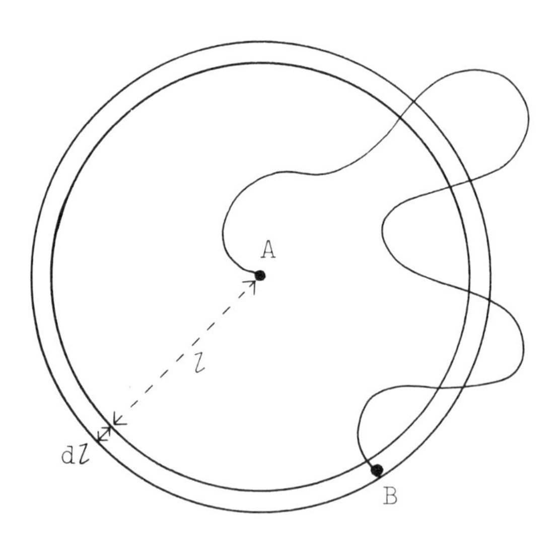

Fig. 2.8 – Positions of a chain with an end-to-end distance l.

A detailed discussion of chain statistics will be left until the next chapter, because it is rather long and complicated and would cause too much interruption in our discussion of polymer states. However, anticipating what will be proved later, Fig. 2.9(a) shows the variation in the probability of finding chain ends separated by a given distance in a given direction: this is the same as plotting probability density, which is the probability of finding a chain end within a *given* volume at a distance from the other end. Fig. 2.9(b) shows the total probability for a given separation irrespective of direction: this function is obtained by multiplying the probability density by l^2, and normalising.

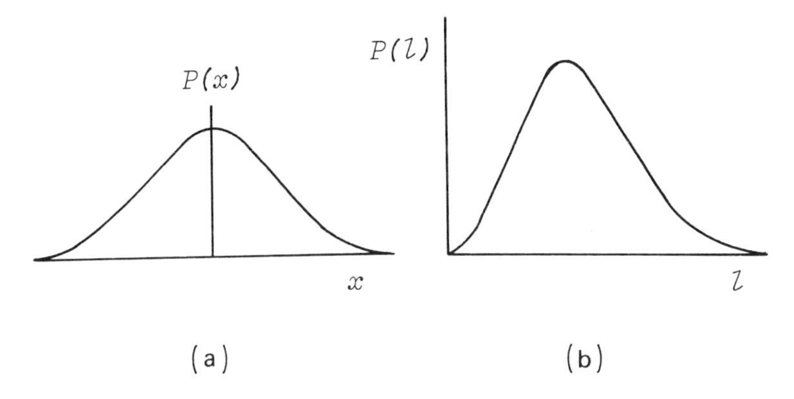

Fig. 2.9 – (a) Probability $P(x)$ of finding one chain end displaced by a distance x in a given direction (irrespective of displacements y and z) from the other end; or probability of finding the end in a particular volume in a given direction x. Note that $P(x)$ is maximum for $x = 0$. (b) Probability of finding one chain end displaced by a distance l in any direction from the other end.

Other units in a free chain will be distributed in a spherically symmetrical distribution around the centre of mass. Clearly there are many possible conformations. An alternative method of obtaining the ideal flexible chain (2.20) depends on determining the number of ways of obtaining a given end-to-end distance; hence finding the entropy; and then using equation (A.3) to find the force. An outline of this method and an indication of the approximations involved will be given in the next chapter. Methods of deriving other parameters of chain conformations will also be described.

2.2.7 Root-mean-square length of a chain

There is one particularly useful parameter which defines dimensions actually taken up by a chain, and for which an independent theoretical derivation can be given. Strictly the end-to-end distance of a chain is a vector quantity $\mathbf{l}$, which needs to be specified both in magnitude l and in direction χ. The simplest scalar quantity, which can be derived from this vector in order to give a measure of magnitude alone, is its scalar product with itself:

$$\mathbf{l}.\mathbf{l} = l^2 \qquad (2.22)$$

For an assembly of n chains – or for n instantaneous conformations of a single chain – we can calculate the mean square value of the chain length:

$$\overline{l^2} = \sum_1^n l^2/n = \sum_1^n \mathbf{l}.\mathbf{l}/n = \overline{\mathbf{l}.\mathbf{l}} \tag{2.23}$$

In order to get back to the dimensions of length, we define the useful parameter:

$$\text{root-mean-square-length} = r = (\overline{l^2})^{\frac{1}{2}} = \left(\sum_1^n l^2/n\right)^{\frac{1}{2}} \tag{2.24}$$

Fig. 2.10 shows a schematic representation of part of one chain of total length L, made up of N links of length a. We can assign a vector displacement

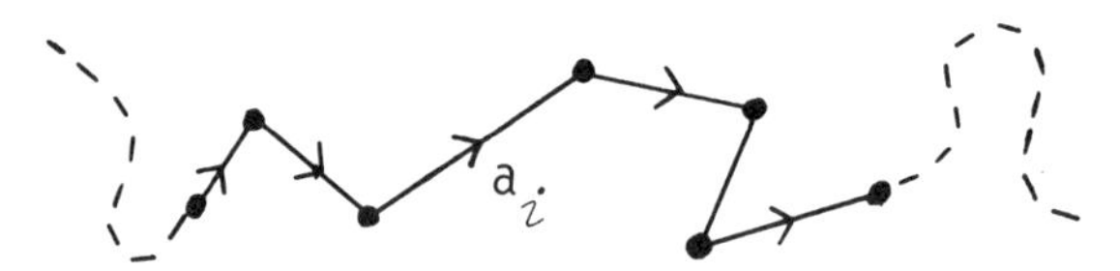

Fig. 2.10 – Part of a polymer chain with a typical link $\mathbf{a}_i$.

– $\mathbf{a}_i$ to the ith link in the chain. In order to find the vector length of the whole chain of N links, we must add the individual displacements:

$$\mathbf{l} = \sum_{i=1}^{N} \mathbf{a}_i \tag{2.25}$$

Hence:

$$\mathbf{l}.\mathbf{l} = \left(\sum_{i=1}^{N} \mathbf{a}_i\right)^2 \tag{2.25}$$

$$\begin{aligned} &= \mathbf{a}_1^2 + \mathbf{a}_2^2 + \mathbf{a}_3^2 + \ldots\ldots\ldots + \mathbf{a}_N^2 \\ &\quad + 2\mathbf{a}_1.\mathbf{a}_2 + 2\mathbf{a}_1.\mathbf{a}_3 + \ldots + 2\mathbf{a}_1.\mathbf{a}_N \\ &\quad\quad + 2\mathbf{a}_2.\mathbf{a}_3 + \ldots + 2\mathbf{a}_2.\mathbf{a}_N \\ &\quad\quad + \ldots\ldots\ldots\ldots \\ &\quad\quad\quad \ldots\ldots\ldots\ldots \\ &\quad\quad\quad\quad + 2\mathbf{a}_{N-1}.\mathbf{a}_N \end{aligned} \tag{2.26}$$

In order to obtain the mean-square-length, we need the average value taken over many possible conformations. Any individual scalar product such as $\mathbf{a}_1.\mathbf{a}_2$ equals $a^2 \cos\alpha$, where α is the angle between the positive directions of the vectors $\mathbf{a}_1$ and $\mathbf{a}_2$. For a random chain, there is no correlation between the directions of any links: consequently all values of α are equally likely, and $\cos\alpha$ is equally likely to be positive or negative. The sum over a large number of such products must be zero. This argument will apply to all the products in (2.26) except the squared terms in the first line. For the squared terms, we have:

$$\mathbf{a}_1^2 = \mathbf{a}_2^2 = \mathbf{a}_3^2 \ldots = \mathbf{a}_N^2 = \mathbf{a}^2 \tag{2.27}$$

Consequently, since we only have to add up the contribution from the first line:

$$\overline{l^2} = \overline{\mathbf{l}.\mathbf{l}} = Na^2 \tag{2.28}$$

$$\text{r.m.s. length} = r = (\overline{\mathbf{l}^2})^{\frac{1}{2}} = N^{\frac{1}{2}} a = L/N^{\frac{1}{2}} \tag{2.29}$$

We note that this length, which is clearly a measure of the extent to which the polymer chain spreads through space, increases less rapidly than the total chain length: it increases only as the square root of the number of links. The more links there are in the chain, the more compact are the conformations taken up.

2.2.8 Real polymer chains

As has been indicated, real polymer chains differ from the ideal flexible chain in various ways. This leads to the need to introduce several corrections. Thus, as with real gases, the effects of the finite volume of the units and of attractive intermolecular forces between them must be brought in: the former reduces the number of possible conformations and the latter introduces internal energy terms into the analysis.

More important, real chains are by no means made up of inextensible, freely jointed links. In one way, they are less rigid than the ideal chain since some change in the length of bonds will occur when the chain is under stress. However, this change is very small and the assumption of inextensibility is usually valid. But the angles between successive links are also difficult to change: they too suffer only a slight elastic deformation under normal forces. The only easy degree of freedom is in the angle of rotation of one bond about the direction of the previous bond, maintaining the bond lengths and the bond angles constant. This restricted degree of flexibility is indicated for a polyethylene chain, in Fig. 2.11. The chain is obviously less flexible than the freely-orienting chain: its vibrations will be less intense; the number of possible conformations will be smaller; and consequently the theoretical predictions so far given will not be valid. In writing down (2.17) many combinations of values of x will not be allowed, and they will all have different *a priori* probabilities. In applying (2.26) all nearest neighbour links will contribute equally and positvely to the sum since $\mathbf{a}_i.\mathbf{a}_{i+1}$ will always equal $a^2 \cos\theta$ where θ is now the fixed angle between neighbouring links; there will also be some correlation, and so some finite contribution, between next nearest links and so on.

However we note that if we remove our attention from a single unit of the chain, and consider instead the chain as made of longer groups of units, then we can regard the chain as being completely flexible. With a sufficient number of units taken together in a group, all correlation between the direction of opposite ends of the group is effectively lost – if one direction is fixed, the other can still be twisted into any direction. It therefore appears that it would be reasonable to describe a real chain as a flexible chain with an effective length of link greater than that indicated by the chemical formula. We shall show, in the next chapter, that for a polyethylene chain, with free rotation round cones of semi-angle 70°, the flexible link length is about 3 carbon–carbon bonds.

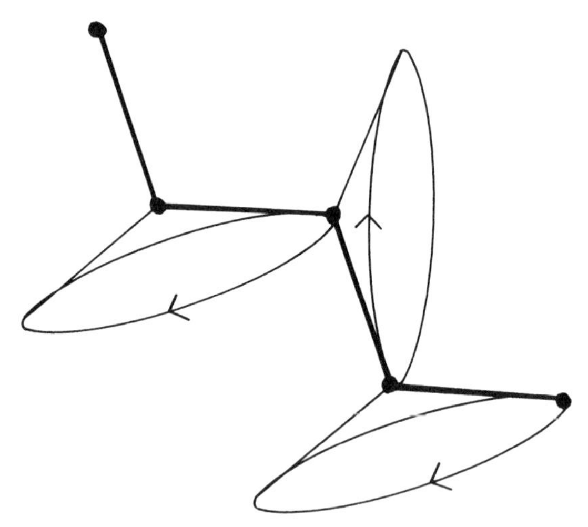

Fig. 2.11 – Freedom of rotation around a cone at a fixed bond angle at each link in a polyethylene chain.

Formally, an effective length of link a_e can be defined as the value which leads to the correct prediction of the root-mean-square-distance between chain ends:

$$r = (\overline{l^2})^{\frac{1}{2}} = N_e^{\frac{1}{2}} a_e \quad \text{or} \quad \overline{l^2} = N_e a_e^2 \tag{2.30}$$

where N_e is the effective number of links of length a_e.

Equation (2.30) could obviously be satisfied by many pairs of values of N_e and a_e. We can however add the further restriction that:

$$\text{extended chain length} = L = N_e a_e \tag{2.31}$$

Combining these equations, we find:

$$a_e = \overline{l^2}/L \tag{2.32}$$

$$N_e = L^2/\overline{l^2} \tag{2.33}$$

We also get the important generalisation of (2.29)

$$\text{r.m.s. length} = r = N_e^{\frac{1}{2}} a_e \tag{2.34}$$

The exact extent of stiffening of a chain will depend on the chemical structure of the chain – it will be influenced by the nature of the successive bonds, and by the presence of side groups. Thus a_e is a measure of chain stiffness.

Even the assumption of free rotation about bonds may not be justified. In many molecules there is steric hindrance. The typical pattern described in section 1.3.4 is one in which certain positions are favoured, and these are separated by energy barriers. Provided jumps over the energy barriers are relatively easy, the concept of an effective random link remains valid. Obviously the chemical nature of the chain will play a large part in determining the length of the random link. Since the ease of overcoming barriers depends on the temperture, the length of the random link will decrease as the temperature rises and the hindrances to rotation are effectively removed.

Quantitatively, the description of chains in terms of effective random links can only give exact values for the r.m.s. length and the fully extended length. However the qualitative predictions of the behaviour of ideal flexible chains can be expected to apply to real polymer chains, and values of other parameters calculated by using the effective random link will give at least a good approximation to the properties of the chains.

So far, we have been considering ways in which the theory of the behaviour of ideal flexible chains can be modified to take account of the behaviour of real chains by the introduction of comparatively minor quantitative corrections to account for excluded volume effects, intermolecular force effects and restrictions on flexibility. But if the temperature falls low enough, the intermolecular attractive forces and the barriers to free rotation will have more severe effects: drastic qualitative changes in behaviour – similar to the changes from gas to liquid or liquid to solid – will occur. We must go on to consider these effects. A brief account of an alternative approach to the problem of flexible chains will serve both to round off the present topic and introduce the next.

2.2.9 The concept of rotation-isomerism

The above discussion of the behaviour of flexible chains has followed the classical line, starting with a discussion of the ideal freely orienting chain, and going on to introduce the effect of restrictions on flexibility. There is another complementary approach.

This starts from the view that, due to the energy barriers to free rotation, a polymer chain can take up only a limited number of preferred conformations. A possible model of a flexible chain is then one in which there are frequent and easy jumps from one preferred conformation to another. It may be tacitly assumed that the time spent in a jump is infinitesimal, so that any instantaneous view would always show each link in the chain in a preferred position. This leads to an elegant method of counting the number of possible conformations. Thus if (as often happens) there are three preferred directions for each bond, relative to previous bonds, then the number of ways of adding another link to a chain is three. There are also three ways of adding the next link, giving a total of 3^2 conformations for two added links; and so on. Consequently, for a chain of N

links, the number of conformations is 3^N: this will be an enormous number if N is large.† Fig. 2.12 illustrates this for a short chain.

The Russian scientists, Volkenstein, Birshtein, and Ptitsyn, made extensive developments in the theoretical physics of this approach. They considered in detail the energy barriers to free rotation and the way in which they will be surmounted. The naive concept given above must be replaced by one in which there are many energy levels capable of being occupied, and in which the values of the main minima and maxima vary.

In addition, there will be correlation between the preferred positions of sequences of neighbouring bonds, since a rotation around one bond will alter the positions of the energy levels on the next bond. The polymer chain should therefore be considered as a one-dimensional, co-operative system. Birshtein and Ptitsyn describe a matrix method of studying the problem based on Ising's model of ferromagnetism. As they point out, earlier methods, such as were used in writing (2.17) are based on the assumption that the free energy A of the polymer molecule can be expressed as a sum of free energies of the individual chain units:

$$A = \sum_{i-1}^{N} A(\Omega_i)$$

where $A(\Omega_i)$ is the energy of the ith unit, dependent only on its conformation Ω_i.

Whereas, in reality, they say that the effect of interactions should be taken into account. For example, if nearest neighbour interactions are brought in:

$$A = \sum_{i=1}^{N} A(\Omega_{i-1}, \Omega_i) \tag{2.36}$$

If necessary the individual energy terms can be made dependent on the conformations of a sequence of a number of neighbouring units.

The detailed analysis by this method is unfortunately too difficult to be included in the present book. It is however worth noting that the Langevin approximation is found to be valid, with modified values of parameters. It should also be added that, despite the claims of its proponents and the undoubted value of the method, the model assumed is not perfect, any more than the freely orienting chain is a valid model. Where the jumps are relatively infrequent, the concept of rotation-isomerism will be most useful. However while a bond rotation is in progress, the height of the energy barrier for rotation of neighbouring bonds will be reduced: once again a coupling of motions causes error.

† If conformations resulting from a mere rotation of the chain as a whole are regarded as identical this reduces to 3^{N-2}; or to 3^{N-3} when N is defined strictly as the number of monomer units rather than the number of junctions. But when N is large, this correction makes no appreciable difference to the value of S, and would in any case only alter the zero position of S, which is immaterial.

When the jumps become very frequent, so that the chain is in continuous motion, the concept of barriers to rotation between isomeric states breaks down completely, and the freely orienting chain is a more reasonable model. Both methods help to illuminate our understanding.

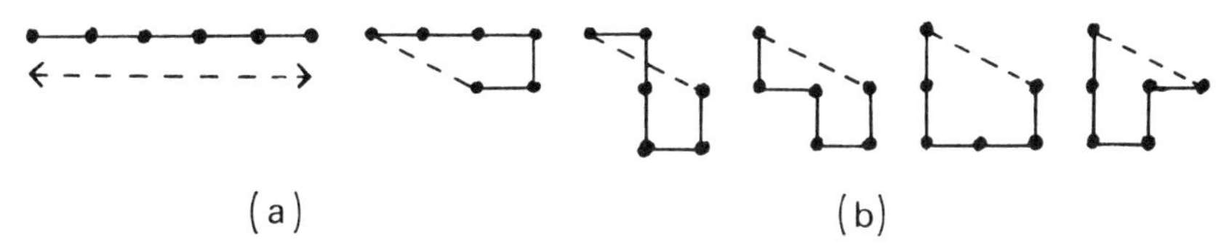

Fig. 2.12 – (a) Single conformation of a straight 6-unit chain at maximum length. (b) Increased number of conformations at reduced end-to-end distance, with three choices at each junction.

2.3 CHAIN STIFFENING

2.3.1 The random glassy chain: elastic behaviour

At low enough temperatures, the isolated chain will not possess sufficient energy to overcome the potential energy barriers to internal rotation. Viewed from the concept of rotation-isomerism, jumps between the different favoured conformations will be very rare. Instead of behaving like a lively wriggling worm, taking up all manner of conformations in rapid succession, the chain will behave like a rigid glass rod. The only remaining major degrees of freedom left to our hypothetical isolated molecule will be those of translation and rotation of the molecule as a whole. The only internal changes of conformation will be due to small elastic vibrations about the basic conformations.

If two ends of the chain are subject to small tension, as indicated in Fig. 2.13(a), the deformation will be due to the slight straining of bond lengths and angles. In terms of internal energy, as indicated in Fig. 2.13(b), this corresponds to small departures from the minimum energy position. For a given displacement of the chain ends, the deformation will be distributed among all the possible changes of bond lengths and bond angles in such a way as to give the smallest change in internal energy: this is a consequence of the minimum energy principles of thermodynamics and mechanics. When the applied force is removed, the molecule will return to its equilibrium conformation: the extension is recoverable, or to use the correct technical term, elastic.

Formally the above argument may be put in the following terms. If x is the change of a bond parameter for an individual element of the chain from its equilibrium position, the change in energy U, can be expressed by an equation of the form:

$$U = \tfrac{1}{2}[\alpha x^2 + \beta x^3 + \gamma x^4 + \dots] \tag{2.37}$$

The lower terms do not appear because we have taken the origin at the minimum energy point with $U = 0$ and $\mathrm{d}U/\mathrm{d}x = 0$. The factor $\frac{1}{2}$ is introduced for reasons which will be apparent later. It must be appreciated that the change x may be a change in bond length, a, in angle between neighbouring bonds, θ, or in bond rotation angle, ϕ.

If the deformations concerned are small, we can neglect the higher terms and put:

$$U = \tfrac{1}{2}\alpha x^2 \tag{2.38}$$

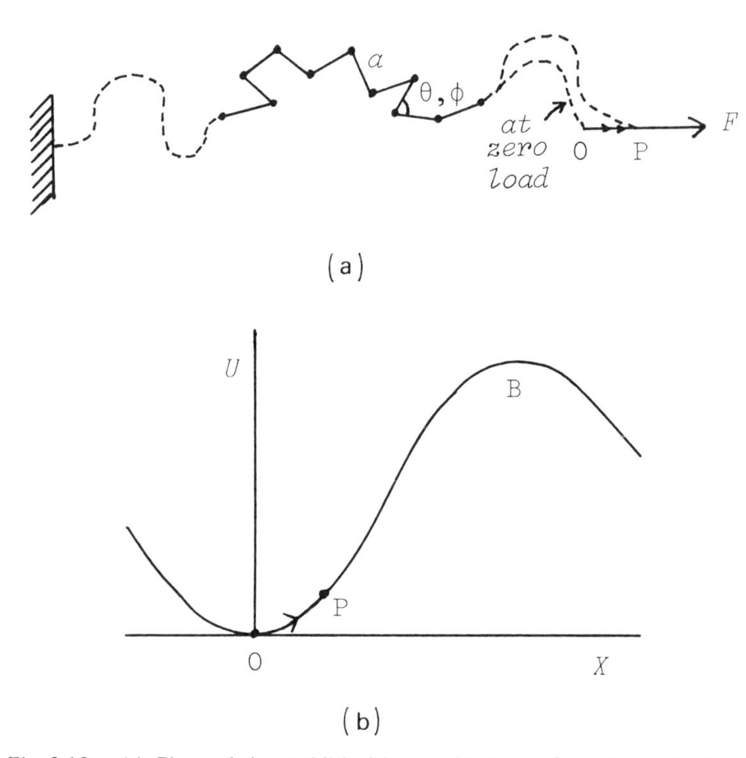

Fig. 2.13 – (a) Glassy chain established in a particular conformation and subject to tension, F, which alters bond lengths (a) and bond angles (θ,ϕ). (b) Change in internal energy U from 0 to P due to length increase X under tension F – much less than the high energy barrier B for change of conformation.

We may note that this approximation is often very good near the origin since if the minimum energy position is symmetrical the second coefficient β is zero. At some distance from the origin, the whole basis of the approximation breaks down because the curve bends over. However, provided the strains are small enough, (2.38) is valid.

If the deformation x is caused by a force p, we have, by conservation of energy during a small increase of x:

$$p\mathrm{d}x = \mathrm{d}U \tag{2.39}$$

$$p = \mathrm{d}U/\mathrm{d}x = \alpha x \tag{2.40}$$

Thus α is the spring constant for the particular mode of deformation of the elements. The terms, force and deformation, are being used generally here: if the spacings between atoms are changing, then the quantities are strictly force and distance displaced, but, if bond angles are changing the corresponding quantities would be torque and angular displacement.

For convenience, we shall now limit the discussion to the problem of a rigid polymer chain subject to an increase X in its end-to-end length, as indicated in Fig. 2.13, though the argument is easily generalized to cover other modes of deformation. If there are a total of ν individual local modes of deformation of the chain, represented by changes in bond lengths and bond angles, we must have:

$$U = \sum_{i=1}^{\nu} [\alpha_i x_i^2] \tag{2.41}$$

where α_i = spring constant for i^{th} mode of deformation; x_i = deformation in i^{th} mode.

The distribution of values of x_i must be made such that U has the smallest possible value, subject to the condition that the combination of all the individual deformations x_i gives the total deformation X. For small strains, it follows that if all the values of x_i are doubled, the value of X will be doubled and the energy terms will be quadrupled – this will apply to any geometrically permissible distribution of x_i values, and hence the form of the distribution giving the minimum total energy will be unchanged. We can therefore put:

$$\begin{aligned} U &= \sum_{i=1}^{\nu} [\tfrac{1}{2}\alpha_i (b_i X)^2] \\ &= \left[\sum_{i=1}^{\nu} \tfrac{1}{2}\alpha_i b_i^2\right] X^2 \end{aligned} \tag{2.42}$$

where the b_i values are coefficients which determine the relation between local deformations and the external deformation, as governed by the minimum energy requirement (or, alternatively, by conditions of internal equilibrium of forces).

By differentiation, as before, we obtain the force – extension equation of the chain:

$$\begin{aligned} F &= \mathrm{d}U/\mathrm{d}X \\ &= \left[\sum_{i=1}^{\nu} \alpha_i b_i^2\right] X \end{aligned} \tag{2.43}$$

Equation (2.43) can be converted into a stress-strain relation by multiplying by the appropriate dimensional parameters. The form of the equation shows that the chain will extend according to Hooke's Law, with a spring constant given by an appropriate combination of the spring constants of all the individual elements.

Clearly the coefficients b_i, which determine the contributions of the individual elements will have to be calculated specially for any particular conformation. Such calculations have been carried out for some simple regular conformations, which occur in crystalline polymers.

For example, estimates of the modulus of an extended polyethylene chain, such as the one in Fig. 2.14(a), have been made. Values of the order of 1 GPa are obtained.

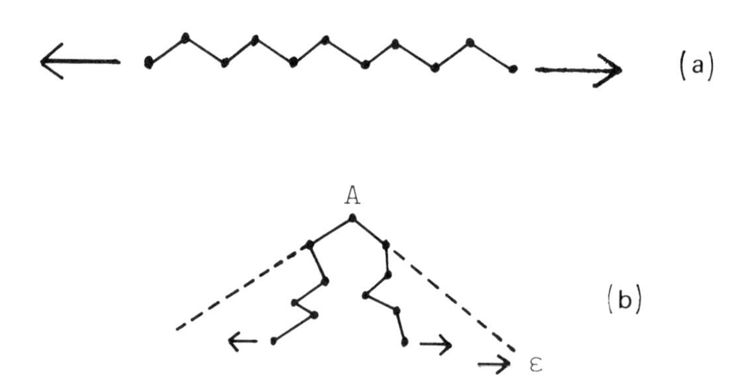

Fig. 2.14 – (a) Planar zig-zag chain under tension. (b) Short portion of random chain under tension.

The detailed behaviour of a typical random chain such as the one shown in Fig. 2.14(b) has not been worked out. We can however bring out an important general point by a crude argument. When links which are far from the centre line undergo a bond angle change, they will cause an appreciable movement apart of the chain ends even if the rest of the chain is completely rigid. Indeed it is quite easy to visualise a typical link, such as the one at A in Fig. 2.14(b), causing a strain ϵ in the whole chain of the same magnitude as the strain ϵ between the ends of the individual pairs of links. Conversely, if we distribute the whole strain ϵ among N such typical links, the strain on an individual link will be ϵ/N. On the other hand, in the fully extended chain of Fig. 2.14(a) the strain in each unit must be ϵ and for the whole chain of N units, the elastic energy must be given by:

$$U_e = N \times \tfrac{1}{2}\alpha\epsilon^2 \tag{2.44}$$

For the random chain on the above argument, the total stored energy will be:

$$U_r = N \times \tfrac{1}{2}\alpha(\epsilon/N)^2 = U_e/N^2 \tag{2.45}$$

Since the amount of stored energy, for a given strain, is proportional to the modulus, we see that the modulus of a random chain would be lower than that of an extended chain by a factor of the order of $(1/N^2)$. This drastic reduction may seem surprising at first, but becomes less so when one considers the change in moduli of wires when they are coiled into springs of various shapes. The differences may be even greater, than indicated by (2.42) if easy bond rotations can contribute to deformation of the random chain while the extended chain deformation is limited to changes in bond length and the angles between successive bonds. In real solid glassy polymers, as we shall discuss later, the situation is very different from that of the isolated chain discussed here: nevertheless the lessons learnt are valuable – they suggest that the effective length of chains free to deform independently in the solid must be small.

As the initial separation between the ends of the glassy chain increases, the links will move closer to the line joining the end-points. Consequently when the chain is strained, larger and larger strains will have to be taken by each link, until in the limit the chain behaves like the fully extended chain. The chain modulus will thus increase greatly as the initial orientation of the chain increases.

2.3.2 Plastic flow of a glassy chain

A corollary of the argument given in the last section is that, even for rather small applied tensions, the bending moments on some individual links may be high. It will therefore be relatively easy to exceed the limit of elastic deformation. Equation (2.38) will cease to be valid. The load-elongation behaviour will become non-linear.

In ordinary substances, this would coincide with the occurrence of some form of failure, such as fracture or yield, as energy barriers were surmounted and elastic deformation ceased. In polymers, the easiest barriers to overcome are the barriers to rotation around bonds. The 'failure' is therefore likely to consist of pulling the chain out of one isomeric conformation into another. This will be a non-recoverable plastic mode of deformation since, on removing the force, there will be no mechanism tending to return the link from its new equally stable, minimum energy position back to the old one. The chain has permanently increased in length, or, in a phrase used for bulk materials, it has been drawn.

The discussion in section A.4.4 shows that there will be time dependence associated with the passing of energy barriers due to the influence of thermal vibrations: consequently the yield stress will increase with increased rate of loading.

Summarising, we can say that a glassy chain when put under load will show a small immediate elastic deformation, perhaps some immediate plastic deformation if some energy barriers are completely overcome, and then a viscous deformation (or secondary, non-recoverable creep) continuing in time. This is one form of visco-elastic behaviour. Conversely, if the chain is subject to rapid extension, the instantaneously developed force will depend on the elastic response, but this can subsequently be relieved by the plastic, viscous flow: stress relaxation will take place. So far we have been assuming that stress will be relieved by rotation around bonds as in Fig. 2.15A, from one isomeric conformation of the chain to another, which can occur without destroying the polymer chain. However high stresses will also develop at some points in the chain in such a way that change in bond angles, as is Fig. 2.15B, occurs to a considerable extent, or, taking a different view of the same situation, such a bond is subject to a high shear force. Under these circumstances, if the load is large enough to exceed the region of elastic deformation, the chain will break. Whether the chain yields by bond rotation or ruptures by bond shear will depend on the relative magnitudes of the barriers of rotation and shear, and on the conformation of the chain. Ultimately, for a fully extended chain under very high tensions, bond rupture must occur anyway due to the stretching and shearing of bonds.

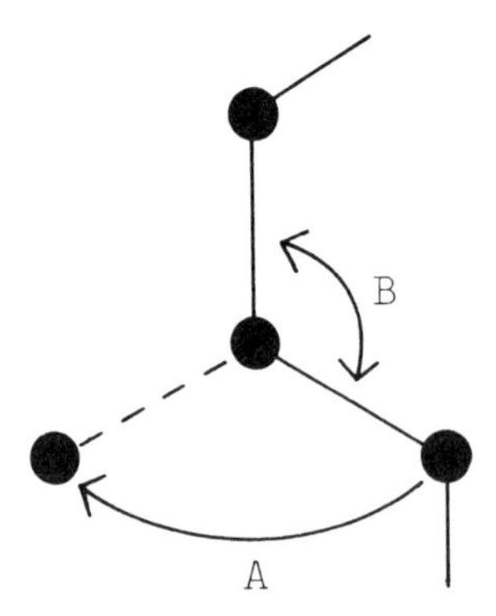

Fig. 2.15 – Bond rotation A over an energy barrier in a glassy chain merely changes conformation, but excessive change in bond angle B will snap the chain.

2.3.3 The transition from glassy to rubbery chain

As the temperature rises, the number of jumps between one isomeric conformation and another increases. The polymer molecule is transformed from a rigid glassy chain to a lively, flexible rubbery chain. We can, if we wish, envisage three regions of behaviour:

temperature rising ↓
- A glassy region – unchanging chain conformation
- B transition region – rate of jumps increasing as temperature rises
- C rubbery region – freely rotating chain

It is instructive to view the nature of the change as the temperature is reduced. We start with a free rubbery chain in a random conformation, or rather in a rapidly changing sequence of random conformations. As the temperature falls, the chain becomes more and more sluggish, since the jumps over the energy barriers become more and more difficult. Eventually, one of the random conformations is finally frozen in; the jumps are so rare that we can regard them as non-existent; the chain is glassy.

What sort of changes will occur on going through the transition? There is no change in structure: an instantaneous photograph would show identical chains in the rubbery and glassy states. There is only a change in the responses of the structure. There is no change in internal energy, for the idealized system, and consequently there can be no latent heat involved; but there is a freezing out of one degree of freedom – the freedom of rotation – for each unit of the chain. So there will be a change in specific heat. The modulus of elasticity will change from the very low value characteristic of a freely orienting chain to the much higher value characteristic of a rigid chain.

If a chain is cooled down while in an extended state under tension, this conformation will be frozen in the glassy state, and will remain (save for a small elastic recovery) when the tension is removed. This is one way of orienting a glassy polymer chain. The same effect can be brought about by brute force, as described in the last section, if big enough loads are applied, or if smaller loads are applied for a long time. In either situation, when the chain is heated up, it will contract to its equilibrium random length on heating to the rubbery state. This is therefore one way in which polymer chains can be temporarily set in special conformations. On heating through the transition temperature, the set is lost.

We shall find that transitions exhibiting these and other related properties are an important feature of the behaviour of real polymer materials. They are said to occur at the glass transition temperature.

Thermodynamically, phase transitions which show a change in specific heat, but no latent heat, are known as second-order transitions. This name is also in common use to describe the changes in polymers. It is however an open question whether the change can be regarded as a true phase transition. The description given above suggests that it is purely a time-dependent process. There are however some convincing counter arguments which suggest that there could be a sharp phase change in an isolated chain molecule. The increase in thermal vibrations in a glassy chain as temperature rises will eventually lead to transition over energy barriers; but, during the jump the barriers to transitions in neighbouring links will be lowered. Consequently, other jumps will occur: the change will be cooperative, which is the requirement for a phase change.

If there are different chemical units in a single chain, several transitions at different temperatures may occur: these can be regarded as successive reductions in the length of an effective random link.

2.3.4 Energy losses: a crude model of a chain in a viscous medium

When the isolated polymer molecule is close to the temperature at which there is a change from the glassy to the rubbery state, jumps over energy barriers will occur infrequently as sudden localised disturbances of the chain. Such disturbances will generate random vibrations which will be dissipated as heat to the surroundings. But this is a classical description of viscosity. Consequently it becomes reasonable to consider the polymer chain as moving in a viscous medium. A model of this sort would, of course, be even more relevant to a condensed system where there are many neighbouring molecules to interfere with the chain motion; but it may be valid even for an isolated chain as a means of working out the changes in the mechanical properties of the polymer chain as it goes through a transition. In addition to the internal viscosity associated with barriers to rotation, there may be effects due to collision of different parts of the molecule.

Qualitatively, we can say that at high enough temperatures the thermal vibrations of the chain are so vigorous that they are unimpeded by the viscosity, and the chain shows rubbery elasticity as already described; but as the temperature falls, the thermal energy of vibration becomes less and the viscosity causes a sluggishness of response. Finally the whole chain becomes rigid and glassy, when the delay in the response becomes much greater than experimental times. This is the classic statement of glass formation as due to a rise in viscosity to effectively infinite values.

In the intermediate region, the viscosity will cause a time-lag in approach to equilibrium. This shows up in real polymer materials in various ways – as creep under a steady load, as stress relaxation at a given deformation, as a change in stress-strain curves with rate of extension, as energy loss and a phase lag in dynamic tests, and so on. The important changes in these properties can be studied either by observing the effect of change in temperature on experiments at a constant rate, as implied above, or by experiments at different rates at a given temperature. The effects of viscosity will be similar in a fast experiment at high temperture and in a slow experiment at low temperature.

A complete study of this problem would clearly be extremely complicated physically and intractable mathematically. We shall therefore again adopt the device of examining a very crude model in order to bring out some of the essential physical features.

Consider a chain of N units as shown in Fig. 2.16(a). We have shown, in previous sections, that in the rubbery state it will have an elastic restoring force F_e, given by (2.19).

In addition there will be a viscous resistance, due to the causes mentioned above, which will be dependent on the rate of deformation and can be regarded as giving a force F_v. The chain in a viscous medium can be represented as behaving like an elastic spring and a viscous dashpot in parallel, as in Fig. 2.16(b). If there is an applied force F, the system will obey the equation:

$$F = F_e + F_v \tag{2.46}$$

If we limit attention to small deformations, then we need only take the linear term in the expansion of the inverse Langevin function, and can put

$F_e = E_r l$, where $E_r = 3kT/Na^2$. Similarly we make the simplifying assumption that the viscous element is linear, so that we have $F_v = \eta(\mathrm{d}l/\mathrm{d}t)$. Equation (2.46) then becomes:

$$F = E_r l + \eta(\mathrm{d}l/\mathrm{d}t) \tag{2.47}$$

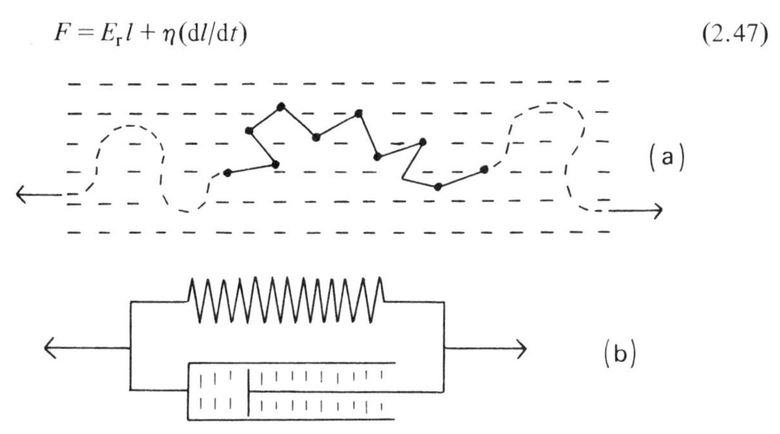

Fig. 2.16 – (a) Polymer chain in a viscous medium. (b) Elastic spring and viscous dashpot model.

This is the equation for a damped simple harmonic motion, and the responses to many different imposed conditions have been worked out. Let us take the behaviour under a constant load F_c applied at time $t = 0$. The equation may be rearranged as:

$$\eta \mathrm{d}l/(F_c - E_r l) = \mathrm{d}t \tag{2.48}$$

On integration, remembering that in the absence of load the equilibrium separation of the ends of the chain segment is zero (so $l = 0$ at $t = 0$), we get:

$$-(\eta/E_r) \log \exp [(F_c - E_r l)/F_c] = t$$

or

$$l/F_c = E_r^{-1}[1 - \exp(-E_r t/\eta)] = E_r^{-1}[1 - \exp(-t/\tau)] \tag{2.49}$$

This is a typical creep response illustrated in Fig. 2.17. The system – as a result of the assumptions introduced – is linear in its reponse to stress, and so a

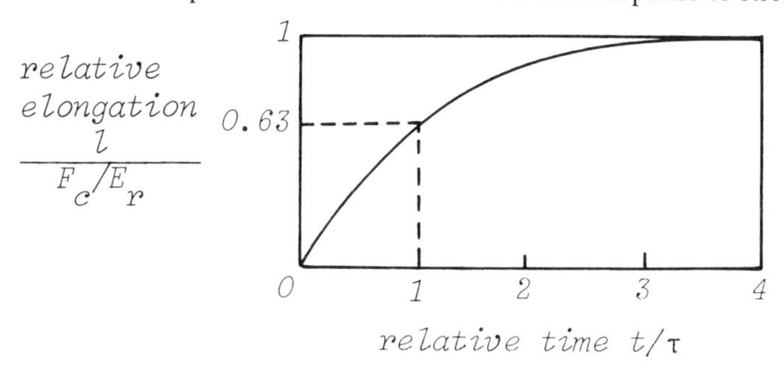

Fig. 2.17 Creep response of simple model of rubbery polymer in a vicous medium.

normalised quantity $E_r l/F_c$ analogous to a creep compliance can be plotted as a representation of behaviour at all stresses. Fig. 2.17 can be regarded in two ways – either as a direct representation of creep behaviour, or as a plot of compliance against the time at which strain was measured. If we were to plot the reciprocal (F_c/l) we would have, crudely, a plot of modulus variation with time, or for dynamic tests with the period of vibration (reciprocal of frequency). However one correction should be introduced if the viscosity is associated with the rubbery jumps from one conformation to another. In addition to the mechanism of chain extension due to rotation around bonds, there will also be an elastic deformation due to changes in bond length and bond angle. This can be added in as additional extension, with an effective compliance E_g^{-1}, and (2.49) then becomes:

$$(l/F_c) = E_r^{-1}[1 - \exp(-t/\tau)] + E_g^{-1} \tag{2.50}$$

As shown in Fig. 2.18(a), this has a marked effect at short times where the modulus then tends to a value E_g, which we see must be the modulus in the glassy state, determined by the arguments discussed in section 2.3.1.

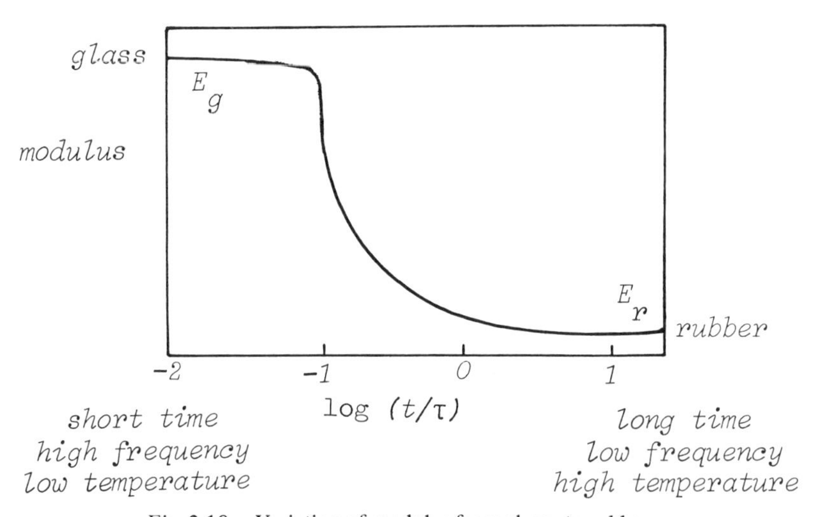

Fig. 2.18 – Variation of modulus from glassy to rubbery.

Looking at the whole form of Fig. 2.18 we see that at low frequencies (long times) the system shows the low modulus (high compliance) of a rubbery chain. But when a frequency of the order of τ^{-1} is approached, the chain begins to stiffen up, and would finally reach the stiffness of the glassy chain.

As usual in a damped simple harmonic motion, there will be a large energy loss when vibrations are imposed at frequencies of the order of τ^{-1}. For a given amplitude of applied force, we have:

$$\text{energy loss} \propto \eta\omega/(E_r^2 + \eta^2\omega^2)$$

where ω radians/sec is the frequency of vibration ($\omega/2\pi$ cycles/sec).

As shown in Fig. 2.19, the energy loss thus shows a peak at a frequency close to $\omega/2\pi = \tau^{-1}$. Thus the transition is characterised by a decrease in modulus and a peak in energy loss (or in associated quantities such as dissipation factor $\tan \delta$ or the imaginary part E'' of a complex modulus) as we pass from the stiff glassy response at high frequencies to the flexible rubbery response at low frequencies.

What will be the value of τ?

Substituting for E_r we get:

$$\tau = \eta N a^2 / 3kT \tag{2.51}$$

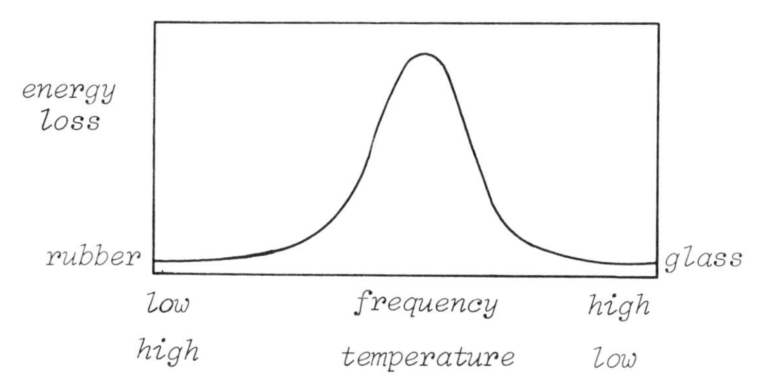

Fig. 2.19 – Energy loss peak.

Since we would not usually expect the viscosity coefficient η to rise with temperature, this shows that, due to the increasing intensity of thermal vibrations of the chain, the characteristic retardation time τ will fall as temperature rises. This confirms what was qualitatively stated above, that a change in temperature at constant experimental time will show similar effects to a change in rate at a given temperature.

Can we estimate a value for the viscous constant η? We would expect this to increase as the number of units in the chain segment increases and to be proportional to the viscosity η_s of the medium (or the equivalent medium) in which they moved. The simplest form would thus be:

$$\eta = C' \eta_s^N \tag{2.52}$$

where C' will be constant of proportionality dependent on the form and dimensions of the chain. This assumes that we can just add up the separate viscous drag on each chain unit. Substituting in (2.51) we then get:

$$\tau = C \eta_s N^2 / kT \tag{2.53}$$

where the other constant terms have been absorbed in C.

The values of C and η_s need to be determined, but the dependence on the length of the chain segment, and part of the dependence on temperature is indicated by this equation.

The crude treatment given above may be developed and corrected in various ways.

(a) The treatment is much oversimplified in that it considers only one mode of vibration of the polymer molecule, whereas in reality all the atoms in the chain are engaged in a myriad collection of Brownian motions. In fact, after replacing the chain by a spring and dashpot we have taken only the fundamental mode of vibration of the system. But if we regard the model, or the polymer chain, as made of discrete segments, as in Fig. 2.16(a), then many higher modes of vibration are possible. The effect of this is to replace the single retardation time τ, by a spectrum of retardation times. One form of development of the theory is concerned with calculating this spectrum, but as this is better discussed in relation to real polymer systems, further analysis will be left until later chapters. It should be noted that once multiple spring and dashpot units are admitted, these can be connected up in various ways: this is another complication which has to be examined later. The general effect of having a range of retardation times is to make the transition less sharp.

(b) The calculation of the constants η_s and C or η can be followed up further and related to molecular mechanisms. Once more, this can only usefully be discussed in detail in relation to real polymer systems, and not the isolated polymer chain which is the subject of this chapter. We can however make the general comment that if the viscous mechanisms are analagous to those in liquids, namely associated with jumps over barriers, η_s will decrease as the temperture rises, thus making the decrease in τ with rise in temperture somewhat more rapid than appears from (2.53). On the other hand in the hypothetical case of a polymer chain subject to gaseous viscosity η_s would rise with temperture. The analysis could be carried further to include the consequences of non-linear viscosity or elasticity.

(c) The formal analysis of the models can be taken much futher by the use of the mathematical theory of viscoelasticity. Even for the simple model of Fig. 2.16(b), there is the problem of the relation between different forms of imposed deformation. And once a multiple collection of units is introduced, the problem becomes more complicated and it becomes necessary to make use of the many interrelations which can be deduced.

It must also be remembered that there are several different forms of experiment which can be used to characterise viscoelastic behaviour, and while these often show up generally similar features there are some significant differences between them.

2.3.5 Regular conformation of a single chain

In the discussion of the irregular glassy chain given above, it has been tacitly assumed that all the isomeric conformations are equivalent or, in other words, that the energy diagram for rotation shows several (usually three) identical maximum and minimum values as in Fig. 2.20(a). The chain is then equally likely to be frozen in any conformation in the glassy state.

But suppose the energy diagram shows different minimum energy levels. In particular, suppose one conformation of the links has an energy much lower than any other, as in Fig. 2.20(b). As the chain cools down, it becomes more and more likely that the links will take up this particular conformation: there may be a substantial thermal fluctuation in the deep trough, without any jump over the barriers to another conformation. The most stable form of the chain, which will be the true equilibrium form at very low temperatures will have every link in the same conformation – the chain will be completely regular all along its length. We may note furthermore that the final formation of such a chain will be somewhat analagous to crystallisation. Jumps from the higher energy minima will occur more rapidly than the reverse jumps and so the regular pattern of the chain will gradually grow in the same way as the rate of attraction of molecules to sites on a growing crystal exceeds the rate of loss. Accompanying these jumps to favoured positions there will be a lowering of internal energy and so there will be a latent heat. The regular chain will have a different structure to the random chain.

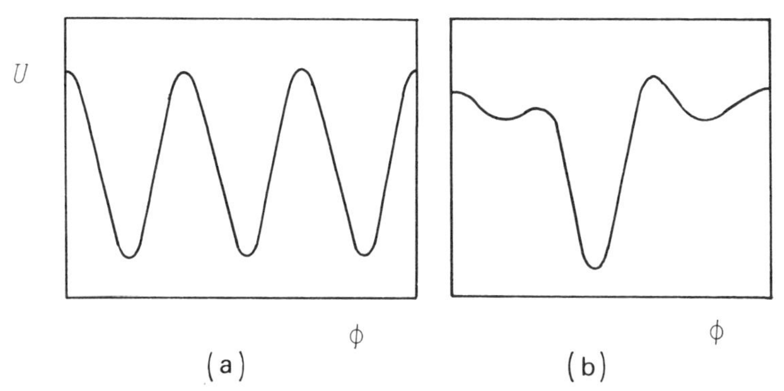

Fig. 2.20 – (a) Energy diagram with equal minima. (b) Variable minima.

Whether or not a regular chain will be formed when a rubbery chain is cooled depends on the magnitude of the differences between the energy minima and on the rate of cooling. If the differences are small, then, at any temperature at which there is an appreciable rate of jumping from one to another, there will be a statistical distribution between the various conformations: thus it becomes impossible to achieve an adequate rate of approach to equilibrium at any temperature at which the lower energy form would be predominant at equilibrium.

Consequently if the differences are small enough, an almost completely random form must be frozen in at any finite rate of cooling. If the differences are somewhat larger, then there are likely to be sequences of units in the regular form with occasional units in a less favourable conformation at the temperature at which the rate of change becomes vanishingly small. For larger differences still, the regular form will be able to be formed almost perfectly by a period of annealing at a suitable temperature. However even when the differences are large, it will be possible to freeze in irregular conformations if the rate of cooling is sufficiently rapid: this process is analagous to the formation of an inorganic glass by cooling a substance under conditions in which it is not able to crystallise.

The particular form of regular chain will depend on the particular form of the minimum energy conformation. It may be an extended chain configuration, as in polyethylene, or it may be a helical form. The chemical reasons for this were discussed in section 1.3.4. These forms will be the stable forms of the molecule at low temperatures (subject to some other factors mentioned later in this Chapter). When the chain is heated up it will 'melt' and the rubbery random chain will be formed. There will usually be a contraction in length when this happens.

We have assumed here that the preferred minimum energy conformation will be a geometrically regular and repetitive chain. This will be so if the constitution of the polymer molecule is regular and repetitive, both chemically and stereochemically. But if it is an irregular chain, such as an atactic polymer molecule, the preferred conformation will be irregular: it will be a geometrically random chain because of the random chemical constitution of the molecule. The formation of a random chain in this way is quite different from our earlier discussion of the formation of a glassy chain. The glassy chain was one in which any random conformation could be frozen in and would be equally stable. The random chain we have now described is a particular preferred conformation, more stable than any other.

2.4 CONDENSED STATES OF A SINGLE MOLECULE

2.4.1 The effect of attractive forces

The sequence of argument about polymer behaviour in this chapter started with a discussion of ideal flexible chains. Now we are in the midst of discussing the effects of various limitations on the freedom of individual chains. In the last section, we were mainly concerned with what happened when the freedom of internal rotation of the chain was impeded – when the chain itself became stiff – but (except partly in section 2.3.4) we did not discard the assumption that there were no effective 'intermolecular' forces between remote parts of the chains. We ignored van der Waals' and similar forces. The only interactions between the units of the chain were taken to be the direct bonds between the atoms, together with some influence from very near neighbours; these are all effects which are within an effective repeat unit of the chain; there are analagous effects within small molecules. But just as we have to bring in the influence of attractive, van der Waals' forces between small molecules to explain liquifaction and crystallisation, so must we take into account the forces between those remote parts of a chain which happen to come close together as the chain takes up its many conformations in space.

If the temperature of the polymer molecule is high enough, and provided it has not chemically disintegrated, the attractive forces will be overcome by the vibrations of the system, and the molecule will behave like a rubbery gas in the manner already described. Thermodynamically, the advantage of the rise in entropy coming from the greater freedom of the molecule when it is in a gaseous state more than counterbalances the rise in potential energy due to breaking the attractive forces. But if the temperature is lower or the attractive forces are higher, the molecule will be pulled together under the influence of the attractive forces. It will then form a condensed state of the single molecule.

2.4.2 The rubbery liquid molecule

We first discuss a molecule which has condensed under the influence of van der Waal's forces while still retaining chain flexibility. The process is analagous to the condensation from vapour to liquid. Instead of taking up all possible conformations in space, the chain will have condensed into a limited volume. This is not quite the densest possible packing of the chain units, since, as in a liquid, there will be sufficient free volume to allow the units to slide past one another. The exact extent of this freedom will depend on the balance between the attractive forces, determining the internal energy, and the forces of disorder, determining the entropy. As the temperature falls, the entropy term (TS) in the free energy diminishes, and so the freedom of movement becomes less: the viscosity rises. In addition to the overall reduction in volume brought about by the attractive forces, it is likely that these forces will also cause some local parallelism of the chains.

In the absence of any applied forces, the condensed molecule will take up a spherical form for two reasons. Firstly, as with any liquid, this is the form with minimum surface energy. But secondly, for long-chain molecules, it is the form of greatest entropy since it includes the greatest number of possible conformations. Any distortion of the molecule, for instance into a elongated shape, will reduce the number of possible conformations. There will always be elastic restoring forces resisting any change from the spherical shape: these will be larger than in simple liquids. There will also, of course, be elastic forces resisting any change of volume: these will be of the same order of magnitude as in simple liquids.

In constrast to simple liquids, we can consider a hypothetical experiment in which the chain ends are pulled apart. This will lower the entropy, because of the reduction in the number of possible conformations. As in the rubbery gas, there will be a load-elongation relation for the pulling apart of chain ends. Qualitatively, the behaviour will be the same: zero mean tension when the chain ends are held coincident, and a rising tension as the separation increases. To go further, we make the assumption that the relative effect of the condensation in reducing the number of available conformations is independent of the separation l of the chain ends. In other words, for all values of l the ratio W_l/W_g is constant, where W_l is the number of possible conformations in the rubbery-liquid state and W_g is the number in the number in the rubbery-gas state. While this assumption has not been rigorously justified, it represents a plausible approximation. For two different values of l we thererfore have:

$$W_{l,1}/W_{l,2} = W_{g,1}/W_{g,2} \tag{2.54}$$

Since the entropy is a function of W_1/W_2, there will be the same entropy changes in the two states for the same change in chain length. Consequently, the stress-strain curves, derived from the entropy changes, will be identical. Equations (2.19) and (2.20) will also apply to condensed chain molecules, provided that they are freely orienting and are therefore in the ideal rubbery-liquid state. The effects of restriction on freedom of orientation and the concept of the effective random link would also be expected to apply in the condensed state.

There is however one important difference. The viscous forces between different parts of the chain must be taken into account, since in pulling the chain ends apart in the rubbery-liquid molecule, it will be necessary to drag individual chain units past one another. The molecule will in fact behave like a spring placed in parallel with a viscous dashpot, as discussed in section 2.2.4, and will display primary creep. The deformation will occur slowly in time due to the viscous resistance, but, on release of the force, the elastic recovery mechanisms will be effective and so creep recovery will occur.

Not only will the elastic behaviour of the chain – its entropy-dependent spring characteristics – be modified by the viscous drag; the viscous behaviour itself will be modified by the polymeric form of the molecules. Suppose a chain is placed in a tube with a pressure difference causing flow. The units cannot move past one another independently: they must drag other units with them and pull portions of the chain into extended conformations, thus giving rise to elastic forces. And if the tube is very fine, the elastic forces set up, as described above. by the overall elongation of the chain may generate sufficient frictional resistance to prevent flow.

We must also note that there will be a whole spectrum of characteristic times associated with the response of the system. Suppose a condensed molecule is subjected to a change of shape. Instantaneously (or rather, for the smallest possible time in which a deformation can be imposed at all) the response can only occur due to a change of bond spacings and bond angles within the region of small elastic deformations. Large stresses will result. Then jumps between different favoured conformations will occur and the stress will be relieved:

stress relaxation will be occurring. Due to the random nature of jumps between states, this will necessarily be time dependent. Furthermore, the relieving of local stress by one jump will usually put more stress on other links, thus increasing the time-dependence by introducing a sequential set of delayed jumps. At first, the jumps will involve only very localised rearrangements of position, but, as time goes on, further relieving of stress can only occur by large sections of chain moving so that more favourable (more random) chain conformations are assumed. These bigger movements will be slowed down by the viscous drag, and will thus be characterised by longer relaxation times.

Both mechanisms mentioned in the last paragraph are associated with jumps over energy barriers, and so will obey relations similar to those discussed in section A.4.3. But, while the former is related to barriers to internal rotation, the latter is related to barriers to the flow of units past one another over barriers due to the attractive van der Waals' forces. With appropriate definitions of the parameters for single or composite units of the chain, equation (2.52) would be expected to give the right form of behaviour for the dashpot in Fig. 2.16, though because of the variety of mechanisms the single dashpot should be replaced by a whole series with different parameters.

2.4.3 The condensed glassy state: two causes

On cooling, the rubbery-liquid state described in the previous section can 'seize-up', become hard, rigid and glassy, by either of two mechanisms – chain stiffening or cross-linking – or by a combination of the two. Because of its origin in a condensed rubbery state, the chain will retain a disordered, entangled, but dense state, with some degree of correlation, or local parallelism, between the paths of neighbouring chains, and possibly some local variations in packing from place to place.

If the chain flexibility is lost, then the individual chains will become glassy. Their rubbery elasticity will be lost, and so will the possibility of simple viscous flow since the chains will be rigidly interlocked and entangled with one another. The molecule will thus behave like a globule of glass. The cause and effects of this mechanism are likely to be generally similar to those described in sections 2.3.1 and 2.3.2 modified because of the condensed, solid state of the system. Under load, the initial response will be stiff elastic deformation with a high modulus: insofar as a number of links in the chain can deform effectively independently of neighbouring chains, the modulus will be lowered by the magnification of forces on elements remote from the line joining the ends of the effective free length of chain, as described for the isolated chain. A complete prediction would require an analysis of the exact distribution of stresses through the network and the way in which it results in changes in bond lengths and angles. If the load is appreciable, and the temperature is not too low, some units will yield by jumping from one conformation to another: plastic flow, or secondary creep, will occur according to the prediction of equation (A.32).

On raising the temperature, jumps between different conformations will become easier, and the globule will develop transitional properties, ultimately changing to the rubber state. The characteristics of this transition will be as described in section 2.3.3 for the single chain. It will be accompanied by changes in expansion coefficients, specific heat and so on, and will be able to be used to give a temporary set.

The second cause of glass formation is that, at lower temperatures, the relative movement of neighbouring units becomes infrequent and sluggish. The attractive forces between chains cannot be overcome: this will be particularly likely in a material like polyacrylonitrite with its highly polar -C≡N side groups giving strong interactions. The viscosity becomes so high that no appreciable deformation is possible in a finite time. This is exactly the same mechanism as occurs in conventional glass formation. It is true that there is some academic discussion over whether the causes of vitrification are purely kinetic or not – whether ordinary glass which is a viscous liquid, on our time scale, at over 1000°C would seem like a viscous liquid on a time scale of millions of years at 0°C, or whether below some critical temperature the viscous flow mechanisms would be inoperative even at infinite times. From a practical point of view, this argument is unimportant. It is also sometimes not very important in practice in polymers whether the transition to the glassy state is caused by chain stiffness or chain attractions: both mechanisms can operate together to provide barriers, which below the glass transition temperture cannot be overcome by the tendency to disorder and vibration. There is often a very strong co-operative effect, with the inhibitions to movement in one unit increasing the resistance to movement of neighbouring units, and, as a result a single transition will occur over a fairly narrow range of temperature. The consequences are similar as a result of both causes, although from the view-point of chemical structure one may be able to allocate more of the source of stiffness to one cause or the other.

With some materials however, the structure may loosen up in a variety of ways at different temperatures due to different causes, and there will be several transitions. Thus side-chains may loosen up at one temperature; main-chain links at another temperature, or at several temperatures if there are different main chain units; and inter-molecular forces at another temperature.

One particular example of the last effect occurs when there are some specific cross-links of intermediate strengths, such as the hydrogen bonds which form between -CO.NH- groups at intervals along a nylon molecule. At low temperatures, these will effectively cross-link the network and will stiffen it up even though there are flexible $-CH_2-$ chain segments between the cross-links. The structure is illustrated in Fig. 2.21. At higher temperatures, the vibrations of the molecule overcome the cross-links: the units joined by hydrogen bonds can be regarded as dissociating, and the modulus falls. Because each cross-link is almost independent of the others, there is little co-operative effect, and so the dissociation is spread over a rather wide range of temperature, in the same way as the dissociation of diatomic molecules into monatomic ones. At a much lower temperature, the chains between cross-links may become rigid in another quite separate transition. The 'glass transition' is split into two separate parts, one occurring at about −100°C and the other at about 50°C.

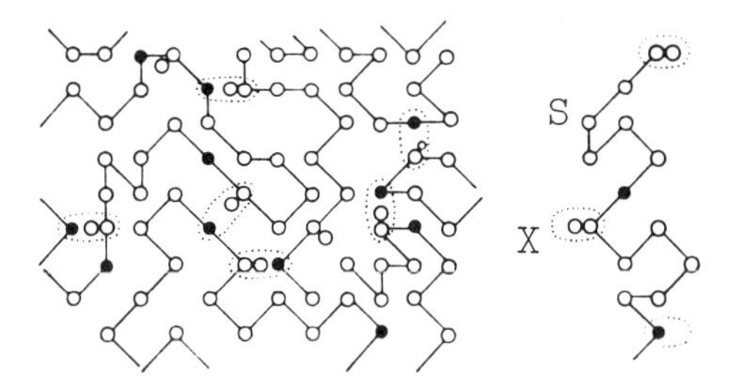

Fig. 2.21 – Schematic view of amorphous polymer structure in a chain such as nylon, which has chain segments, S, which becomes soft and flexible above a low temperature transition, and chain units X, which form cross-links dissociating at a higher transition temperature. (Hearle, *J.Pol.Sci.*, C, **20**, 215, 1963.)

2.4.4 Crystallisation

While some substances composed of small molecules vitrify, a much commoner effect is for the molecules to pack together in the regular lattice of a crystal when the vibrations become insufficiently energetic to maintain the mobility of the liquid state. The same thing can happen with polymers. As indicated in Fig. 2.22(a) sections of chain can come together and form a stable crystal lattice. What is more a single polymer molecule can crystallise by folding back on itself, as in Fig. 2.22(b). Just as an ordinary crystal can form either from the liquid or vapour phases, so could a single molecule crystallise on cooling either from the rubbery-gas state in which the chain is taking up all possible conformations in space or from the condensed rubbery-liquid state.

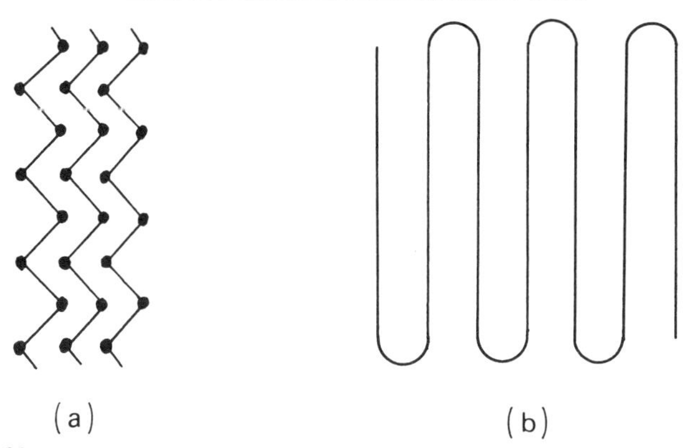

Fig. 2.22 – (a) Packing of segments of a polymer molecule in a crystal lattice. (b) Crystallisation of a single long-chain polymer molecule by folding back and forth.

There are several factors which help to promote crystallisation in polymers. Firstly, the chain should be regular, both chemically and stereochemically, so that if neighbouring units fit together in one place, then they must also do so all along the chains. Secondly, the attractive forces should be strong, so that crystallisation will be helped, for example, by hydrogen bonding. Thirdly, if the attractive forces are inherently only weak Van der Waals' forces, they will be most effective if the shape of the chain is such that good close-packing is possible. Ideally, when the individual chain is in a preferred minimum energy conformation, the chains should pack closely together in order to minimise the inter-chain energy; a low value of energy of chain packing should be accommodated with little distortion of the chain itself. Fourthly, if the chain itself is fairly stiff then the tendency to vibration, leading to break-up of the crystal lattice, will be weaker. Fifthly, if the chain has itself settled into a regular conformation of minimum energy, then all the attractive forces have to do is to bring the chains together. It is not necessary for all these conditions to be perfectly satisfied: a strong tendency to crystallisation for one reason will make up for a weaker tendency in other respects.

The detail crystallography is one area of study in polymer physics. The perfect lattice may be regarded as made up of infinitely long chains packed side-by-side. Under these circumstances, the conformation of each individual chain is often very close to the preferred conformation which it would take up in isolation, with only slight distortion due to intermolecular forces. The intermolecular forces then mainly determine the way in which the chains pack together. Sometimes intermolecular forces are more important; and some polymers show different crystal forms under different conditions. For instance, extended chain lattices with attractive forces satisfied between neighbouring chains and helical chain lattices with attractive forces satisfied within a chain may be stable at different temperatures; the extended chain lattice may also be induced by tension. We shall return to these considerations in a later chapter.

2.4.5 Equilibrium crystalline form of a single chain

In order to see how a single chain might crystallise, we will assume for convenience that the chain has one strongly preferred regular extended conformation, as in Fig. 2.23(a). The argument is easily modified to make it more general. We will take the form of Fig. 2.23(a) without thermal vibration as an arbitrary zero of free energy. Under other conditions the free energy A will be given by $(U\text{-}TS)$, where U is the internal energy difference and S is the entropy difference from the chosen reference state. At high temperatures the entropy contribition will be dominant, and will overcome the increased internal energy of disordered chain forms. But at lower temperatures, the internal energy effect will be relatively more important, and we assume that a regular conformation is preferred to a disordered one.

The chain may now lower its free energy by folding back on itself as in Fig. 2.23(b). The internal energy is reduced by satisfying all the attractive forces between the two portions of the chain which are in contact.† But the fold itself will be a less favoured form, and so will cause some rise in internal energy. The net effect is that:

$$\text{change in internal energy as a result of one fold} = -(L/2)u_c + u_f$$

where L = length of chain; u_c = energy required to separate unit lengths of two chains; u_f = increase of energy due to one fold.

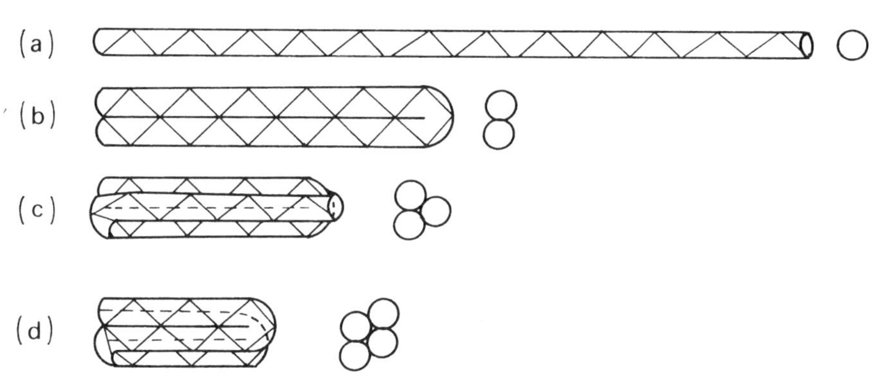

Fig. 2.23 – (a) Regular extended chain conformation, length L. (b) One fold, length $L/2$, one contact; (c) Two folds, length $L/3$, three contact. (d) Three folds, length $L/4$, five contacts.

If $(L/2)u_c > u_f$, the chain energy will decrease on forming one fold, and so this will tend to occur. In other words if the chain has a length greater than $2u_f/u_c$, folding will be favoured.

The chain could bend back on itself a second time as in Fig. 2.23(c). Assuming what is not necessarily true, namely that the attractions at the three chain contacts and the fold energy at two positions are all the same as before, we get:

$$\text{change in internal energy on forming two folds} = 3(L/3)u_c + 2u_f$$
$$= -Lu_c + 2u_f \quad (2.56)$$

Both terms here are double those in (2.57): if the tendency to form one fold was favourable, then two folds must be twice as favourable. What about three folds, as in Fig. 2.23(d)? There are then five lines of contact, and we have:

$$\text{change of internal energy on forming three folds} = -5(L/4)u_c + 3u_f$$
$$= -1.25Lu_c + 3u_f$$
$$(2.57)$$

† Instead of considering the favourable lowering of energy due to satisfying attractive forces, it is common to consider its converse the unfavourable surface energy – in other words the reference state is taken as one in which there is no free surface.

In this expression the first term is relatively smaller than it was in the previous expressions, and so three folds will not necessarily be preferred over two folds. The exact condition for preferring three folds is:

$$(-1.25Lu_c + 3u_f) < (-Lu_c + 2u_f)$$

or

$$(L/4)u_c > u_f \;;\quad L > 4u_f/u_c \tag{2.58}$$

In summary, on this simple argument it is possible to predict that chains of length less than $2u_f/u_c$ will prefer not to fold – the advantage of satisfying the attractive forces will not overcome the inconvenience of the fold; chains between lengths $2u_f/u_c$ and $4u_f/u_c$ will prefer to form two folds; chains longer than $4u_f/u_c$ will form three or more folds. It is possible to continue this argument indefinitely to predict the number of folds which will be formed. If we were to follow a sequence of packing, we should find that certain numbers of folds will be less favourable, because they lead to the formation of a new layer with few interchain contracts. To do the analysis properly, as was done by Lindenmeyer and his colleagues, it is necessary to consider the precise details of the interactions between chains and the precise details of fold forms. Thus certain lines of contact between chains will be more favourable than others, so that the values of u_c will not be identical for all lines of contact, and the fold energy u_f will be influenced by the direction of folding.

For a very long chain, which can form many folds, the following approximate argument may be used. Let the number of folds be $(n - 1)$, so that there are n chain segments. If the area occupied by a single chain is a square of side z, then the area of cross-section of the whole crystal is nz^2. If the crystal itself is square in cross-section, as in Fig. 2.24 its perimeter will be $4\sqrt{n}z$, and the number of molecules on the perimeter will be $4\sqrt{n}$. We assume that unit length of chain in the interior of the crystal has a reduction in internal energy of $-u_i$, while unit length on the perimeter has a reduction of only $-(1-p)u_i$: p is the fraction of the chain which is exposed at the surface and could be put equal to 1/4. We then have:

change in internal energy due to crystallisation with folds

$$= U = \underbrace{-n(L/n)u_i}_{\text{all units}} + \underbrace{4\sqrt{n}(L/n)pu_i}_{\text{correction for surface units}} + \underbrace{(n-1)u_f}_{\text{folds}}$$

$$= -Lu_i + 4Ln^{-1/2}pu_i + (n-1)u_f \tag{2.59}$$

At equilibrium, we must have:

$$\mathrm{d}U/\mathrm{d}n = -2Ln^{-3/2}pu_i + u_f = 0 \tag{2.60}$$

$$n = (2Lpu_i/u_f)^{2/3} \tag{2.61}$$

If $p = 1/4$,

$$\text{fold length} = L/n = (4Lu_f^2/u_i^2)^{1/3} \tag{2.62}$$

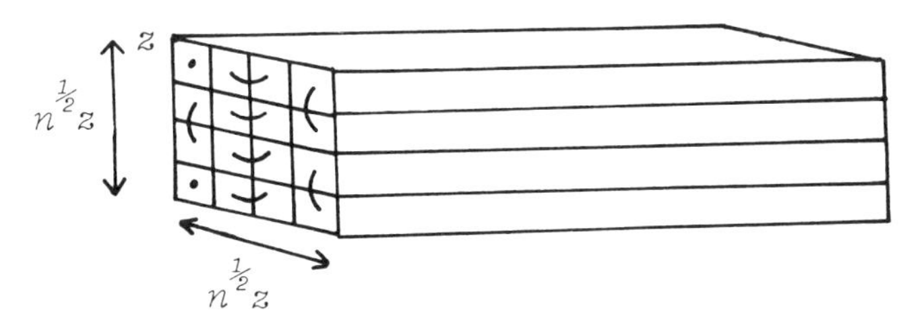

Fig. 2.24 – Model of chain folded crystalline polymer chain.

Obviously this expression will be modified in its numerical factors if the geometry of the crystals is different, and will need correcting in other ways to account for specific interactions. Nevertheless it does give a general indication of the form of the arguments which can be used to predict fold lengths, and of the way in which the equilibrium fold length for a single molecule will rise as the chain length rises and as the fold energy increases, and decrease as the energy of interaction increases.

There are other aspects of the crystallisation of a single chain which are relevant to the way in which real polymer systems will behave. For instance, folds are likely to be somewhat bulky due to the less favourable chain conformations. They are therefore very likely to pack together in a staggered form as in Fig. 2.25. These forms are observed in polymer single crystals, and will be discussed in a later Chapter.

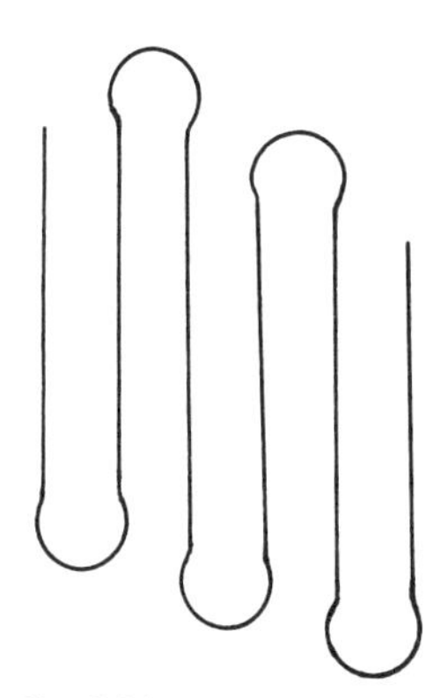

Fig. 2.25 – Bulky folds will lead to staggered arrangement.

So far we have ignored the effects of entropy. The quantities u_c, u_i and u_f in the above analysis should really have been the free energy differences and not the internal energy differences from the fully extended reference state. While the internal energy effects do dominate the situation and give rise to the major features of behaviour, there are some secondary aspects which are associated with entropy changes. Thus entropy differences may swing the balance from one form of crystal lattice (or isolated chain formation) to another of slightly different internal energy: since the entropy contribution $-TS$ becomes more important at higher temperatures, a first order transition in crystal form may result.

Furthermore we see that if u_i and u_f are modified to include even fairly small but different entropy contributions, then the ratio u_f/u_i will vary with temperature. Consequently the equilibrium fold length will also be a function of temperature. These matters have been the subject of considerable theoretical and experimental investigation, to which we shall return later. It has been suggested that the freedom for certain types of oscillation within a polymer crystal depends on the length of the chain in the crystal: this further complicates the problem by making the entropy a function of the fold length itself.

We must also appreciate the general point that, except at very low temperatures, entropy effects will cause some disturbance and vibration within the crystal, subject to the restrictions of the internal energy forces. In crystals of small molecules, entropy also dictates that a fraction of the total number of molecules in a closed system should be present as vapour, continually evaporating and condensing in a dynamic equilibrium. This cannot happen with a single polymer molecule: the chain units are all linked together. Nevertheless it does seem possible that at the surface there may be some increased disorder and looseness of packing, or some short chain end segments projecting in disorder from the surface. There may also be defects moving through the crystal, raising the entropy without too much disturbance of the potential energy, and contributing to the mobility of whole chains. It is important to note that the relative movement of two sections of chain in a crystal is inherently unlikely because it would necessitate the co-operative upsetting of the intermolecular attractions of many chain units: but a defect, like the one in Fig. 2.26 can move easily, since each individual step concerns only a few chain units. A succession of defect motions will move a whole section of chain an appreciable distance, just as a carpet is more easily moved along the floor by pushing ridges across it than by

Fig. 2.26 – A defect, such as one due to an extra unit, within a crystal lattice may be mobile.

dragging it all together. Defect mobility is thus a way in which a polymer crystal can most easily reach its true equilibrium state, or in which it can respond to applied forces.

2.4.6 Kinetic effects in crystallisation: metastable equilibria

In the previous section, we were concerned with the prediction of the true equilibrium crystalline state of a single polymer molecule; but, in order to reach such a state, there must be a possible route, and the system must move along it at an adequate rate. Many states of truly low free energy will be isolated by such high energy barriers that they can never be reached, at least in any measurable time. The starting point and other external influences operating during the approach to equilibrium will also play a part: any system will follow a route leading locally to a lower free energy – this may well lead away from, rather than towards, the true position of minimum energy, and the system will finish up in some deep trough of metastable equilibrium.

A geographical analogy may help. The sources of the Don and the Volga are not far apart, but because of the topography, their waters finish up in different and separate troughs of minimum gravitational potential energy. The Don leads to the Black Sea, which is connected through the Mediterranean to the energy level of the oceans. The Volga flows to the Caspian Sea in an inland depression below sea level. The oceans are, in fact, in a metastable state with respect to the Caspian Sea, but the waters cannot flow over the mountains in order to establish true equilibrium.

On this continental scale, and on the scale of everyday things, we are used to the effective permanence of metastable states. What is an usual in polymers is to find so many and such important metastable states at the fine scale of molecular dimensions. Because of the way in which all the units are linked together, it is very difficult for a polymer molecule to escape from a deep state of metastable equilibrium: the molecule is imprisoned in its own chains!

To come down to details, let us consider a very long single polymer molecule which has condensed to a globule in the rubbery-liquid form. As the globule is cooled, conditions will become favourable for crystallisation. At various points within the whole system, sections of chain may be attracted together or may fold back on themselves giving the nuclei of crystals. From each nucleus growth will take place, until finally there is such interference that no further change is possible: a state which is strictly metastable, but, on any realistic basis, is truly stable will have been reached. The exact form of this final state will clearly depend on the detailed form of nucleation and on the pattern of growth, influenced as this will be by temperature and stress differences consequent upon the crystallisation or imposed externally. These are matters which we shall take up in detail when we discuss polymer crystallisation in Chapters 7 and 8. The detailed morphology is too complicated to discuss here, determined as it is by kinetic effects, by the competing rates of different processes acting simultaneously.

The final state is far from being a perfect single crystal. At the time of reaching effective stability the structure will be polycrystalline with a considerable degree of disorder trapped within the system. There are many ways of looking at this: the disorder may be regarded as due to an accumulation of crystal defects, or if appreciable lengths of chains are left trapped between separate crystals they may be better regarded as regions of amorphous material. Because of the diversity of form, this aspect of the subject is better discussed in relation to real crystalline polymers.

Finally we may point out that kinetic effects may prevent crystallisation completely. If a molecule is very rapidly cooled, it may be set into a glassy state before there is time for crystallisation to occur. We can therefore recognise two different glassy states of polymers: some polymers become truly glassy because there is no ordered state of lower free energy, while others reach a glassy state because the rate of crystallisation was not fast enough to be effective in the imposed circumstances. The latter materials will be able to be crystallised by annealing. Similarly any particular pattern of crystallisation may be changed by subsequent annealing at a higher temperature.

2.5 ASSEMBLIES OF POLYMER MOLECULES

2.5.1 The polymer gas and polymer solutions

Apart from a few passing comments, the whole discussion so far in this chapter has been about behaviour of single polymer molecules – albeit so long that the single molecule is truly a massive system. In principle, there is no reason why such systems should not be realised experimentally; but, in practice, real linear polymer systems are assemblies of many molecules of limited length. With three-dimensional network polymers, it is, of course, easy to obtain a real massive single molecule. In order to link this chapter to the rest of the book, we shall now discuss briefly other factors which must be taken into account in these real systems.

The simplest system theoretically would be a polymer gas, with the molecules scattered in a large volume at low pressure and interacting only slightly with each other as in Fig. 2.27(a). At a high temperature, the molecules would possess translational and rotational energy as a whole, and, in addition have many modes of internal vibration. As the temperature fell, these internal vibrations would gradually reduce in intensity, and particular modes of vibration would become more important. Some of these interactions would be qualitatively different to the interactions of small molecules: the chain molecules could tangle up, as in Fig. 2.27(b), and thus temporarily link two molecules together as one. With still more entanglement, larger clusters of molecules would become common; ultimately, all the molecules would be linked and the system would no longer be gaseous.

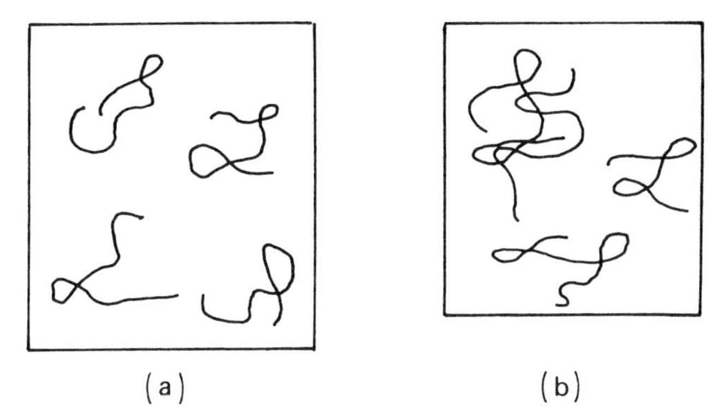

Fig. 2.27 – (a) A polymer gas, or dilute solution. (b) At a greater concentration, chains may become entangled.

The polymer gas is not a system for practical study, but effects very similar to those described in the last paragraph can be obtained in polymer solutions. The only difference is that instead of existing in empty space the chain molecules are dispersed in an environment of solvent molecules: clearly, the interactions with the solvent must be taken into account, but conditions can be found in which the chain molecules take up conformations similar to those which would be predicted for the gaseous state. In very dilute solution, we can therefore examine polymer conformations experimentally. As the solutions become more concentrated, the effects of entanglements between separate molecules will become more important. It is easy to see that the viscosity of a solution will be affected by the volume occupied by a polymer molecule and by the entanglements. Furthermore, viscous drag will tend to pull chains out into rather extended conformations. But as we have shown, there is an elastic force between the ends of an extended chain: consequently, the behaviour of polymer solutions must demonstrate another manifestation of visco-elasticity. In very dilute solutions the viscous effects will be dominant: in concentrated solutions, the elastic effects become more important. We shall take up the study of polymer solutions in the next chapter.

2.5.2 Condensed state: the influence of chain ends

If we were able to examine in detail a small region of a condensed polymer system, whether in the rubbery-liquid, glassy, or crystalline state, the odds are that we could not tell whether this was a region of a single molecule or of a multi-molecular system. Only if more than two chain ends were observed simultaneously, could we be certain that there was more than one chain molecule! More generally, we can say that in a multi-molecular system, a certain finite fraction of regions of a given size would contain chain ends. What effect will these chain ends have?

In the rubbery-liquid state, with weak interactions and flexible chains, flow will be easy from chain ends and whole molecules will be able to slide past one another. If a force is suddenly applied to the system, its reaction can be regarded as occurring in three ways. Firstly, there will be an elastic deformation with no relative movement of portions of the chains past one another. This may be represented by the spring A of Fig. 2.28. Secondly, the molecules will deform into more extended conformation with both elastic and viscous resistance, as in the unit B of Fig. 2.28. Thirdly, whole chains can slide past one another: this

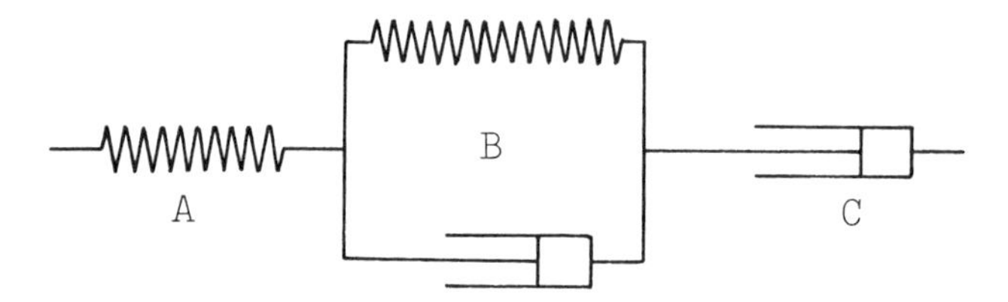

Fig. 2.28 – Representation of viscoelastic behaviour of an assembly of rubbery liquid chain molecules.

gives the viscous element C in Fig. 2.28. If the deformation stops, the force will remain because of the elastic elements, but it will gradually fall as the flow of chains past one another enables each molecule to take up a random conformation with zero elastic stress. The relative magnitudes of the various viscous and elastic effects will determine the character of the material. If the viscosity of the element C is low, due to mobility of whole chains, then flow will occur rapidly and the material will clearly be a liquid; but if the viscosity of C is very high, then little flow of chains will occur in a finite time and the elastic character will be more important. Neither of the first two mechanisms, represented by units A and B, depend on chain ends: they occur also with single polymer molecules in the rubbery liquid state as discussed in section 2.3.4. The chain ends lead to the introduction of the viscous element C, which is a representation of secondary creep in the material.

Fig. 2.18 can now be extended for an amorphous assembly of polymer chains by the addition of two more regions of mechanical behaviour, as shown in Fig. 2.29 (reversed to be expressed as a compliance instead of modulus). At low temperature, the material is stiff, glassy and elastic, but as chain segments develop some greater freedom of motion the transition region of viscoelasticity appears, and leads to the rubbery elastic region. Then there is another region of viscoelasticity as whole chain flow starts, and finally there is the fully molten polymer which is almost purely viscous in behaviour. This ideal behaviour may be modified in real polymer systems in various ways. Firstly, if the chains are short, the two transition regions may merge together, since there will be no clear distinction between a region of segment mobility and a region of whole chain mobility: the rubbery region will then be absent. Secondly, if the chains are cross-linked the final regions associated with flow of whole chains will be absent: the material will remain rubbery and elastic until the temperature is so high that it degrades chemically. Thirdly, if there are multiple transitions there may be several separate rubbery elastic plateaus separated by transition visco-elastic regions: the multiple transitions may be due either to additional freedom of chain rotation changing the length of an effective random link or to a break-down of temporary cross-links, such as hydrogen bonds, which limit the length of chain segments. Some lesser transitions may be apparent if side-chains become mobile; these would have an influence on the effective viscosity of the material.

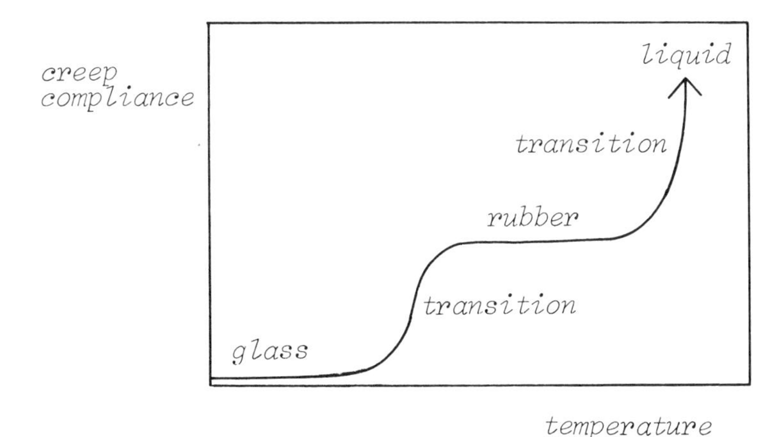

Fig. 2.29 – Creep compliance (extension in given time under given force) of an amorphous polymer as a function of temperature.

In the glassy state, viscous flow of one chain past another cannot occur either because the chains are rigid and entangled with one another, or because the attractive forces between chains are so strong. However chain ends will be points past which stress cannot be transmitted: they will act as flaws in the material and will lead to associated local stress concentrations. Yield and fracture may therefore occur more readily near chain ends. Values of modulus will also be lowered because portions of chains near to chain ends will not be contributing fully in resistance to deformation.

In the crystalline state, chain ends will also occur as defects. Recently, rather extensive theoretical studies of the crystal dislocations which could result from chain ends have been made: the movement of these dislocations can be used to explain various features of the deformation of crystalline polymers.

In discussing crystallisation of single polymer molecules, we saw that chain length played a significant part. This will be reflected in the behaviour of real polymer systems. The crystallisation pattern will be influenced by chain lengths and if there is a distribution of chain lengths some segregation may occur.

In conclusion, we can say that while many properties of polymer systems could be explained in terms of the behaviour of single molecules, there are some modifying effects which arise when chain ends are present.

2.5.3 A simple theory of slip from chain ends or folds

The general consequences of slippage at chain ends can be roughly quantified by the following simple argument, which uses a model consisting of a straight polymer chain of length l aligned in the direction of the applied tension and terminating in free ends as shown in Fig. 2.30(a). There must be zero tension at the ends. We assume that, if there was no slip, the deformation would lead to a stress σ.

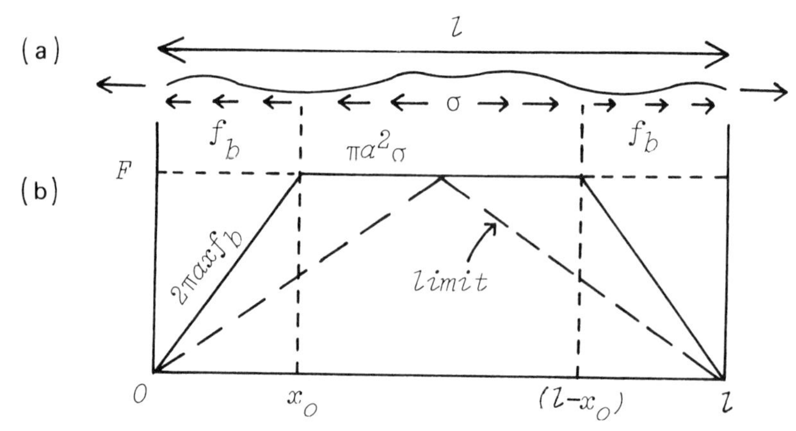

Fig. 2.30 – (a) Hypothetical straight polymer chain of length l under tension. (b) Variation of tension along chain.

The corresponding force F_0 would be given by

$$F_0 = \pi a^2 \sigma \tag{2.63}$$

where a is the effective radius of the chain.

If the maximum shear stress bonding the molecule to its neighbours is f_b, then the maximum force F which can be exerted on a length x without pulling it through the material by inter-molecular slippage is:

$$F = 2\pi a x f_b \tag{2.64}$$

The change from a slipping region to the non-slip region will occur at x_0 given by:

$$F_0 = 2\pi a x_0 f_b = \pi a^2 \sigma \tag{2.65}$$

$$x_0 = \tfrac{1}{2} a\sigma / f_b \tag{2.66}$$

Consequently the variation of tension along the length is that shown in Fig. 2.30(b). If we define a slippage factor S as the ratio of tension without slip to tension with slip, then simple geometry gives the relation:

$$S = (l - x_0)/l = 1 - x_0/l$$
$$= 1 - \tfrac{1}{2}a\sigma/lf_b \quad (2.67)$$

We now put $K = \sigma/f_b$ and assume that the repeat units of the polymer chain have a ratio of length to width β so that:

$$\text{number of repeat units in chain} = N = 1/\beta a$$

This leads to:

$$S = 1 - \tfrac{1}{2}K/\beta N \quad (2.68)$$

Fig. 2.31 shows the form of this relation. For low values of $\beta N/K$, the value of S is negative, which is clearly incorrect; but in these circumstances the simple geometric argument based on Fig. 2.30 will have ceased to be applicable since there will be slip over the whole chain length, when $S < 0.5$. The dotted line indicates a more realistic form of behaviour.

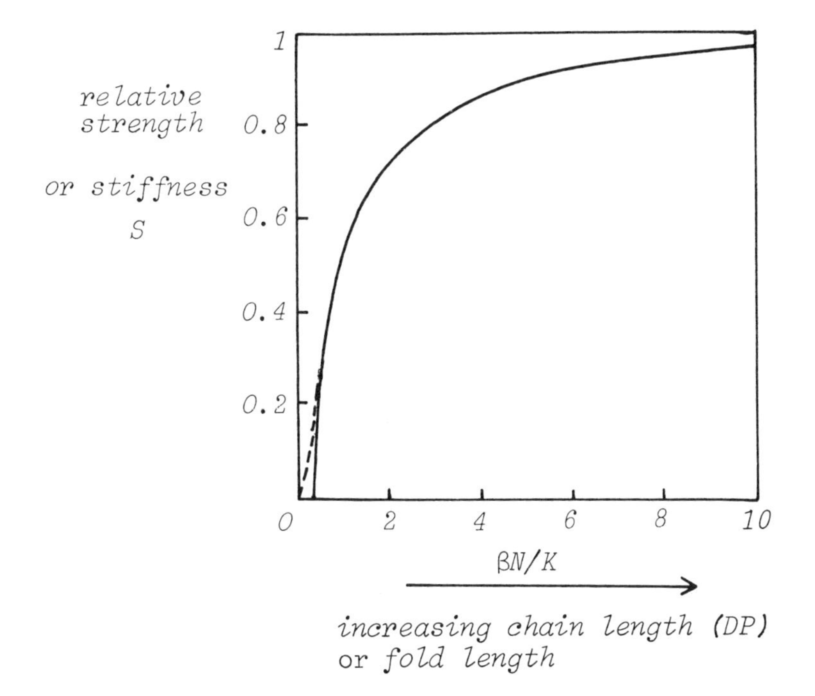

Fig. 2.31 – Effect of slippage from chain ends to resistance to deformation or rupture.

The parameter K can be interpreted in two ways. If we are concerned with failure then it will be related to the ratio of axial strength along the molecules to shear strength between the molecules. If, on the other hand, we are concerned with modulus (resistance to deformation) then K will be related to the ratio of axial stiffness of the molecules to cohesion between molecules.

The treatment was based on a straight chain aligned in the direction of applied force, but one can expect that the same general argument would apply to chains in other directions bearing only a component of load. Furthermore some deviation from the perfectly straight path would lead to effects which could be averaged out and give the same form of expression for the slippage factor.

However if the chain folds back on itself, then, unless the fold is trapped in an entanglement, it will act like a false end and will not be able to transmit load. Consequently the length l, and the corresponding quantity N, should really be taken as the mean effective length of chain segments between either folds or free ends, that is to a quantity which may be termed the molecular extent.

Fig. 2.32(a) shows an idealised form of a folded chain molecule with total

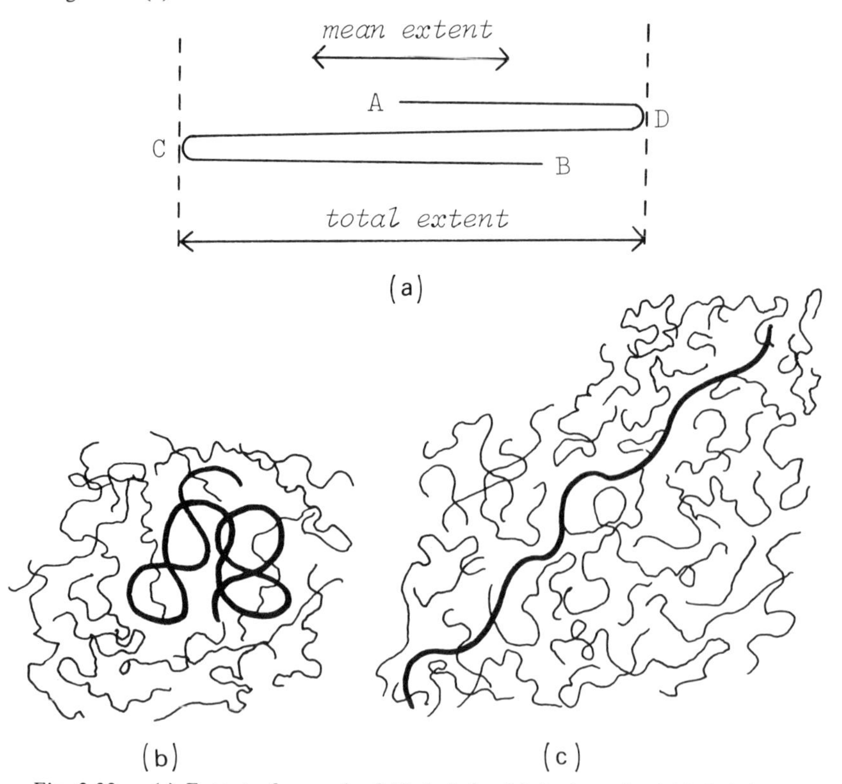

Fig. 2.32 – (a) Extent of a regular folded chain. (b) An irregular folded chain. (c) A generally extended chain.

extent and mean extent indicated. Slippage will occur from the ends A and B and the false ends C and D. Figs. 2.32(b) and (c) show chain molecules of similar length, respectively with many folds giving a low chain extent, or no folds and a chain only slightly less than the fully extended chain length.

When the degree of polymerisation is low, or when there is a very high degree of chain folding, the ability of the polymer material to resist stress will be much reduced.

2.5.4 Cross-linking and gelation

The converse of splitting up a single very long molecule into chains of limited length is to tie separate chains together by cross-links. We have already discussed the influence of lesser attractions between different parts of the same chain: now we must consider strong covalent bonds between different chains. Weaker attractive forces, such as hydrogen bonds, between different chains will have similar effects, except that these lesser bonds may easily be broken and re-formed. In some polymers, covalent bonds may also change position as a result of chemical exchange reactions.

It is simplest to consider first the effect of cross-linking in solution. A few cross-links, like a few entanglements, will merely link the chains into small clusters. This will change the viscosity and other properties, but will not qualitatively alter the nature of the system. A dramatic change will come when the cross linking is so prevalent that a single network extends throughout the whole system. This is known as the point of gelation. It will be a point at which separate polymer molecules can no longer flow past one another, even as clusters of chains: the material will be 'jelly-like'. It is possible to calculate how much cross-linking is necessary for this to occur.

Suppose we have a set of chains, each containing N monomer units, in which a fraction q of monomer units are cross-linked. On the average, there will then be qN cross-links per chain; or, if $qN \ll 1$, then the probability of a chain containing a cross-link is qN. The link from one chain will lead to a particular unit on another chain: there is however a probability of $q(N - 1)$ that one of the remaining $(N - 1)$ units is also cross-linked to another chain. So the probability of finding a chain with two other chains attached through cross-links is $qN \times q(N - 1)$, which is approximately $(qN)^2$ if N is large. The same argument can be applied to the remaining $(N - 1)$ units in the third chain, and so on. We see therefore that the probability of finding a chain with n further units attached through a cross-link is approximately $(qN)^n$. Crudely, we can say that the probability of n chains being linked together is $(qN)^n$.

This is exact enough for our purposes, although in a rigorous analysis of the situation there would be other points to be considered: the exact definitions; the fact that the first chain may also be cross-linked elsewhere; and, most important, the fact that links may lead back to the same chain.

The above analysis shows that if $qN \ll 1$, the clusters will be very small. But as qN approaches 1 then large clusters will become common. In fact, on the simple treatment, when $qN = 1$, all the possibilities are simultaneously certain; and for $qN > 1$, they are more than certain, with the higher values of n being more certain than the lower values! The statements in the last sentence are nonsensical, because the approximations of the simple analysis are no longer justified and one must take account of multiple linkages. We can however see that near $qN = 1$, there will be a very rapid change from a situation in which the chains are in small clusters to a situation in which there is a finite probability that a chain will be attached to a network which extends indefinitely throughout the system. This is known as an infinite network, although it is, of course, limited by the finite size of the system. When $qN > 1$, there is a high probability that the system will contain a single network molecule.

Although the gel point is thus given by $qN = 1$, it does not follow that, at this point, all the chains are linked into the one network: it merely implies that there is one infinite network present, though there may be other single molecules or clusters present with the system. As qN becomes larger still, more chains will join on to the infinite network, and the degree of internal cross-linking within the single network molecule will increase, further restricting the freedom of deformation and movement. The jelly will get stiffer.

Summarising the above discussion, we note that a polymer solution in which cross-linking has occurred has three components:

A – solvent molecules.
B – single molecules and small clusters.
C – an infinite network.

Before the gel-point, C is absent; just beyond the gel-point, the component B is negligible and C is becoming more heavily linked internally.

Similar effects can occur in the absence of solvent. Without cross-links, a molten polymer or rubber consists of chains free to flow over one another; and this flow remains possible, though with greater difficulty as some cross-links form. However once the gel point has been passed, a deformation of the material must lead to an elastic deformation of the whole single network molecule. It is then impossible to relieve this completely by viscous flow: the viscous element C in Fig. 2.28 has been eliminated. We shall be much concerned with this type of behaviour in the Chapter on rubber elasticity. As the number of cross-links increases, the rubber becomes more difficult to deform: it becomes stiffer. Finally, at very high degrees of cross-linking, freedom is so restricted that the material is a hard rigid solid capable of undergoing only a very small elastic deformation.

2.5.5 Linking by entanglement

In the absence of chemical cross-links, entanglement between chains will perform a similar function. The chains can be regarded as split up into separate segments, which are interlocked where they pass round one another. The difference from a chemical cross-link is that an entanglement can move; high tensions in one segment can be relieved to some extent by slip from a neighbouring segment past the entanglement point. If there are free chain ends, then the molecules may pull completely apart and disentangle from one another.

But as with gelation, there can be a very dramatic change of properties when the incidence of entanglement changes from less than one per chain to more than one per chain. In the former, the molecules will be effectively separate except for a limited number of clusters of two or three chains, but in the latter all, or almost all, the chains will be entangled together in an infinite system.

The condition in which this change-over occurs is influenced by a number of factors.

(a) *Concentration of solution.* In a dilute solution, the chains will be far apart. In a concentrated solution, they will mingle with one another, and so, provided the other factors are satisfied, there will be a critical concentration at which the change-over will occur. A considerable change in the variation of viscosity with concentration can be expected at this point. In a bulk polymer, the chains must be intermingled, though they will only be effectively entangled if the other factors are strong enough.

(b) *Molecular weight.* The longer the chains, the greater is the chance of effective entanglement. In a bulk polymer this leads to a change from a material which acts as an assembly of small molecules, to one which behaves as a polymeric system.

(c) *Chain conformation.* Tightly bunched chain conformations will entangle less than more open conformations.

(d) *Time and temperature.* Entanglements will only be effective if they hold for the period of the experiment. With long times or with higher temperatures (leading to more rapid thermal fluctuations) the chains will be able to slide away from one another and act independently.

2.6 SUMMARY: POLYMERIC STATES

2.6.1 The states of a single molecule

The substance of this chapter can usefully be summarised by considering the various states in which polymers can exist. For a single molecule these are indicated in Fig. 2.33. Which of the states can be reached will depend on the nature of the polymer; for example cellulose degrades chemically before it melts, so that the liquid state is inaccessible. Which state the polymer molecule is in at any time will depend on the conditions of temperature and pressure, and on the previous history of the molecule.

If the polymer molecule is regular in form, the true equilibrium state at very low temperatures will be a single crystal. However, in real situations, meta-stable imperfect polycrystalline or glassy states may occur. The condensed glassy state will be the true equilibrium form at low temperatures when the molecule is so irregular that no crystal lattice is possible. At intermediate conditions a condensed rubbery-liquid state will occur, softening to a liquid at a higher temperature. At high enough temperatures and very low pressures the chain would take the open conformations of the rubbery gas. At lower temperatures irregular open glassy chains or regular extended single chains may occur, either as true equilibrium or metastable states. Finally there are states in which depolymerisation has occurred by chemical degradation, or increased polymerisation through cross-linking of reactive chains to give rubbers or rigid thermoset plastics.

2.6.2 The states of condensed systems of many polymer molecules

The isolated polymer molecule is a fiction, though perhaps not a complete fiction, introduced at the beginning of this chapter as an aid to fundamental thought. In reality we are concerned with condensed systems composed of a large number of polymer molecules.

Except for the top left compartment, which is the polymer gas, the states shown in Fig. 2.33 all occur in practice, although some of the compartments

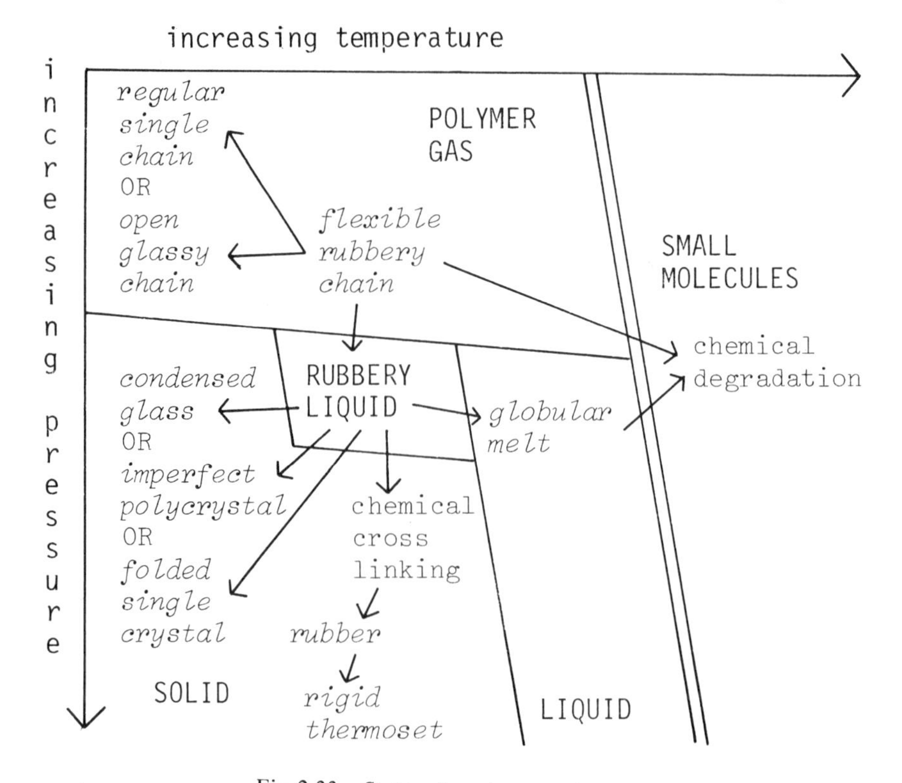

Fig. 2.33 – States of a polymer molecule.

may be missing with particular types of polymer. The behaviour of rubbers, glassy polymers and crystalline polymers will be taken up in later chapters. Rigid thermosets and melts will receive a briefer mention.

However, there is another liquid system which is important: polymer solutions, where the polymer chains are separated by solvent molecules, and thus behave in a way which is similar to what would be expected in a hypothetical polymer gas. This is therefore the appropriate next step in our study.

CHAPTER 3

Chain Statistics and Rheology: Polymer Solutions and Melts

'**Rheology** is the science of **flow** and **deformation** of matter. Flow and deformation are the results of **movements of the particles** of a body, relative to one another. The branch of physics which deals with movements of material bodies is called **mechanics** . . . . When the movements of the planets around the sun are dealt with and the planets are considered as material points, what is relevant is only the mass M of each planet. But it is irrelevant whether the planet consists of water, rubber or jam. However the rheological properties of these materials are very different: water flows, rubber is elastic, jam is plastic.

Markus Reiner, *Lectures on Theoretical Rheology*

3.1 CONFORMATIONS OF A FREELY ORIENTING CHAIN

3.1.1 Distribution of chain lengths

Equation (2.19), derived in the previous chapter, is an expression for the mean tension exerted in a freely orienting chain when its ends are held fixed. From this equation, it is possible to work out the distribution of end-to-end distances in a chain whose ends are free. Although an isolated, free chain is not a practical system for experimental study, the results will be applicable – with some modification – to the behaviour of polymers in solution. Paradoxically, although the mean force is zero when the chain ends are coincident, the most probable end-to-end distance is not zero.

It is easiest to approach the problem by considering one end A of a chain of N links to be fixed in space, with the other end B at a distance l as shown in Fig. 3.1. The *a priori* probability that B lies within a particular thin, spherical shell, so that l has a value between l and $(l + \mathrm{d}l)$ is proportional to the volume $4\pi l^2 \mathrm{d}l$ lying between radii l and $(l + \mathrm{d}l)$. The increase in this *a priori* probability more than offsets the increase in free energy A_l with increase of l. If we now apply the Maxwell–Boltzmann law we get:

$$\begin{aligned}&\text{probability that } l \text{ has a value between } l \text{ and } (l + \mathrm{d}l) =\\&= P(l)\mathrm{d}l\\&= \text{constant} \times 4\pi l^2 \mathrm{d}l \exp(-A_l/kT) \qquad (3.1)\end{aligned}$$

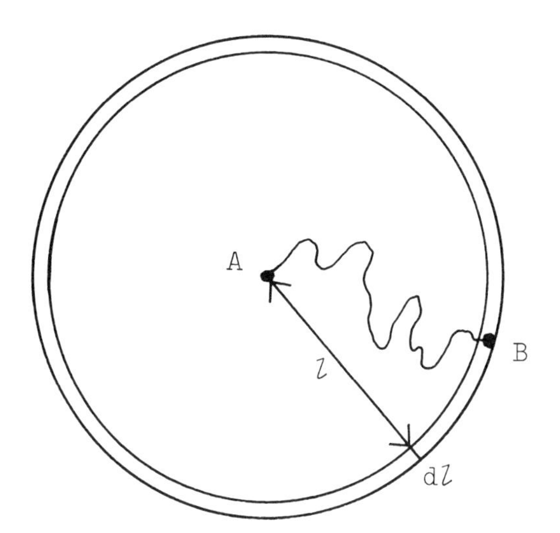

Fig. 3.1 – Freely joined chain with ends at a distance between l and $(l + \mathrm{d}l)$.

The free energy A_l of the molecule with an end-to-end distance l is given by:

$$A_l = \int_0^l F\mathrm{d}l \tag{3.2}$$

where F is the tension in the molecule at a length l.

Equations (3.1) and (3.2) are quite general, and may be applied to any polymer chains. Substituting from (2.19), as appropriate for a freely jointed chain, we obtain:

$$A_l = \int_0^l (kT/a)\mathcal{L}^{-1}(l/L)\mathrm{d}l \tag{3.3}$$

where a is the length of a link, and L is the total length of the chain.

If we insert this expression in (3.1), collect together the various constant factors, and put $a = L/N$, we get:

$$P(l)\mathrm{d}l = \text{constant} \times l^2 \exp\left[-\int_0^l (N/L)\mathcal{L}^{-1}(l/L)\mathrm{d}l\right]\mathrm{d}l \tag{3.4}$$

Since the end of B of the molecule must lie somewhere within a distance L of the fixed end A, it follows that:

$$\int_0^L P(l)\mathrm{d}l = 1 \tag{3.5}$$

Consequently the constant in (3.4) can only be a normalising factor, and the equation becomes:

$$P(l) = \frac{l^2 \exp\left[-\int_0^l (N/L)\mathcal{L}^{-1}(l/L)\mathrm{d}l\right]}{\int_0^L l^2 \exp\left[-\int_0^l (N/L)\mathcal{L}^{-1}(l/L)\mathrm{d}l\right]\mathrm{d}l} \tag{3.6}$$

If the integration in (3.4) is carried out, we obtain:

$$P(l) = \text{constant} \times l^2 \exp\left[-N\frac{l}{L}\mathcal{L}^{-1}\left(\frac{l}{L}\right) + \log_e \frac{\mathcal{L}^{-1}(l/L)}{\sinh \mathcal{L}^{-1}(l/L)}\right] \tag{3.7}$$

Alternatively, expansion in series can be shown to give:

$$P(l) = \text{constant} \times l^2 \exp\left\{-N\left[\frac{3}{2}\left(\frac{l}{L}\right)^2 + \frac{9}{20}\left(\frac{l}{L}\right)^4 + \frac{99}{350}\left(\frac{l}{L}\right)^6 + \ldots\right]\right\} \tag{3.8}$$

Fig. 3.2(a) shows the distribution of end-to-end lengths of a freely jointed polymer chain as predicted by the above equations. For a reference system with an origin located at one end of the chain, it represents a spherically symmetrical distribution for the location of the other end of the chain, with the probability that the chain shall be at any particular distance from the origin rising to a maximum and then falling.

The argument so far has been concerned with the distribution of end-to-end lengths irrespective of direction. However we may instead be interested in the probability that the chain end lies within a given small volume at a certain

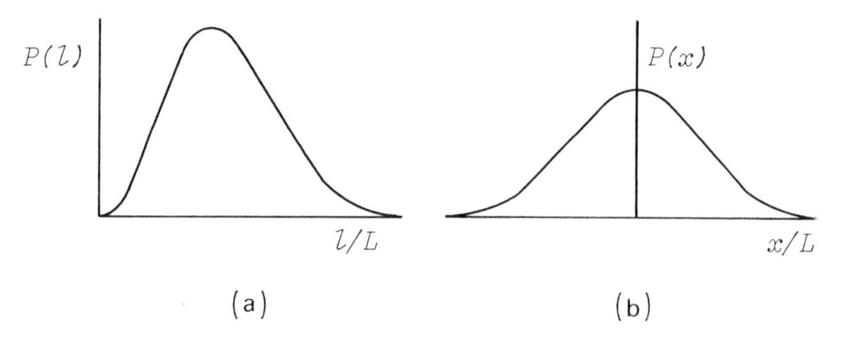

Fig. 3.2 – (a) Probability distribution of chain end separation l in any direction, for a chain of 100 freely jointed links. (b) Probability distribution for separation by a distance x in a given direction. Note that for a chain of 10,000 freely jointed links the peak in (a) and the point, of inflexion in (b) would occur when l/L is about 0.01; negligibly few chains are extended by more than a few per cent.

distance x in a particular direction. Under these circumstances, the *a priori* probability is the same for all values of x. Consequently, there would be no factor corresponding to $4\pi l^2$ in (3.1) and we get the probability that x has a value between x and $(x + dx)$ given by:

$$P(x)dx = \text{constant} \times \exp(-A_l/kT)dx \tag{3.9}$$

The remaining analysis is similar giving:

$$P(x) = \text{constant} \times \exp\left[-\int_0^x (N/L)\mathcal{L}^{-1}(x/L)dx\right] \tag{3.10}$$

This distribution function is shown in Fig. 3.2(b). It shows, as would be expected, a maximum value at $x = 0$ and is symmetrical about this maximum. This resolves the paradox mentioned earlier: it shows that the probability of finding the chain end B in a given volume element is greatest when this element is located coincident with the end A; but there are many more similar elements at a greater distance from A.

In calculating end-to-end distances in a polymer chain it is perfectly valid to use a reference system based on an origin at one end of the chain. However, it would be physically more instructive, though mathematically more difficult, to use a reference system with an origin at the centre of mass of the chain. Both ends would then take up a distribution of locations in space: intuitively we can see that these would both be identical, spherically symmetrical distributions with a maximum probability at some particular distance from the centre of mass, and zero probability both at zero distance (due to the shrinkage of the available volume to zero) and at a distance $L/2$ equal to half the fully extended chain length. We may note incidentally that the latter state can only occur when the chain is fully extended in a straight line in both directions from the centre of mass.

A more complete statistical analysis would yield expressions for the distribution of the other units in the chain as well as of the end-points, and give formulae for other dimensional characteristics of the free chain. Derivations of some of these quantities, based on another approximate form of chain statistics, are given later.

3.1.2 An alternative form of derivation

It is worth noting that equations identical with those obtained in the previous section can be derived by quite different methods. The method used by Kuhn and Grün is based on the number of conformations that a freely orienting chain can take up. The method of analysis is somewhat complicated and only an outline will be given here.

For any given end-to-end length, many distributions of the orientations of chain links are possible. Therefore, what should be done is to calculate the *total* number of ways in which the links can orient themselves to give any particular value of l. This has not been done, because of the mathematical difficulties.

Instead, the link direction is defined by the angle θ which it makes with a line joining the chain ends, as in Fig. 3.3. The *a priori* probability that a link

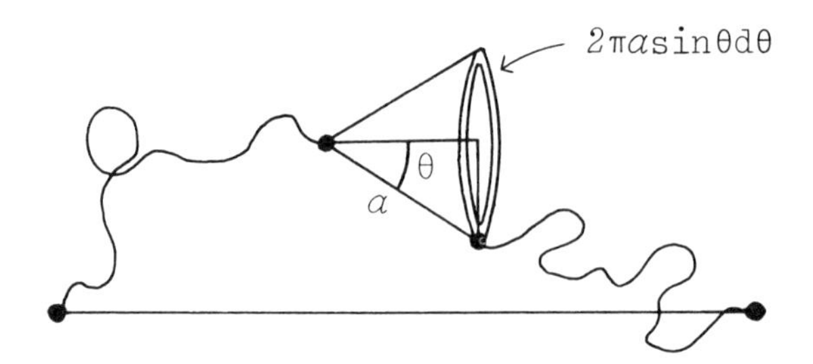

Fig. 3.3 – An individual link in a freely jointed chain, making an angle θ with the line joining the chain ends.

makes an angle between θ and $(\theta + d\theta)$ is proportional to the solid angle within these limits, namely to $\frac{1}{2}\sin\theta\, d\theta$. The probability P of obtaining a distribution with n_1 links making angles between θ_1 and $(\theta_1 + d\theta)$, n_2 between θ_2 and $(\theta_2 + d\theta)$, and so on, arranged in any order, is:

$$P = \frac{[(\frac{1}{2}\sin\theta_1)^{n_1}(\frac{1}{2}\sin\theta_2)^{n_2}\ldots]N!}{[(n_1!)(n_2!)\ldots]} \tag{3.11}$$

For a particular chain with a particular end-to-end length, the possible values of the distribution are restricted by the conditions:

$$\text{number of links} = N = \Sigma n \tag{3.12}$$

$$\text{chain end separation} = x = \Sigma n a \cos\theta \tag{3.13}$$

The most probable distribution of link directions for a given chain length, can be found by maximising the function P, subject to the restricting conditions. There are standard procedures for carrying out this operation, and the result is that the *most probable* distribution is defined by the equation:

$$dn = \tfrac{1}{2}[N\mathcal{L}^{-1}(x/L)\sinh\mathcal{L}^{-1}(x/L)]\exp[\mathcal{L}^{-1}(x/L)\cos\theta]\sin\theta\, d\theta \tag{3.14}$$

where dn is the number of links between θ and $(\theta + d\theta)$.

The probability P_m of this most probable distribution is given by substitution in (3.11), and, with the use of Stirling's approximation for the factorials, gives the result:

$$P_m = \text{constant} \times \left[-N\frac{x}{L}\mathcal{L}^{-1}\left(\frac{x}{L}\right) + \log_e \frac{\mathcal{L}^{-1}(x/L)}{\sinh\mathcal{L}^{-1}(x/L)}\right] \tag{3.15}$$

It is then assumed, as is commonly done in statistical mechanics, that the probability of this most probable value is proportional to the required total probability of achieving a given end-to-end displacement. This is equivalent to taking the peak value of the distribution curve as a measure of the total area under the curve – a procedure which is only strictly valid if all distribution curves are of the same shape. Evaluation in this way thus yields an equation similar to (3.7) which was derived by a different method. The l^2 factor is missing because the angle θ was defined above in relation to a particular direction: the expression gives a value for $P(x)$, corresponding to (3.10), which has been decuced by the alternative method.

In addition to yielding expressions for $P(l)$ or $P(x)$, the analysis gives the most probable distributions of link angles for any given end-to-end length. Fig. 3.4 shows the polar distribution of these angles for various values of l/L. As would be expected, when $l/L = 0$, all angles are equally probable; but, as l/L increases, the distribution becomes increasingly asymmetrical with more and more chains pointing in directions close to the line joining the chain ends.

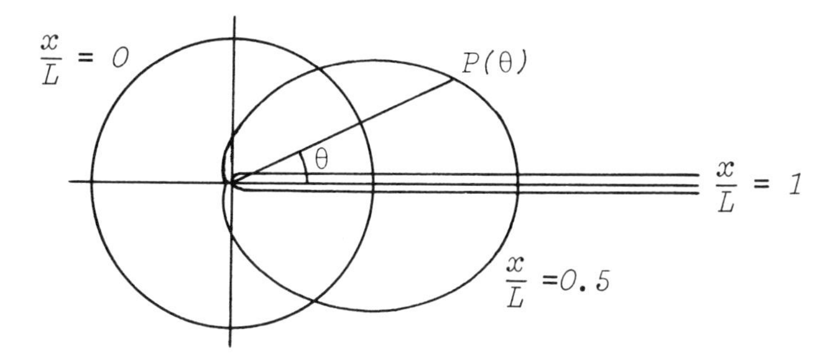

Fig. 3.4 – Polar plot of probability distribution of angles of links for chain ends separations of 0, 0.5, and 1.

Finally, it should be noted that, despite appearances, neither method of deriving the Langevin approximation for $P(l)$ is completely rigorous. Both methods involve certain improper averaging procedures, although the effects of these are probably small. Certainly other aspects of behaviour of real polymer chains will cause greater differences from the theoretical predictions.

3.1.3 The Gaussian approximation

The method, adopted in this book, of deriving the force-extension relation for a rubbery chain, and then obtaining the chain statistics, does not follow the historical line of thought. Chronologically, a more approximate form of chain statistics, known as Gaussian statistics, came first. Because of the greater mathematical simplicity of the final equation, the Gaussian form is still worth using.

The Gaussian expression turns out to be identical with the first term of the series expansion of the Langevin function form, omitting all the higher powers: it may be regarded solely in this way, but the independent method of derivation does add to our qualitative understanding of chain statistics, and its basis – though not all the mathematical details – will be given here.

We consider a chain of N links in which each link is free to orient itself in any direction. The chain conformation is thus equivalent to a random flight in three dimensions, where each step is of the same length but may be taken in any direction. Expressed formally, if each link is regarded as a vector of length **a** directed outwards from a point O, their ends would be randomly distributed over a sphere of radius **a**. In actuality, each vector link proceeds from the end-point of the previous chain, as indicated in Fig. 3.5 where the chain form is related to fixed axes with one end one of the chain at the origin O. With the situation defined in this way, it is then necessary to calculate from statistical arguments the probability $P(x,y,z)\mathrm{d}x\,\mathrm{d}y\,\mathrm{d}x$ that the other end of the chain lies within a volume element $\mathrm{d}x\,\mathrm{d}y\,\mathrm{d}z$ located around the point (x,y,z) at distance l from 0.

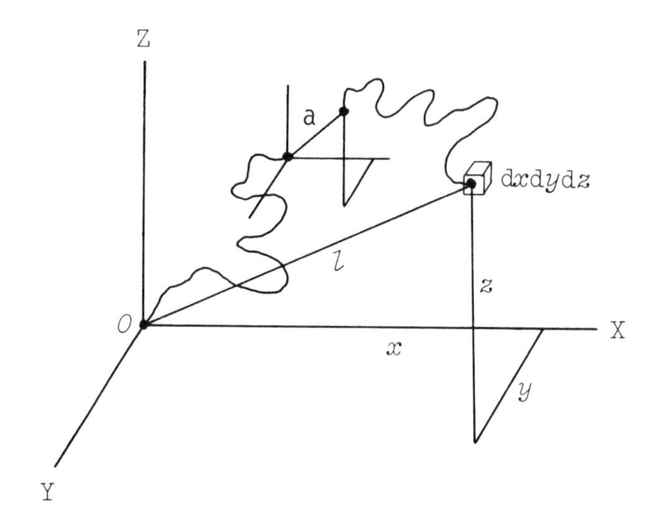

Fig. 3.5 – Freely jointed chain with ends at 0 and x, y, z.

Details of the derivation are given in many other books, but since the method is physically less exact and mathematically more complicated than the methods given in previous sections, they will not be repeated here. The approximation is made that the probability function for the value of the co-ordinate x has the Gaussian error form, $\exp(-x^2)$, which occurs in many statistical problems governed by the combination of random effects. Introduction of the same form for the other co-ordinates, y and z, leads to the expression:

$$P(x,y,z)\mathrm{d}x\,\mathrm{d}y\,\mathrm{d}z = (b^3/\pi^{3/2})\exp[-b^2(x^2+y^2+z^2)]\,\mathrm{d}x\,\mathrm{d}y\,\mathrm{d}z$$

where $b^2 = 3/2\,Na^2$. (3.16)

If we now consider a given direction, and transpose the axes so that x and l are identical, and y and z are zero, we find:

$$P(x) = (b^3/\pi^{3/2})\exp(-b^2x^2) \qquad (3.17)$$

However if we include all directions as possible, then we have:

$$l^2 = x^2 + y^2 + z^2 \qquad (3.18)$$

Since the volume enclosed between distances l and $(l + \mathrm{d}l)$ is $4\pi l^2\mathrm{d}l$, we obtain:

$$P(l)\mathrm{d}l = (b^3/\pi^{3/2})\exp(-b^2l^2) \times 4\pi^2 l^2\mathrm{d}l$$

$$= (4b^3/\pi^{1/2})\,l^2\exp(-b^2l^2)\mathrm{d}l \qquad (3.19)$$

Alternatively by integrating over all possible values of y and z, it can be shown that the probability that x has a value between x and $(x + \mathrm{d}x)$ irrespective of the values of y and z is given by:

$$P_x(x)\mathrm{d}x = (b/\pi^{1/2})\exp(-b^2x^2)\mathrm{d}x \qquad (3.20)$$

The distribution functions represented by these equations are shown in Fig. 3.6. Incidentally it may be noted that (3.17) and (3.20) are of the same form, but with different constants: the former represents the probability of

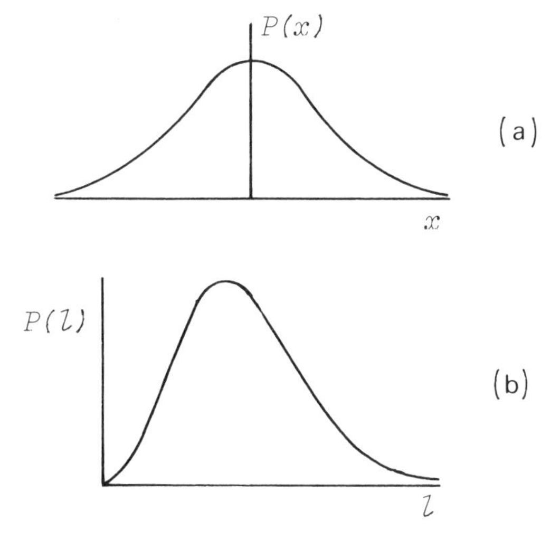

Fig. 3.6 – (a) Form of (3.18): probability $P(x)$ of chain end separation, x, in a given direction, for Gaussian approximation. (b) Form of (3.70) probability $P(l)$ of chain end distance l, in any direction; for Gaussian approximation.

finding a chain end in a given volume $dx\,dy\,dz$ at a given distance in a particular direction; but the latter is the probability of occurrence of a given displacement in the given direction – in other words, of the chain end being located within a given small distance dx of a plane perpendicular to the given direction at the given distance.

The Gaussian forms are clearly incorrect for highly extended chains, since they show a finite probability that the end-to-end distance is greater than the chain length! This fault derives from approximations, of which two are noted here. Firstly Stirling's approximation is used to evaluate the factorials which occur in the calculation of the number of ways of reaching a given end-point: this approximation is only valid for large numbers. Secondly, it is assumed that chain displacements in the directions x, y and z can be treated independently of one another: this is clearly not completely true. If displacement is used up in one direction, less is available for other directions. In the extreme instance of a chain which is fully extended in one direction, say x, there can be no displacement in the other directions, y and z. These two approximations do not cause any appreciable error in estimating the numbers of highly coiled configurations of chains with many links. Serious errors are introduced however when more extended chain configurations are being counted. If the chain is not under load there will be very few extended configurations, and so the error in using the Gaussian approximation will be small.

Equation (3.19) may be rewritten in the form:

$$P(l) = \text{constant} \times l^2 \exp[-(3N/2)(l/L)^2] \qquad (3.21)$$

This form is identical with the first term in the series expansion of the Langevin approximation given in (3.8) – the constant, in both equations, must be adjusted to be the appropriate normalising factor. This confirms that we can view the Gaussian approximation as derived from the Langevin expression by omitting all except the first term in the series. Clearly this will not cause any error for conformations in which $(l/L) \ll 1$, but will lead to error when l/L is close to 1.

3.1.4 An exact derivation

Both the Gaussian and the Langevin function approximations depend on the assumption that the chain contains many links. Using nothing but rigorous geometrical and statistical arguments, Treloar has obtained an exact expression for the for the distribution of end-to-end distances in a chain of N links each of length a:

$$P(l)\mathrm{d}l = \frac{l}{2a^2}\frac{N^{N-2}}{(N-2)!}\sum_{S=0}^{S=KN}(-1)^s\binom{N}{S}\left(m-\frac{S}{N}\right)^{N-2}\mathrm{d}l \qquad (3.22)$$

where $m = \frac{1}{2}(1 - l/Na)$; K is an integer, such that $(m-1) \leqslant K \leqslant m$; and

$$\binom{N}{S} = \frac{N!}{(N-S)!S!}$$

The comparison given in Fig. 3.7(a) shows that the Langevin approximation is not bad for a chain of 25 links, and is good for chains of 100 links or more. The plot given is for $P(x)$, proportional to $(1/l^2)P(l)$, and is in logarithmic form in order to emphasise the differences at high chain extensions. It should be noted that the logarithmic scale of relative probability means that there is a rapid reduction of probability as end-to-end distance increases, and also that multiplication by $(1/l^2)$, namely substraction of 2 log l, would make little difference except when l is very small. The Gaussian approximation diverges from the exact expression as the end-to-end distance increases, becoming seriously in error when the chain is extended to more than about a third of its fully extended length. For the 6-link random chain, shown in Fig. 3.7(b), there is

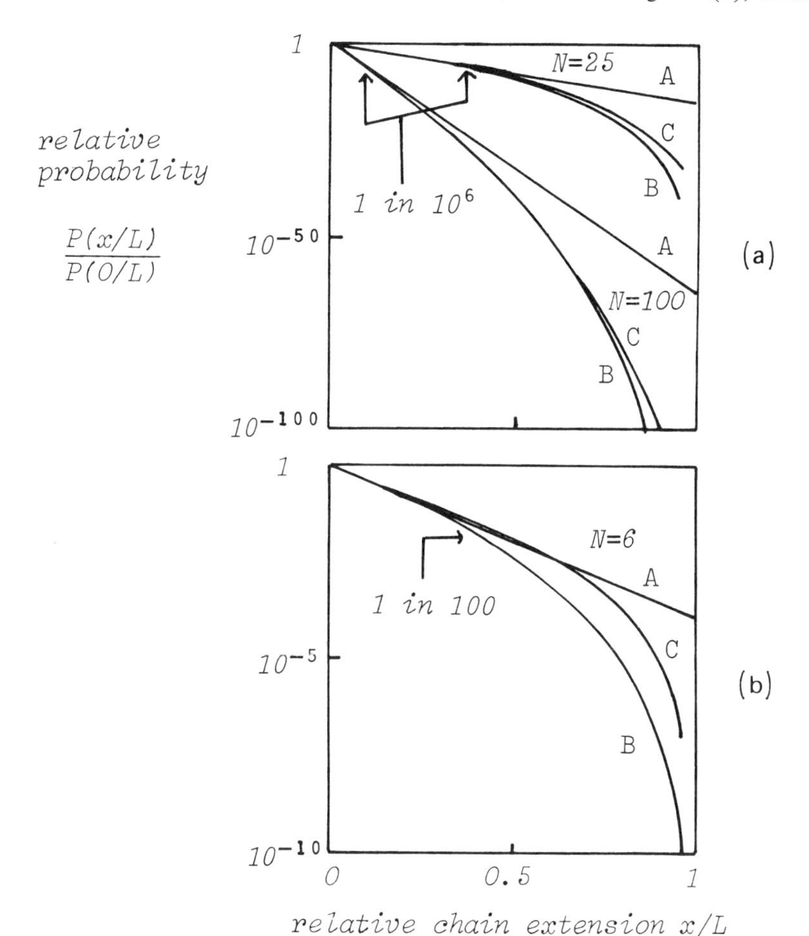

Fig. 3.7 – Comparison of distributions of chain end separation in a given direction predicted by (A) Gaussian, (B) Langevin function, and (C) exact forms for (a) $N = 25, 100$ (b) $N = 6$.

appreciable error in the Langevin approximation, and the Gaussian approximation is rather better except at very high chain extensions.

3.1.5 Derivations from chain statistics

There are a number of specific parameters which can be deduced from chain statistics.

From any distribution of end-to-end distances, the most probable end-to-end distance can be obtained by solving the equation:

$$\frac{\mathrm{d}P(l)}{\mathrm{d}l} = 0 \tag{3.23}$$

For a free random chain, using the Gaussian approximation, differentiating (3.19), we get:

$$\frac{\mathrm{d}P(l)}{\mathrm{d}l} = \frac{8b^3}{\pi^{\frac{1}{2}}}\, l\,(\exp(-b^2l^2))(1-b^2l^2) = 0 \tag{3.24}$$

The solutions of this equation are:

$$l = 0; \quad l - \infty; \quad l = \pm(1/b)$$

The first two are obviously the minima at the ends of the distribution, and so we see that $1/b$ is the required most probable value of the end-to-end distance:

$$\begin{aligned}\hat{l} &= 1/b = (2N/3)^{1/2}a \\ &= (2/3N^{1/2})L\end{aligned} \tag{3.25}$$

When N is large the most probable values will be only a small fraction of the fully extended length L and so the use of Gaussian statistics is justified.

A mean end-to-end distance could be calculated from the expression:

$$\bar{l} = \int_0^L lP(l)\,\mathrm{d}l \tag{3.26a}$$

However it turns out that a root mean square value is more useful, and the general expression for this is:

$$r_r = (\overline{l^2})^{1/2} = \left[\int_0^L l^2P(l)\,\mathrm{d}l\right]^{1/2} \tag{3.26b}$$

Substitution of the Gaussian expression (with a change in the upper limit to ∞) leads to:

$$r_r = (\overline{l^2})^{1/2} = N^{1/2}a = L/N^{1/2} \tag{3.27}$$

This equation is identical with (2.29) previously derived by a general argument in section 2.2.7. The same equation can be obtained by substituting the Langevin approximation, (3.7).

We note that both the most probable length and the root mean square length are proportional to the square root of the number of links in the chain, and so increase less rapidly than the fully extended length.

So far we have considered only the relative positions of the end-points of the chains. It is also useful to know how the intervening portions are distributed. On symmetry grounds, it is obvious that the chain elements of a long chain will be symmetrically distributed around the centre of mass. The exact form of the distribution is a complicated function, for which the integrations have not been solved in a closed form, though Debye and Bueche have obtained numerical solutions using a series method based on Gaussian statistics.

If $P_m(r)\,\mathrm{d}r$ is the probability of finding an element at a distance between r and $(r + \mathrm{d}r)$ from the centre of mass, an approximate form of distribution valid except at large values of r, is

$$P_m(r) = (9/\pi Na^2)^{3/2} \exp(-9r^2/Na^2) \tag{3.28}$$

The quantity $P_m(r)$ is, of course, proportional to the density of chain elements at any position. Fig. 3.8 shows how the total relative density $r^2P_m(r)$ varies with distance from the centre of mass.

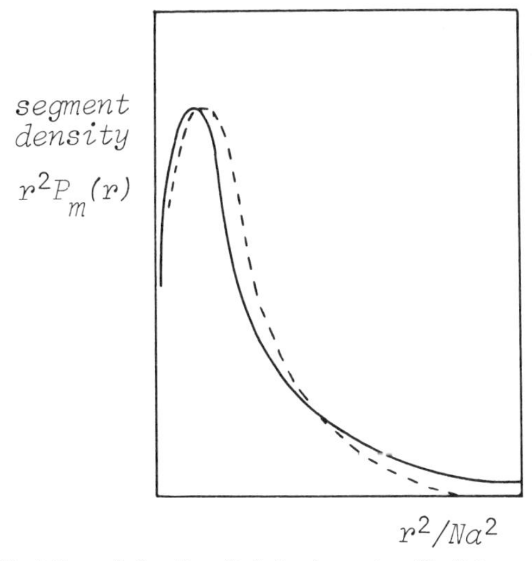

Fig. 3.8 – Variation of density of chain elements with distance from centre of mass, solid line from (3.28) and dotted line from an exact Gaussian form.

An average value for the distribution about the centre of mass is given by the radius of gyration r_g, defined by:

$$r_g^2 = (1/N)\sum_{i=1}^{N} r_i^2 = \int_0^{L/2} r^2P_m(r)\,\mathrm{d}r \tag{3.29}$$

where r_i is the distance of the ith element from the centre of mass.

Bueche has shown by a direct argument that, for the random chain, the radius of gyration is related to the mean square end-to-end length by the equation:

$$r_g^2 = \overline{l^2}/6 \tag{3.30}$$

or

$$r_g = r_r/\sqrt{6} = \sqrt{N}a/\sqrt{6} = L/\sqrt{6}\sqrt{N} \tag{3.31}$$

There are other measures of chain dimensions for which expressions are given in the literature.

3.1.6 Derivation of force equations from chain statistics

It is also possible to use chain statistics to calculate stress-strain relations for isolated chains: this is the reverse of the methods used earlier in this book. In general, we see that the number of ways W of obtaining a chain with a given end-to-end displacement is given by

$$W = \text{constant} \times P(x) \tag{3.32}$$

Note that it is necessary to use $P(x)$ and not $P(l)$ because we are concerned with particular relative positions of chain ends, and not with a given length in any direction.

With an appropriate definition for the origin of entropy values, we then get:

$$S = k \log_e W/W_0 = k \log_e P(x) \tag{3.33}$$

Hence, using (A.3) and remembering that there are no internal energy changes, the tension in the chain is given by:

$$\begin{aligned} F &= -T(\partial S/\partial l)_T \\ &= -kT(\partial/\partial l)\,[\log_e P(x)] \end{aligned} \tag{3.34}$$

Consequently, if an expression for $P(x)$ is known, the force can be determined. Substituting the Gaussian expression, (3.17) we get

$$\begin{aligned} F &= -(kT)(\partial/\partial l)\,[\log_e (b^3/\pi^{3/2}) - b^2 l^2] \\ &= 2kTb^2 l \\ &= 3(kT/Na^2)l \\ &= (kT/a)(3l/L) \end{aligned} \tag{3.35}$$

Transforming this in the same way as before into an expression for specific stress, we find:

$$f = (N_0 kT/M_0)(3l/L) \tag{3.36}$$

Equation (3.36) derived from Gaussian statistics, is the same as (2.20) with the expansion of the inverse Langevin function as in (2.21) limited to the first order term in the series.

Generally, we may note that if an expression for $P(l)$ can be found, then the stress-strain curve can be derived from it, just as conversely, knowledge of a stress-strain curve leads to the distribution function by the use of (3.1) and (3.2).

3.1.7 Conformations of real chains

As discussed in section 2.2.8 real chains are not freely orienting. Usually the bond angles are fixed, and only rotation around the bonds is at all easy. However, as we have seen, a real chain is often regarded as replaced by an **effective free chain** having the same total length and the same root-mean-square length as the real chain.

For some molecules, more detailed calculations have been made. In order to calculate the root-mean-square length for a restricted chain, it is necessary to evaluate all the terms in (2.26). As before, the first line of squared terms equals Na^2. We can obtain the required average value $\overline{\mathbf{l}.\mathbf{l}}$ by averaging separately over each of the remaining terms in the sum.

For a chain of identical links, and with no appreciable end effects, all terms such as $\mathbf{a}_1.\mathbf{a}_2$, $\mathbf{a}_2.\mathbf{a}_3$ . . . $\mathbf{a}_i.\mathbf{a}_{i+1}$ must have equal average values. In general, any set of terms of the form $\mathbf{a}_i.\mathbf{a}_{i+m}$ will have equal averages. In a chain of N links there are $(N-1)$ ways of obtaining pairs of form $\mathbf{a}_i.\mathbf{a}_{i+1}$, or, in general, $(N-m)$ ways of obtaining pairs like $\mathbf{a}_i.\mathbf{a}_{i+m}$. Consequently, by adding up all the individual terms in (2.26), we obtain the equation:

$$\text{mean square length} = \overline{l^2} = \overline{\mathbf{l}.\mathbf{l}}$$

$$= Na^2 + 2\,[(N-1)\,\overline{\mathbf{a}_i.\mathbf{a}_{i+1}} + (N-2)\,\overline{\mathbf{a}_i.\mathbf{a}_{i+2}} + \ldots\, \overline{\mathbf{a}_i.\mathbf{a}_{i+N-1}}] \quad (3.37)$$

The single, final term in this series is, of course, the term $\mathbf{a}_1.\mathbf{a}_N$.

Equation (3.37) may be applied to any chain for which the scalar products of the vectors can be evaluated. Consider a simple chain, with free rotation about a fixed bond angle θ. Each successive bond can take up a position around a cone of semi-angle α equal to $(\pi-\theta)$, as shown in Fig. 3.9. We are immediately that all adjacent links make an angle α with one another, and so the terms $\mathbf{a}_i.\mathbf{a}_{i+1}$ are constant at $a^2\cos\alpha$. If we examine all the positions which the vector $\mathbf{a}_{i+2}$ can take up, we see that their average will be a vector of magnitude $a\cos\alpha$ in the direction of the vector $\mathbf{a}_{i+1}$, namely at an angle α to the vector $\mathbf{a}_i$. Consequently:

$$\overline{\mathbf{a}_i.\mathbf{a}_{i+2}} = a^2\cos^2\alpha \quad (3.38)$$

This argument may be repeated, and it follows that:

$$\overline{\mathbf{a}_i.\mathbf{a}_{i+m}} = a^2\cos^m\alpha \quad (3.39)$$

Except for the special cases of $\alpha = 0$ (when the whole chain would be in a straight line) or $\alpha = \pi$ (when the chain would just be folding back and forth), the values of $\cos^m\alpha$ must tend to zero as m becomes large. In other words, only near neighbours, where there is correlation between chain directions, make any appreciable contribution to the sum. The full expression may be obtained by substitution in (3.37):

$$\overline{l^2} = Na^2 + 2a^2\,[(N-1)\cos\alpha + (N-2)\cos^2\alpha + \ldots + \cos^{N-1}\alpha] \quad (3.40)$$

When N is very large, it can be shown that (3.40) reduces to the form:

$$\overline{l^2} = Na^2\,[(1+\cos\alpha)/(1-\cos\alpha)] \tag{3.41}$$

or

$$r_r = N^{1/2}a\,[(1+\cos\alpha)/(1-\cos\alpha)]^{1/2} \tag{3.42}$$

As with the random chain, the root-mean-square length is proportional to the square root of the number of links. Indeed, it can be shown that all long chains give this proportionality. For a chain of this type, we see from Fig. 3.9(b) that the fully extended chain length equals $Na\cos\frac{1}{2}\alpha$. Substitution in (2.32) and (2.33) then yields the following values for the parameters of the equivalent free chain:

$$a_e = \overline{l^2}/L = a(1+\cos\alpha)/\cos\tfrac{1}{2}\alpha(1-\cos\alpha) \tag{3.43}$$

$$N_e = L^2/\overline{l^2} = N\cos^2\tfrac{1}{2}\alpha(1-\cos\alpha)/(1+\cos\alpha) \tag{3.44}$$

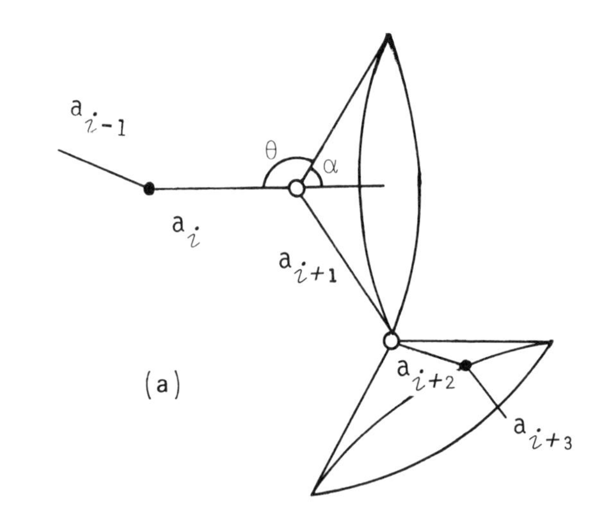

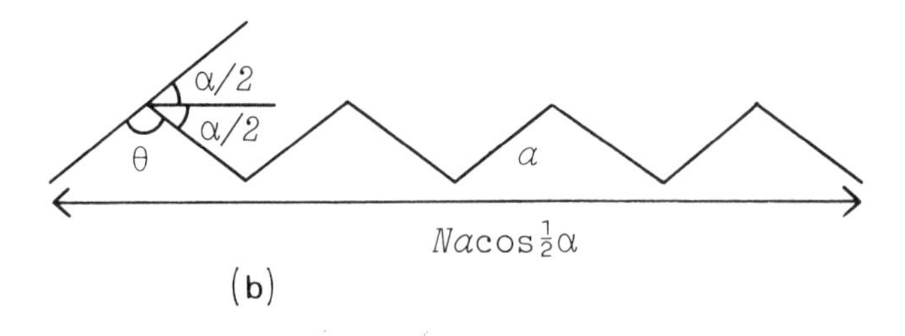

Fig. 3.9 – (a) Successive links of chain with a bond angle θ. (b) Fully extended chain.

A chain of particular interest is the paraffin-type chain of undistorted carbon atoms in which adjacent bonds are directed towards opposite corners of a regular tetrahedron. With this geometry, cos α is exactly 1/3, θ is $109\frac{1}{2}°$, and α is $70\frac{1}{2}°$. Substituting these values we find that $N_e = N/3$, so that three C–C bonds make one effective free link and a_e equals $\sqrt{6}a$, which equals $2.45a$ or 3.8 Å.

As an illustration of an actual conformation which would occur in a random paraffin-type chain, Treloar constructed the model illustrated in Fig. 3.10. Each bond angle was restricted to the valeney angle, but the choice of position around the bond was selected by the throw of a die.

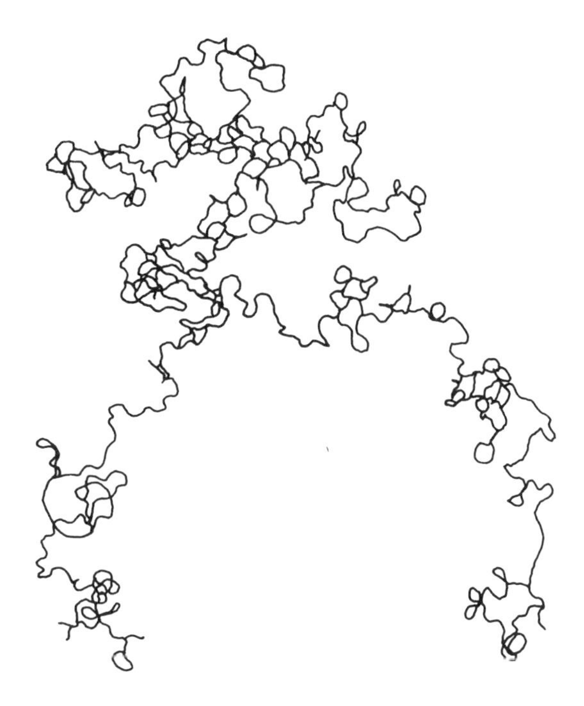

Fig. 3.10 – Model of a chain of 1000 C–C bonds constructed by Treloar with free rotation around the bond ($\alpha = 70\frac{1}{2}°$). In practice the 'free' rotation was a random choice of 6 positions.

Another interesting example is the polyisoprene chain. The statistics are more complicated because there are both single and double bonds along the chain, but parameters can be worked out by a method devised by Wall. Values calculated from these formulae are given in Table 3.1. For *cis*-polyisoprene, which occurs in natural rubber, we see that each isoprene unit, of length 4.60 Å, can be regarded as contributing 1.31 free links from its four bonds.

Table 3.1 – Parameters calculated for a polyisoprene chain.

	cis-polyisoprene (natural rubber)	*trans*-polyisoprene (gutta-percha)
r.m.s. length $(\overline{l^2})^{1/2}$	$2.01N^{1/2}$Å $= 4.02N_i^{1/2}$ Å	$2.90N^{1/2}$Å
equivalent free links, N_e	$0.33N$ $= 1.31N_i$	
length of free link, a_e	3.52Å	

N = number of bonds; N_i = number of isoprene units.

When there are barriers to free rotation around a bond, the chain is still further restricted. The probability of obtaining any particular value of ϕ, the angle of rotation round the bond, will depend on the value of the internal energy $U(\phi)$, as given by diagrams such as Figs. 1.18 to 1.22, and will equal $\exp[-U(\phi)/kT]$. For a chain with a particular distribution of values of ϕ, it has been shown that:

$$\overline{l^2} = Na^2\left(\frac{1+\cos\alpha}{1-\cos\alpha}\right)\left(\frac{1+\psi}{1-\psi}\right) \tag{3.45}$$

where ψ = mean value of $\cos\phi$.

$$= \int_0^{\pi} \exp[-U(\phi)/kT]\cos\phi \, \mathrm{d}\phi \Bigg/ \int_0^{\pi} \exp[-U(\phi)/kT]\mathrm{d}\phi \tag{3.46}$$

Compared with the freely rotating paraffin chain, the root-mean-square length is increased by a factor $[(1+\psi)/(1-\psi)]^{1/2}$. Values of $(1+\psi)/(1-\psi)$, which are naturally a function of temperature, calculated using the internal energy function appropriate to rotation in a $-CH_2-$ chain are plotted in Fig. 3.11. The number of effective free links will be decreased by the factor $(1-\psi)/(1+\psi)$ also plotted in Fig. 3.11.

The behaviour of real chains depends on the exact sequence of bonds and side-groups along the polymer chain, and more detailed discussion of this problem is part of polymer chemistry. Some general ideas about likely behaviour can be predicted from the numerical values of energy barriers.

3.2 POLYMER SOLUTIONS

3.2.1 Physics and physical chemistry

By tradition, the study of solutions is regarded as part of physical chemistry. The justification for this is that solubility and the behaviour of solutions are determined by the chemical interaction between solvent and solute. The 'chemical

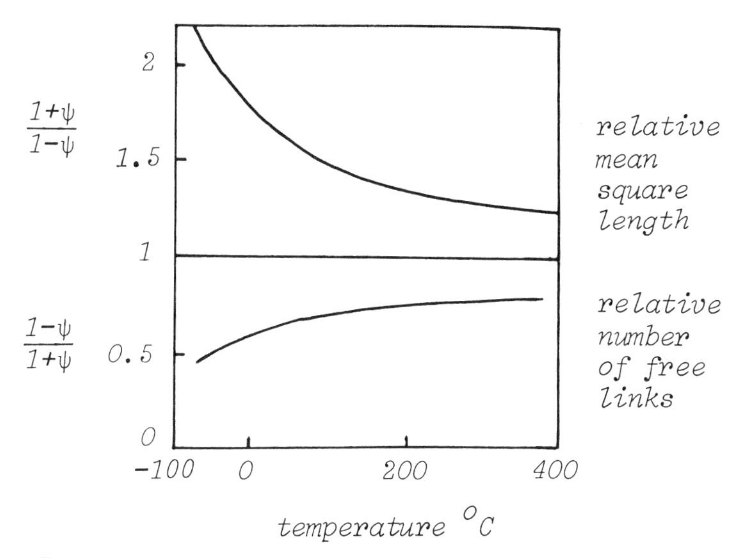

Fig. 3.11 – Effect of temperature on paraffin chain dimensions.

potential' dominates the thermodynamics. It would be inappropriate in this book to go into detail on the thermodynamics and physical-chemical aspects of polymer solutions: these subjects are extensively treated in text-books of polymer chemistry and in monographs.

There are however two areas of interest to the physicist. Firstly, there is experimental and theoretical work on physical properties of solutions, in particular on viscosity and visocoelasticity. Secondly, polymer solutions offer the only easy experimental method of investigating the conformations of individual free polymer molecules.

3.2.2 Interactions and excluded volumes: 'theta' solvents.

The actual conformations taken up by polymer chains in solution will depend on the strength of the interactions between the molecules. In a poor solvent, where the interactions within the polymer chain are strong and the interactions with the solvent are weak, there will be many points of close contacts between different parts of the chain: the molecule will take up a fairly tight conformation, which occupies less volume than a conformation calculated from an ideal theory based on no interaction between different parts of a chain. It is these interactions which lead, in the absence of solvent, to the condensation of polymer molecules: they also result in an error in the theoretical prediction of the behaviour of polymers in the high-temperature, low-pressure gaseous state. On the other hand, when the interactions with the solvent are stronger, the polymer chain will tend to spread out into a more open conformation, so as to be associated with as many solvent molecules as possible. If the two opposing effects could be just balanced, it might appear that the polymer chains would take up the theoretically predicted conformations. The results would be analogous to neutralising the need for the interaction correction $(p + a/v^2)$ in van der Waal's equation for real gases.

There is however another factor which must be taken into account. All the theories are based on the assumption that the polymer chains are made up of links of infinitesimal thickness: the ideal polymers are assumed to be **line** chains, in the same way that ideal gas molecules are regarded as point particles. But in reality, molecules take up space. Many of the theoretical conformations would not be possible, because they would demand the simultaneous occupation of the same space by different parts of the chain. This **excluded volume** effect means that real polymer conformations should be more open than those predicted theoretically, just as the $(v - b)$ term in van der Waals' equation implies that the volume occupied by real gases would be greater than that occupied by ideal gases at the same temperature and pressure.

Flory has shown that it is possible to obtain conditions in solution in which the tendency for the conformation to be more open due to interaction with the solvent just counterbalances the excluded volume effect. In summary, we can describe the various conditions in solution in the way shown in Table 3.2. As the temperature of a solution rises, the increased energy of vibration will alter both the effective volume occupied by parts of a chain and the effect of interactions. Consequently, the balance, which is necessary for the interaction and excluded volume corrections to just neutralise one another, will only occur for a particular solvent at a single temperature. Flory called this the *theta* Θ temperature. Development of the thermodynamics yields relations which can be used to determine the value of Θ and to reduce experimental data to the Θ condition. The temperature Θ is analagous to the temperature in real gases at which the two correction terms

Table 3.2 – Effect of solvent on chain conformations.

Type of solvent	Interactions between		Effect on polymer conformation
	solvent-polymer	polymer-polymer	
Poor	Weaker	Stronger	Tight conformation.
Moderate	Equal	Equal	Close to ideal conformation.
Theta	Slightly stronger	Slightly weaker	Ideal conformation: interaction with solvent just balances excluded volume.
Good	Stronger	Weaker	Open conformations.

in real gases at which the two correction terms in van der Waal's equation just neutralise one another, so that the gas behaves like an ideal gas with $pv = RT$. At the theta temperature which can only be reached with poor solvents, the solution behaviour is similarly ideal.

Another requirement for ideal behaviour in any solution is that it should be dilute. This is particularly true in polymer solutions where inter-actions and entanglements between separate long-chain molecules are very likely to occur. Studies of the thermodynamics of more concentrated polymer solutions have also been made.

3.2.3 Colligative properties

In an ideal solution of simple substances, in the limit as infinite dilution is approached, the colligative properties, such as osmotic pressure, change in vapour pressure, and change in freezing-point, depend on the relative numbers of solvent and solute molecules. Measurement of these properties may therefore be used to determine molecular weights. In the early days of polymer science, it was doubted whether the classical arguments applied to polymers. But the theoretical basis is now well-established, and measurements of molecular weight may be made provided appropriate precautions are taken. The osmotic pressure is the most practical quantity to measure, since the changes in the other quantities are very small. Details of methods are given in many text-books. Any real polymer material will have a distribution of molecular weights, and the particular average value given by the colligative properties is the number-average molecular weight.

From the molecular weight and a knowledge of the chemical constitution of the monomer, the degree of polymerisation, the total number of links in the chain, and the fully extended chain length can be determined.

Experimental studies of the colligative properties of more concentrated solutions give a means of checking the validity of particular theories of the thermodynamic behaviour of polymers in solution.

3.3 VISCOSITY AND RELATED PHENOMENA IN POLYMER SOLUTIONS

3.3.1 Empirical treatment of viscosity without chain entanglement

It is not surprising that when long thread-like molecules are dissolved in a liquid its viscosity should increase: indeed, this happens on a larger scale when fibres are dispersed in water. For dilute solutions, which we can define as those in which each polymer molecule is far from any other, there will be a small increase in viscosity; but, in more concentrated solutions, where interactions and entanglements between polymer molecules become important, the viscosity increases rapidly, until, at quite modest concentrations, the solution becomes very thick.

The viscosity of polymer solutions may be measured by observing the relation between stress and flow in a number of standard forms of apparatus:

capillary viscometers (most commonly used in molecular weight determinations), concentric cylinder viscometers, rotating disc viscometers, and cone-and-plate viscometers. A **coefficient of viscosity** η is defined in the usual way in terms of the relation between shear stress σ and velocity gradient du/dz or rate shear strain $d\gamma/dt$:

$$\sigma = \eta(du/dz) = \eta(d\gamma/dt) \tag{3.47}$$

This defintion is quite general; but, in order to apply it to any particular form of viscometer, it is necessary to integrate over the whole flowing liquid, thus obtaining for example Poisseiulle's equation for flow in a capillary. With a sensibly designed viscometer, there are no problems about this when η is constant; but more detailed analysis and careful experimental design are needed when η is itself a function of the rate of shearing.

As an expression of the results of measurements of viscosity of polymer solutions, the following terminology (due more to physical chemists than physicists) has grown up. A dimensionless quantity, the **specific viscosity** η_{sp} is defined as:

$$\eta_{sp} = (\eta - \eta_0)/\eta_0 \tag{3.48}$$

where η_0 is the viscosity of the solvent.

In simple situations, with no interference between neighbouring polymer chains, one might expect this fractional increase in viscosity due to the polymer to be proportional to the amount of polymer added: in other words, the ratio η_{sp}/c, where c is the concentration, should be constant. In practice, it usually increases as c increases. However, the effects of interference can be removed by extrapolating back to zero concentration, and so an **intrinsic viscosity** or **limiting viscosity number** is defined as the value of η_{sp}/c extrapolated back to $c = 0$. The intrinsic viscosity is denoted by the symbol $[\eta]$.† Effectively this is a limitation to the first term in a power series expression of η_{sp} in terms of c.

Another quantity which it is convenient to introduce is the ratio of η_{sp}/c to $[\eta]$, which may be termed the **reduced specific viscosity**: we note it would remain equal to one if η_{sp} was proportional to c, but changes due to the non-linear dependence of η_{sp} on c.

As an alternative to the specific viscosity, some workers use the **relative viscosity** η_r equal to η/η_0, and thus to $(1 + \eta_{sp})$. Except at very small concentrations, the difference between η_{sp} and η_r will be negligible.

It is reasonable to expect that the longer the polymer chains (for a given weight of polymer present), the greater would be the interference with the flow of the liquid. So the intrinsic viscosity should increase with molecular weight M. An abundance of experimental observation fits the semi-empirical Mark-Houwink Equation:

$$[\eta] = KM^a \tag{3.49}$$

† A disadvantage of the terminology is that η has the dimensions $ML^{-1}T^{-1}$; η_{sp} is dimensionless; and $[\eta]$ has the dimensions of c^{-1}, namely $M^{-1}L^3$.

where K and a are parameters which depend on the particular polymer solvent, and temperature, but are independent of molecular weight.

Consequently viscosity is an arbitrary measure of molecular weight, and in quality control and similar practical operations this is often all that is needed. In order to use viscosity to get actual values of M, it is necessary first to calibrate the system by using fractionated polymer specimens, whose molecular weight can be determined by other methods, in order to find the values of K and a. Once this is done, measurement of viscosity may be the simplest method of molecular weight determination. However, in an ordinary sample with a range of molecular weights, it must be noted that the values obtained from the above equation is the so-called viscosity-average molecular weight, which usually lies between the number-average and mass-average values. The viscosity-average of a given disperse sample is a function of the value of a, and so depends on the conditions of measurement. In a solvent at the theta-temperature, it is found that $a = 0.5$ and K depends on the value of the theta-temperature. In good solvents, a is closer to 1.

The dependence of viscosity on the other parameters of concentration, temperature, nature of solvent, nature of polymer, and (once one moves away from the extrapolation to zero concentration which gives $[\eta]$) molecular weight is complicated and not well understood.

Fig. 3.12 gives raw data on the rapid increase of viscosity with concentration. Various empirical schemes of correlation have been proposed, usually

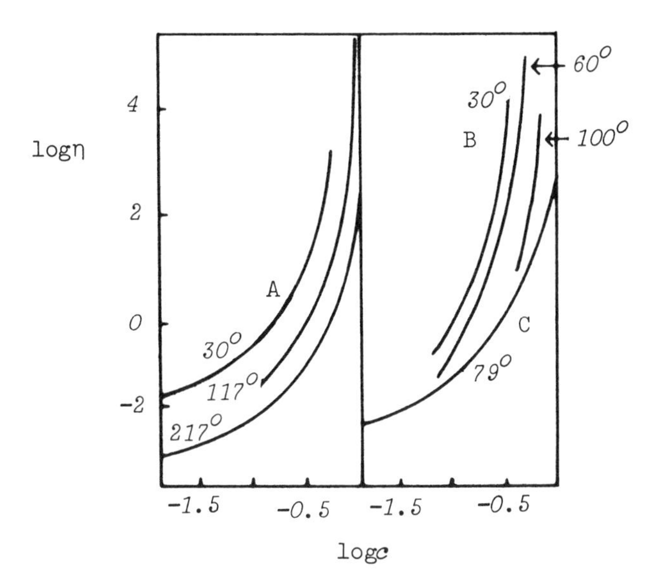

Fig. 3.12 – Dependence of viscosity η on concentration of solution: A = polystyrene in dibenzyl ether; B = PMMA in diethyl phthalate; C = decamethylene adipate in diethyl succinate, at temperatures indicated. After Fox.

with a viscosity variable such as η_r related to a combined parameter of the form $c^x M^y$. For example much data can be fitted to a plot against $cM^{0.68}$, becoming asymptotically proportional to $(cM^{0.68})^5$ or $c^5 M^{3.4}$ above some critical value of concentration. However in many instances this relation does not fit. Fig. 3.13 shows examples where values of $x = 1$ with $y = 0.68$ and 0.5 are the forms to use.

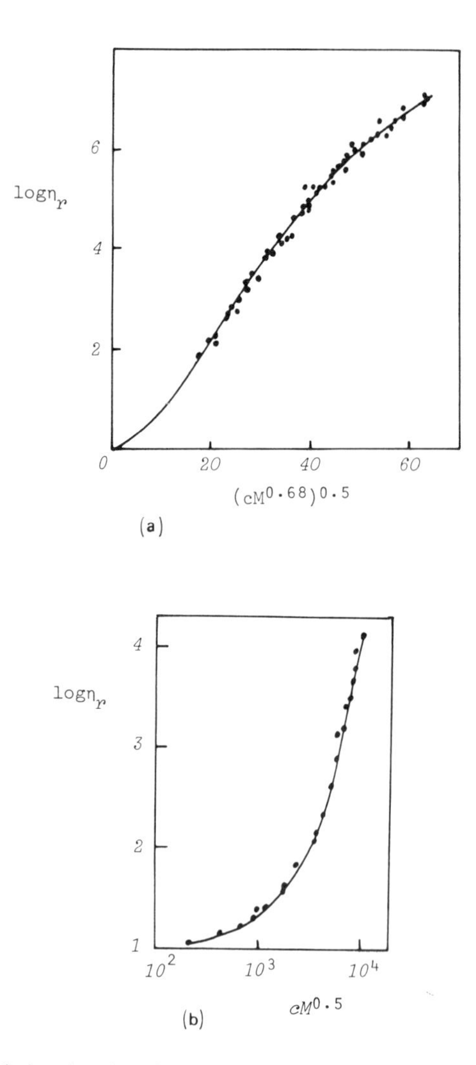

Fig. 3.13 – (a) Relative viscosity of polyisobutylene solutions plotted as a function of $(cM^{0.68})^{1/2}$ for various molecular weights and solvents and concentrations from 5 to 40% by weight. After Ferry. (b) Relative viscosity plotted against $(cM^{0.5})$ for polystyrene in decalin, for two molecular weights. After Gandhi and Williams.

Simha suggests that it is instructive to plot log $(\eta_{sp}/c)/[\eta]$ against c. Fig. 3.14 shows two examples of this form of plot. In all cases, except the non-polar good solvent, there is a change of slope at some concentration, where structure must develop in the solution although the concentration is well below the level of extensive entanglement.

Fig. 3.12 also shows the considerable influence of temperature. Typical examples of the decrease of viscosity with shear rate are shown in Fig. 3.15.

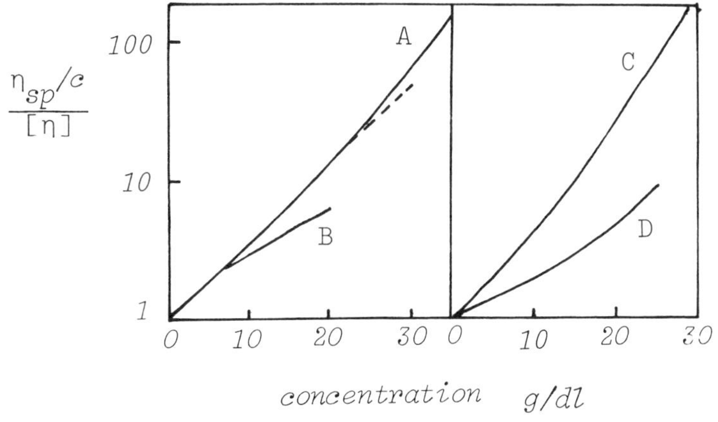

Fig. 3.14 – Plots of log $(\eta_{sp}/C)/[\eta]$ against c for (A) poor solvent with nonpolar interaction, polystyrene in decalin; (B) good solvent with nonpolar interactions, polystyrene in toluene; (C) poor solvent with polar interaction, PMMA in m-xylene; (D) good solvent with polar interactions, PMMA in chlorobenzene. After Gandhi and Williams.

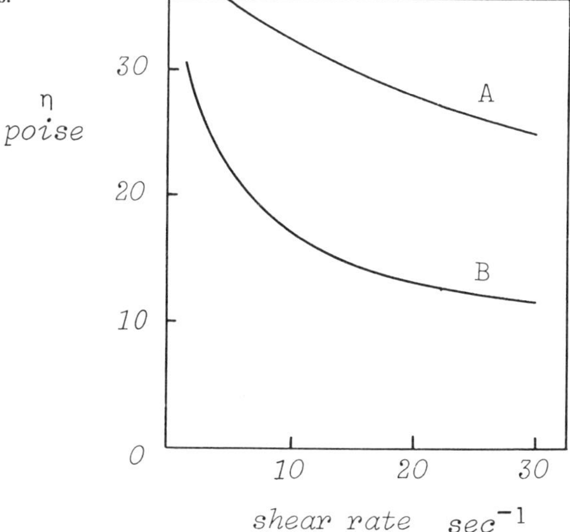

Fig. 3.15 – Viscosity plotted against shear rate for (A) $2\frac{1}{2}$% concentration of PMMA in dimethyl phthalate (B) 6% concentration of polyisobutylene in decalin. After Kaye.

3.3.2 The movement of polymer molecules in dilute solution

The general problem which faces us when we try to develop a theory of viscosity of polymer solutions – even very dilute solutions – is one which will crop up many times in this book. The real situation during flow is of complicated, coupled movements of what is effectively an infinite number of atoms (actually about 10^{24} for one gram-molecule of material). Progress is therefore only possible by introducing a simplified model of the situation. In some instances, for example the regular lattice of a crystalline solid or the random distribution of an ideal gas or a rubbery polymer chain, the model is realistic and gives good predictions. In many other instances, the problems are more difficult: the subject of viscosity, even of simple liquids, is one of these, where various theories, based on crude models, give some understanding but no satisfying complete analysis. It will be even more difficult to develop a full theory of the viscosity of polymer solutions where the forces which must be taken into account are those between solvent and polymer molecules, those between separate polymer chains, those transmitted along polymer chains, and those associated with changes in polymer chain conformations.

Progress has however been made in predicting the changes in solvent viscosity which will occur as a result of the addition of polymer molecules. For dilute solutions, in which the polymer molecules are far apart, a drastic simplification is usually made. We choose to ignore almost completely the molecular nature of the polymer chain, and treat it instead as a macroscopic solid object, through and round which the liquid can flow. This removes the problem from the realm of molecular physics, and makes it a problem – albeit a complicated one – in classical properties of matter (or in other terminology, a problem in continuum mechanics or in applied rheology).

This approach can be applied to several observed phenomena: to viscous flow, where the movement of the solvent will lead to movements of the molecules; to diffusion, where the molecule moves through the solvent as a result of Brownian motion (here the argument goes full circle, since the motion which Brown saw was that of solid particles of pollen, and from this the motion of small molecules was inferred – now we treat the polymer molecules as solid particles); to the fall of a polymer molecule through a solvent under gravitational (or centrifugal) force in a sedimentation experiment; and to the flow of a charged polymer molecule through a solvent in an electric field. The easiest to consider are the last two, where the polymer is moving through the liquid under a clearly defined applied force. Since we are only concerned with the relative motion of polymer molecule and solvent it is permissible, and leads to easier understanding, to switch the system round and regard the polymer molecule as fixed with the solvent flowing past it.

For a spherical solid particle of radius R_s, as in Fig. 3.16(a), Stokes' law will apply (provided the liquid is Newtonian, and the velocity is not too high) and we have:

$$F = 6\pi\eta_0 R_s v \tag{3.50}$$

where F is the viscous force exerted on the particle and v is the relative velocity of particle and fluid (at this point it is simpler to think of the particle moving through the fluid).

More generally, we can define a frictional coefficient ζ between particle and liquid and put:

$$F = \zeta v \tag{3.51}$$

For a polymer chain, we have to try to estimate the radius R_s, equal to $(\zeta/6\pi\eta_0)$, of the hydrodynamically equivalent sphere. Two extreme simplifications come to mind. If the polymer molecule is tightly coiled, as in Fig. 3.16(b), it may be reasonable to regard it as effectively impermeable to solvent and to take R_s as the radius of the circumscribing sphere. On the other hand, if the polymer chain is a very open conformation as in Fig. 3.16(c), the solvent may be regarded as flowing uniformly through it as a free-draining molecule, with the resistance to flow given by the sum of the resistance of all the individual units. If the chain is regarded as made up of N_v units† having a frictional resistance ζ_0 and a hydrodynamically effective radius R_0, we have:

$$\zeta = N_v \zeta_0 = N_v(6\pi\eta_0 R_0) \tag{3.52}$$
$$R_s = N_v R_0 \tag{3.53}$$

However for a typical moderately open chain conformation, as in Fig. 3.16(d), the flow pattern will be intermediate between the two extreme forms.

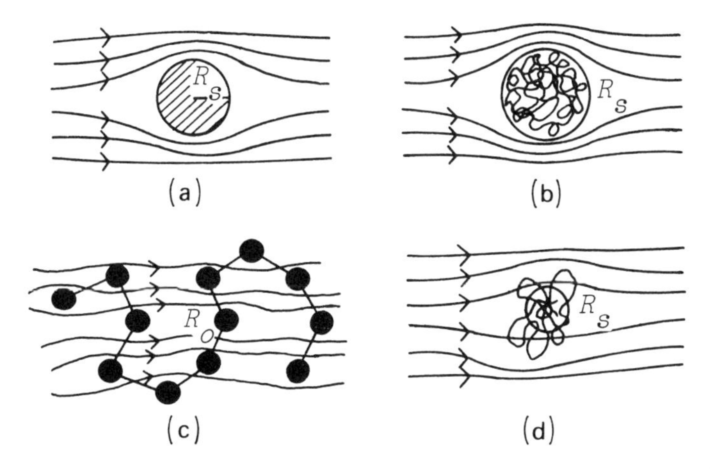

Fig. 3.16 – (a) Spherical solid particle in fluid flow. (b) Densely packed polymer molecule acting as a solid molecule. (c) Flow through an open polymer molecule round spherical units of chain. (d) Mixture of flow round and through a polymer molecule.

† We may choose to identify N_v with N, the number of chemical units in the chain, with N_e, the number of effective random units, or with some other choice of chain unit.

Solvent will flow through the molecule, but at a slower rate than the flow round the molecule. The behaviour of such a system can be inferred from the following suggestive argument. We consider the simplest situation in which the quantity ζ_0/η_0 is large: this means that the frictional resistance to flow through the polymer molecule will be very much greater than the viscous resistance to flow round it. The effect will be particularly marked for the rather densely packed regions near the centre of gravity of the molecule. The central regions can thus be regarded as effectively impermeable to solvent, while the outer regions offer little resistance, as in Fig. 3.16(d). Thus it seems reasonable to regard the molecule as replaced by an effective hydrodynamic sphere whose radius R_s is proportional to $(\bar{s}^2)^{1/2}$ the root-mean-square distance of the chain units from the centre of gravity. Provided ζ_0/η_0 is large enough, one can expect the pattern of flow to be independent of its actual value.

We can thus put:

$$R_s = K_1\,(\bar{s}^2)^{1/2} \tag{3.54}$$

When ζ_0/η_0 is smaller, there will be appreciable flow through the central regions of the molecules, and it has been suggested that R_s should be reduced by some factor. However comparison with experimental results and more detailed analysis of particular models shows that in most instances it is reasonable to assume the form given in (3.54). It is reasonable to assume, at any rate for most common polymer conformations, that K_1 will be a universal constant, dependent on the pattern of flow past the molecule when ζ_0/η_0 is high. A more detailed analysis of a particular model suggests a value of 12.5 for K_1. In practice it is more convenient to express the space occupied by the polymer by r_r the root-mean-square end-to-end length of the chain, and put:

$$R_s = K_2 r_r \tag{3.55}$$

The ratio K_2/K_1 will depend on th nature of the chain conformation, and will be $1/\sqrt{6}$ for a Gaussian chain. Substituting from (2.30) we can therefore write:

$$R_s = K_2 a_e N_e^{1/2} \tag{3.56}$$

or

$$\zeta = 6\pi\eta_0 K_2 a_e N_e^{1/2} \tag{3.57}$$

Once the friction coefficient ζ, or the radius R_s, has been specified, the problem of sedimentation is solved. If we neglect the extent to which it is opposed by Brownian motion, we have:

$$(m_p - m_l)g - \zeta(\mathrm{d}x/\mathrm{d}t) = m_p(\mathrm{d}^2x/\mathrm{d}t^2) \tag{3.58}$$

where m_p is the mass of the particle (polymer molecule); m_l is the mass of displaced liquid; g is the acceleration due to gravity (or the centrifugal acceleration), and $\mathrm{d}x$ is the displacement in time $\mathrm{d}t$.

The limiting velocity, when $\mathrm{d}^2x/\mathrm{d}t^2 = 0$, is given by:

$$(\mathrm{d}x/\mathrm{d}t)_\infty = (m_p - m_e)g/f \tag{3.59}$$

Similarly the flow in an electric current is given by substituting the electrostatic force in (3.50).

The problem of viscosity is more complicated. The polymer molecules move with the flowing liquid, so there is no relative transitional motion, such as we have been discussing. However, as illustrated in Fig. 3.17 the velocity gradient in the liquid means that polymer molecules (or particles) must rotate. It is this rotation which generates additional energy loss in the system, and so leads to the increase of viscosity.

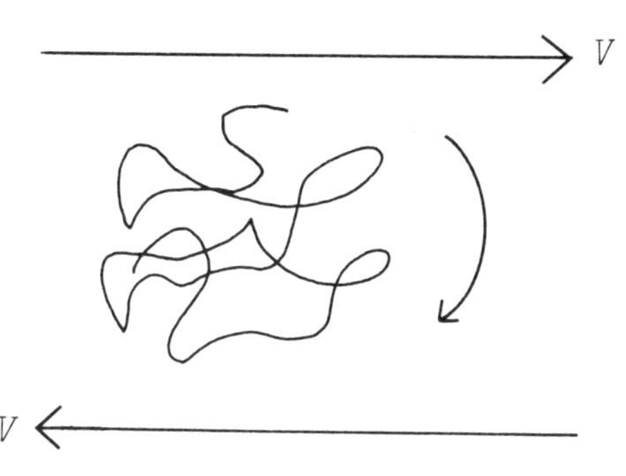

Fig. 3.17 – Rotation of polymer molecule due to velocity gradient.

Einstein showed that the specific viscosity due to rigid spheres suspended far apart in a fluid was given by:

$$\eta_{sp} = (5/2)\phi \tag{3.60}$$

where ϕ is the volume fraction occupied by the particles.

It is interesting to note that the increase in viscosity depends only on the total volume of added particles, and not on their size. Considering the number, size, and mass of the spheres, we have:

$$\phi = (c/m)(4/3\pi R_s^3) \tag{3.61}$$

where c is the concentration of dispersed material, and m is the mass of a single particle.

For polymer molecules in solution, $m = N_e M_e / N_0$ where M_e is the molecular weigth per random link, and N_0 is Avogadro's number. It is also reasonable to expect that an expression for the radius of the effective hydrodynamic sphere for the rotational motion in viscous flow would be of the same form as for the translational motion already discussed. By analogy with (3.56), we therefore put:

$$R_s = K_3 a_e N_e^{1/3} \tag{3.62}$$

where K_3 may or may not equal K_2.

Substitution from the above equations then gives:

$$[\eta] = \eta_{sp}/c = (10\pi N_0 / K_3^3 a_e^3 / 3M_e) N_e^{1/2} = K_4 N^{1/2} \tag{3.63}$$

where N is the number of chemical units in the chain and K_4 is a constant for a given type of polymer.

It will be observed that this equation is of the same form as the empirical (3.49). Its derivation depends on the assumption of Gaussian statistics, and so the exponent a will be expected to be $\frac{1}{2}$ only when this applies, namely at the Θ temperature.

3.3.3 Improvements in the model for dilute solutions

There are obviously many approximations and crude assumptions in the treatment as given in the last section and the subject has been developed further in a number of papers. The topics dealt with include:

(a) a more exact analysis of the behaviour of the free-draining model bringing in the hydrodynamic interaction in the flow between neighbouring chain units; whereas the simple model assumes that the chain units act as a cloud of separate particles, the Kirkwood-Riseman theory takes account of the fact that they are really connected together in chains, so that the flow will be different;

(b) a closer study of the movement of particles during viscous flow, including effects when the diameter of the flow channel is not much greater than the effective radius R_s of the spheres;

(c) consideration of the effects of deviation from Gaussian conformations, either towards more open conformations in good solvents or tighter ones in bad solvents;

(d) the effect of some solvent molecules being effectively immobilised by association with the polymer molecules.

These are all detailed improvements of the crude theory – complicated in mathematics, valuable in enabling more detailed predictions to be made in special cases, but not adding greatly to our physical understanding.

There are however other aspects where a fundamental improvement of the model is required.

The simple model assumes that the polymer molecule can be regarded as a spherical symmetrical, rigid, random coil. But, except, below the glass-transition temperature for an isolated chain in solution, the random coils will be flexible. It is therefore necessary to take account of the internal movements of the chain itself, and of the change of shape of the polymer molecule. Apart from changing the drag, the change of shape will introduce an elastic force, since when the flow stops the chain will revert to its spherical random form. The large scale analogue would be a dispersion of threads which deform elastically under the influence of viscous forces. Polymer solutions will then be viscoelastic as discussed later.

A different situation arises with polymers – particularly proteins – which take up a specific rigid conformation in solution. These may be globular particles or rods, and the viscous drag will then be influenced by the shape of the particle.

The usual procedure is to introduce one additional parameter to characterise shape and regard the molecules as ellipsoidal particles, so that the dimensions of an equivalent hydrodynamic ellipsoid of revolution have to be estimated. The analysis of the flow behaviour of dispersions of ellipsoidal particles can then be applied.

3.3.4 Anistropy in flow

Elongated particles tend to line up in a flowing fluid: they follow the streamlines. Consequently when polymer molecules are inherently anistropic in shape, for example rod-like protein molecules, they will orient and give rise to anistropy in the structure of the flowing solution.

Even where the equilibrium form of the polymer molecule is a spherically symmetrical random coil, an elongated form will be induced by a shear gradient. This is illustrated in Fig. 3.18, where we see that because the solvent near end A is moving faster to the right than that near end B it must lead to an elongation of the molecule. There will also be some other internal deformation of the molecule, resulting from the tendency of segments at C to move faster than those at D. If this is combined with rotation of the molecule as a whole, it will lead to a cyclic deformation of the molecule. However it may be that rotation is impeded, but even if rotation is completely prevented, there will still be viscous drag, because with the molecule moving at a uniform speed, segments near C will be moving slower than the neighbouring liquid, while those near D will be moving faster. The difference in viscous drag will lead to an elastic distortion of the chain, in order to balance the forces.

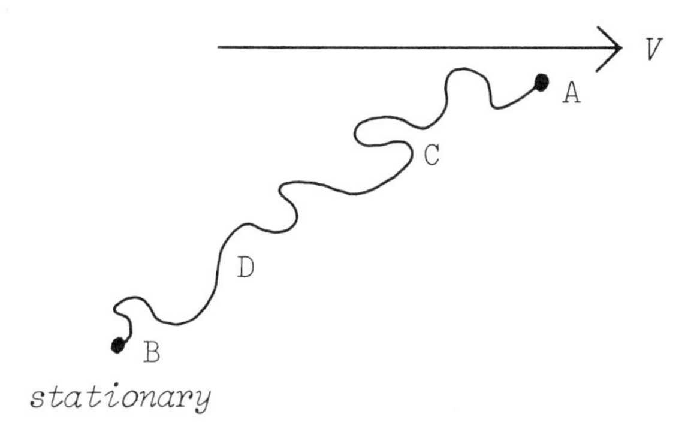

Fig. 3.18 – Elongation of a polymer molecule due to velocity gradient.

There are several effects which should be taken into account in studying the behaviour of flexible chain molecules in solution. Firstly, the flexibility of the molecule means that it will tend to move locally with the liquid, and so will not increase the viscosity as much as a rigid chain would. Secondly, extension of the chain leads to the molecules lying more along the lines of flow and so giving less viscous drag: in real high-speed flow conditions, this can lead to a considerable reduction in viscous drag if it suppresses turbulent flow and leads to stream-line flow, which is the type of flow which we have been considering in this chapter. Thirdly the internal motion of the molecule will contribute to energy loss.

Directly or indirectly these factors will lead to the dependence of viscosity on rate of shear.

The consequences of orientation of polymer chains as a result of flow can be observed optically as flow birefringence. Anisotropy in flow behaviour can also lead to movements perpendicular to the direction of applied stress. Thus if a rod is rotated in a fluid with anisotropic viscosity, the liquid may climb up the rod due to the development of flow perpendicular to the applied stress.

3.3.5 Chain interaction in flow

When the solution is more concentrated, so that motion is transmitted directly from one polymer chain to another, instead of being transmitted only through solvent molecules, the viscosity will become much greater, and the forms of analysis given earlier will no longer be valid. One can envisage three regions of behaviour, though these merge into one another: (a) for very dilute solutions, as in Fig. 3.19(a), there will be virtually no direct interaction between polymer molecules; (b) for more concentrated solutions, as in Fig. 3.19(b) we can consider the polymer molecules as separate entities, but must take account of interaction; (c) for very concentrated solutions, as in Fig. 3.19(c), the molecules are extremely entangled, and flow would be dominated by the movement of molecular segments past one another.

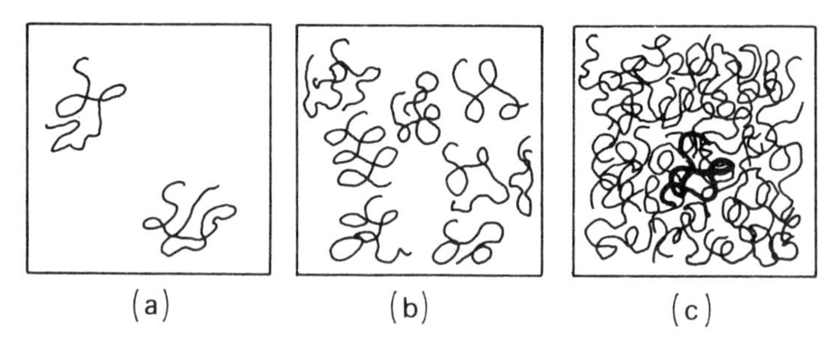

Fig. 3.19 – (a) Very dilute polymer solution. (b) Dilute polymer solution. (c) Concentrated polymer solution.

We can take the effective outer radius r_{ext} of the space occupied by the polymer molecule as about $3r_r$. The spacing S between molecules will be of the order of $n^{-1/3}$ where n is the number of molecules per unit volume. Therefore:

$$\begin{aligned} S/r_{ext} &\cong n^{-1/3}/3r_r \\ &= (N_e M_e / c N_0)^{1/3} / 3 N_e^{1/3} a_e \\ &= (M_e / 27 c N_0 N_e^{1/2} a_e^3)^{1/3} \end{aligned} \tag{3.64}$$

The solution can only be regarded as dilute if $S/r_{ext} \gg 1$, that is if:

$$c \ll M_{\Theta}/27N_0 N_e^{1/2} a_e^3 \qquad (3.65)$$

A typical set of values would be $M_e = 40$, $N_0 = 6 \times 10^{23}$, $N_e = 1000$, $a_e = 10^{-7}$ cm: this would mean that a polymer solution would be effectively dilute only if $c \ll 10^{-4}$ g/cm^3 (0.01%). If $c \sim 10^{-4}$ g/cm^3, there would be appreciable interaction; and if $c \gg 10^{-3}$ the solution will appear highly concentrated with much entanglement of the chains. It is therefore clear why the constant limiting value of η_{sp}/c occurs only at extremely low concentrations.

In highly concentrated solution, the pulling out of chains into oriented conformations will be pronounced. The solution is likely to be highly viscoelastic and to show many anomalous effects resulting from the anisotropy induced by flow.

In the discussion of the behaviour of dilute solutions, it is assumed that the friction coefficient is given by summing the drag on each unit of the chain. In a very dilute solution, where the molecules are widely separated from one another as in Fig. 3.19(a), it is necessary, as we have seen, to take account of the disturbance of the flow path near a molecule regarded as a particle. In a more concentrated solution, where the molecules are close together as in Fig. 3.19(b) the flow path will be reasonably uniform past all the units of each chain. The frictional coefficient per molecule is then given by (3.52):

$$\zeta = N_v \zeta_0$$

However, this situation will no longer be true when the chains are entangled with one another. In these circumstances, one chain will drag along another. If the chains were tied together in groups, each of x chains, the new number of units in the chain would be xN_v. This argument suggests that with entanglement, we might replace N_v by a value N_v^* giving the number of units linked together as an effective single molecule.

If the linking together of chains was by chemical cross-links, the picture would be fairly simple. As discussed in section 2.5.3 on gelation, there is a rapid change from a situation in which few molecules are linked to one in which all the chains are linked together. In the former instance, the viscous behaviour would be very little affected by the cross-linking: in the latter it would form a solid network – a gel – with an infinite viscosity.

But entanglements are weaker links, and the molecules can slide past one another, although this will be inhibited by the entanglement. Bueche has analysed this problem approximately by introducing a slippage factor s which gives the relative viscosity of successive chains. However this treatment also applies, with less complication, to the viscosity of molten polymers, namely at 100% concentration, and will be discussed further in that context in section 3.5.2. There is a change over from a prediction of proportionality to M at moderate concentrations and molecular weight to $M^{3.5}$ when there is a high degree of entanglement and interaction, as shown in Fig. 3.20. In solutions, as distinct from melts, the concentration interacts with the molecular weight in determining the conditions for change from one mode to another.

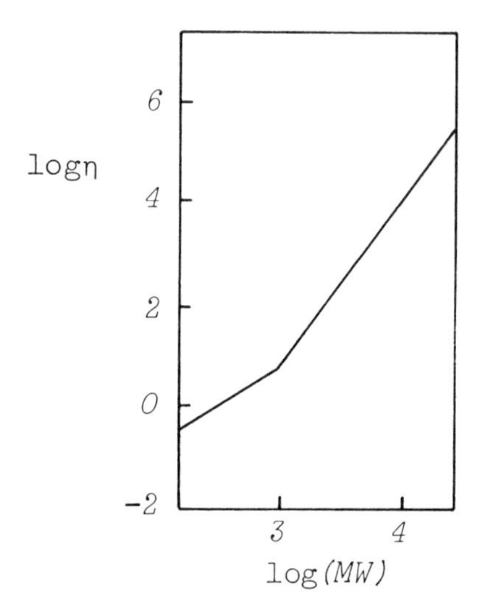

Fig. 3.20 – Plot of viscosity of 25% solution of PMMA in diethyl phthalate at 60°C.

In summary, we now have three relations between viscosity and molecular weight. For very dilute solutions, in solvents at the Θ temperature, we have:

$$\eta \propto N^{1/2} \quad \text{or} \quad M^{1/2}$$

For moderately concentrated solutions, we have:

$$\eta \propto N \quad \text{or} \quad M$$

For highly concentrated solutions, we have:

$$\eta \propto N^{3.5} \quad \text{or} \quad M^{3.5}$$

It must be remembered that the concentration ranges for which these equations apply are themselves dependent on the molecular weight, so that the boundaries between the ranges are surfaces defined by concentration and molecular weight values.

The whole discussion so far has also tacitly assumed that the polymer chains are all of the same length. With real polymer materials, the effects of the dispersion of chain lengths must, of course, be considered. The averaging problems are complicated.

3.4 VISCOELASTICITY OF POLYMER SOLUTIONS

3.4.1 The phenomenon of viscoelasticity

As shown in Chapter 2, the equilibrium state of a polymer chain is a random conformation, although as discussed earlier in this chapter the conformation may be somewhat expanded or contracted. Any distortion away from the random conformation, such as would occur due to viscous drag in a flowing liquid, gives a decrease in entropy and hence generates an elastic force: a polymer solution can be expected to be visoelastic. This is a particular example of the behaviour of a chain in a viscous medium discussed in section 2.3.4.

Since, by definition, we are considering liquid systems in this chapter, the model of behaviour must have the form shown in Fig. 3.21, where the elements may have nonlinear responses. There can be no spring in parallel with the whole system, since this would imply ultimate complete recovery. In appropriate circumstances, the behaviour of such materials can be represented by a complex modulus, and Fig. 3.22 illustrates a typical viscoelastic response for a polymer solution.

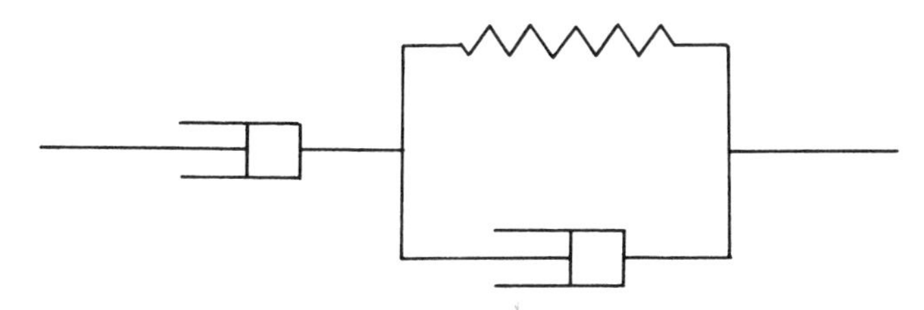

Fig. 3.21 – Basic model of viscoelasticity in a polymer solution.

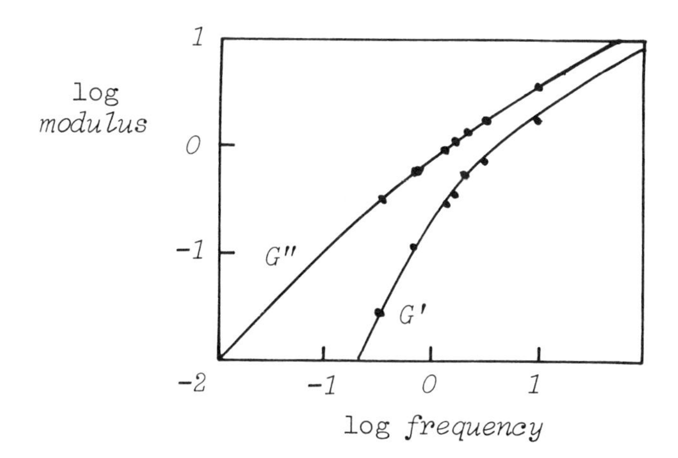

Fig. 3.22 – Viscoelasticity response of poly-α-methyl styrenes in cyclohexane at 39°C, for two molecular weights. The curves are a fit of the general Tschoegl theory. After Ferry.

In more concentrated solutions, the viscoelastic response becomes much more complicated, because of the dominance of the anisotropy of structure induced by flow. We can no longer characterise the behaviour by a constitutive equation (equation of state) which treats stress and strain in one direction effectively as scalar quantities. Instead we must consider the relation between the stress tensor and the strain tensor. Lodge has shown that, for uniform states of flow or changes of shape, this can be analysed by describing the stress and strain in terms of reference vectors embedded in the material: a knowledge of tensor analysis is not necessary until non-uniform states of flow are considered.

This complication leads to curious effects. For example, Fig. 3.23(a) illustrates how a concentrated solution will swell transversely on emerging from a tube as the elastic deformation recovers, and Fig. 3.23(b) shows the climbing of a rod by a viscoelastic solution, known as the Weissenberg effect.

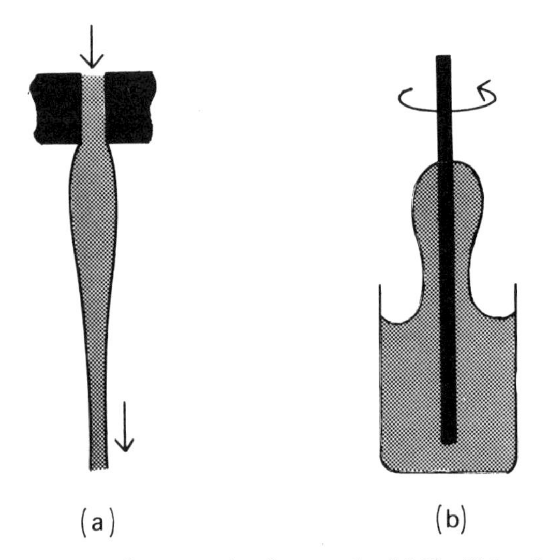

Fig. 3.23 – (a) Die swell on extrusion from a tube. (b) The Weissenberg effect.

3.4.2 Models

Theoretical studies of the viscoelasticity of polymer solutions are based on a development of the bead model for a free-draining molecule already used in our discussion of viscosity. Instead of the single spring and dashpot introduced in the crude discussion of section 2.3.4, the polymer molecule is regarded as made up of a number of sub-units whose mass and bulk is concentrated into the beads, which are linked by elastic strings as illustrated in Fig. 3.24. The problem is then to determine how such a model would behave when immersed in a viscous liquid. In macroscopic terms this is the problem of the vibrations of a loaded elastic string subject to damping. The string could be subject to bending, twisting, or tensile oscillations, and each could take place in many modes.

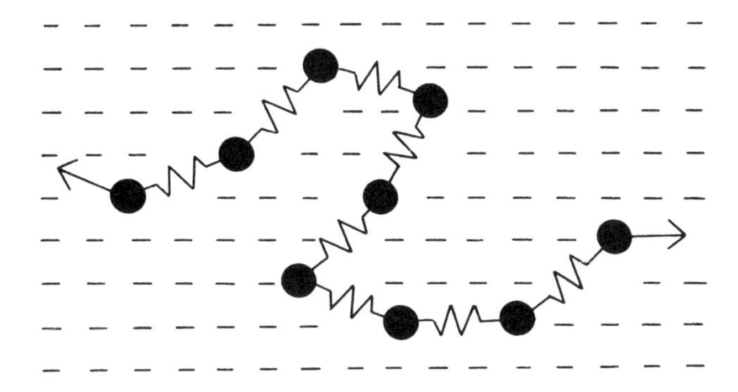

Fig. 3.24 – String and bead molecule of polymer chain in solution.

There have been several different analyses of the problem. Three of these theories, due to Rouse, Bueche, and Blizzard have been shown by Gross to yield mathematically equivalent results. None of them deals explicitly with the detailed geometry of chain deformation; the various possible types of deformation (bending, twisting, extension) are all lumped together in a schematic representation. We can regard each sub-unit of the molecule as composed of q random links, each of length a_1, behaving as a typical entropy-dependent rubbery chain. If we take the simple, linear approximation, this is an elastic spring with a force F_e following the equation:

$$F_e = 3kT/qa_1^2 \delta l \tag{3.66}$$

where δl is the distance between the ends of the sub-unit.

The viscous drag comes from the movement of the beads at a rate dx/dt through the solvent, and will be given by the sum of the resistance to movement of the q monomer units making up the bead. The viscous force F_v will therefore be given by:

$$F_v = q\zeta_0(dx/dt) \tag{3.67}$$

where ζ_0 is the monomeric friction coefficient.

If we regard Stokes' law as valid at the molecular level (although it is macroscopic in concept), we have:

$$\zeta_0 = 6\pi\eta_0 a_2 \tag{3.68}$$

where a_2 is the radius of a monomer unit.

We thus have a model composed of a combination of elastic springs with a force constant $(3kT/qa_1^2)$ and viscous dashpots with viscosity coefficients $(6\pi\eta_0 qa_2)$.

The next problem is to see how the model should be put together. The elastic strings, corresponding to the z sub-units, will obviously be linked in series in order to be equivalent to the whole chain of $N = qz$ random links. Each bead is linked to the surrounding medium (regarded as 'fixed') by the viscous element. We thus get Blizzard's ladder network† as shown in Fig. 3.25. We should note that this is a model for the behaviour of the polymer chains in the viscous liquid, and they have a complete elastic recovery to an equilibrium conformation (or set of conformations), even though there is a viscosity in the system as a whole due to the relative movement of complete molecules. With the formulation of the model, the physics of the analysis is complete. All that remains is to analyse the response of the model to any particular stimulus. The simplest to treat is the behaviour during steady state sinusoidal loading.

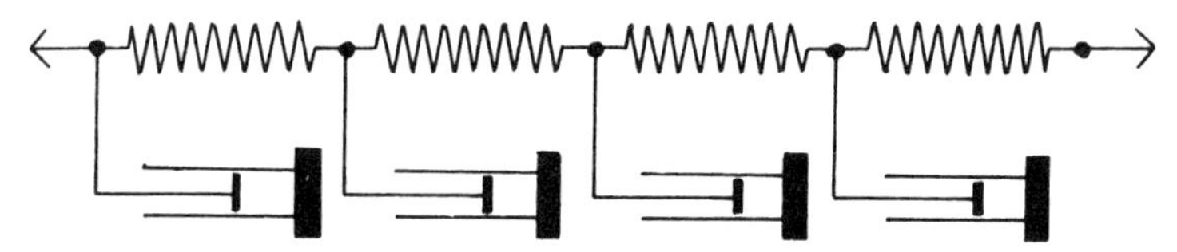

Fig. 3.25 – Blizzard's ladder network.

The response of the system might be thought to depend on the arbitrary choice of q, the size of the sub-unit. However, we can see from Fig. 3.24, that the lower modes of vibration will be independent of the choice of q: the elastic and viscous contribution of neighbouring sub-units can be combined without appreciably altering the behaviour. The low-frequency response will be independent of q. However, as the wavelength of the vibration approaches the length of the sub-unit, the response will depend on the size of the unit: indeed the highest mode of vibration is limited by z the number of sub-units. It would thus seem desirable to make q as small as possible in order to avoid this error in high-frequency response. But this introduces another error: (3.70) is only valid for large values of q, since it depends on the statistical behaviour of a chain of many random links. In summary, one can say that while the model will be valid over a range of frequencies, it must fail at the highest frequencies of dynamic testing.

3.4.3 Alternative approaches to the problem

The treatment given in the last section is really complete in itself. Both the fundamental physical arguments and the basis for calculation of any particular aspect of viscoelastic behaviour are given.

Historically, however, there have been other ways of approaching the problem. As these are superficially different, although physically and mathematically equivalent, and are often described in other books as separate theories,

† In his original treatment, Blizzard used arbitrary constants for the elastic and viscous elements. The molecular interpretation given here is based on the theories of Bueche and Rouse.

it will be useful to describe them. Furthermore one approach may be more useful than another in dealing with more advanced applications.

Bueche, in his approach to the problem, did not explicitly describe a visco-elastic model of springs and dashpots. He viewed the problem directly as one of studying the normal modes of vibration of a series of elastic springs, with force constants given by (3.66), joined by beads which suffered a damping force, given by equation (3.67). A single sub-unit would thus be analogous to a weight oscillating on the end of a spring while immersed in a viscous medium. A series of such units, as shown in Fig. 3.26, would undergo a set of co-operative modes of vibration. If there was no damping, the fundamental mode of vibration, with one node at the centre, would have a frequently ω given by:

$$\omega_1 = (\pi/Na)(3kT/m)^{1/2} \tag{3.69}$$

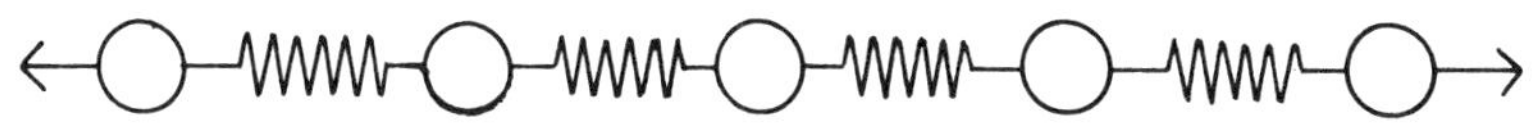

Fig. 3.26 – Bueche's model.

And the higher modes have frequencies given by:

$$\omega_n = n\omega_1 \tag{3.70}$$

where n is the number of nodes.

Clearly with this model it is not possible to define a vibration for which $n > N$, and in practice the model is only useful for $n < N/5$.

The treatment can be generalised to cover the consequences of damping, and linear viscoelasticity theory then enables other responses to be calculated.

Physically, it is clear that the Bueche model is identical with the ladder network though this was not immediately obvious from the way in which the theories were introduced. So it is not surprising that the results prove to be mathematically equivalent.

Rouse, in his analysis went back further towards first principles and worked with the model shown in Fig. 3.27, which is more closely related to the physical representation originally introduced in Fig. 3.24. The entropy of the sequence

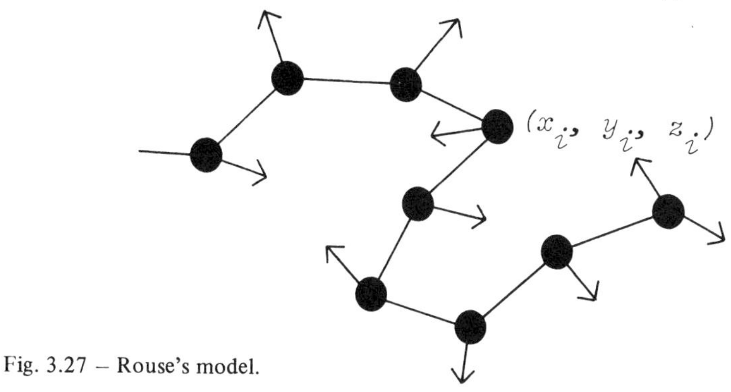

Fig. 3.27 – Rouse's model.

of random links in any sub-unit is given by the relative position of the ends of the sub-units, according to the usual calculation based, in approximation, on Gaussian statistics. Assuming that there are no internal energy changes, the entropy calculations lead directly to the free energy.

If we now consider the motion of one particular bead with co-ordinates (x_i, y_i, z_i), the free energy (potential energy in macroscopic mechanical terms) will come from the sets of random links on either side; and so will be given by the values of (x_i, y_i, z_i), together with the co-ordinates of neighbouring beads $(x_{i-1}, y_{i-1}, z_{i-1})$ and $(x_{i+1}, y_{i+1}, z_{i+1})$. Viscous damping is introduced, as in the other theories, by a friction coefficient $q\zeta_0$. This analysis thus gives a set of equations of motion for the chain, of which one member would be:

$$q\zeta_0 \frac{dx_i}{dt} = \left(\frac{\partial A_i}{\partial x_i}\right) \tag{3.71}$$

where A_i is the free energy associated with position of the ith unit relative to the neighbouring unit.

Values of free energy are then substituted from the statistical thermodynamics of chain conformations. There will be similar equations for all values of i for the x, y, and z co-ordinates.

Once again, the situation is fully defined physically: all that is needed is to solve the set of simulations differential equations. This is what Rouse did: the method is to transform the equations, by a matrix method, into a set of equations based on new co-ordinates – so-called normal co-ordinates – such that the new equations are functions of a single co-ordinate. The results may be expressed in various forms. For example the relaxation modulus is given by:

$$G(t) = NkT \sum_{p=1}^{z} \exp(-t/\tau_p) \tag{3.72}$$

where τ_p is a relaxation time given by:

$$\tau_p = zl^2 q\zeta_0 \, [24kT \sin^2 p\pi/2\,(z+1)\,]^{-1} \tag{3.73}$$

where p may take any integer value for 1 to z.

Once again the limitation on high frequency (short τ_p) response is evident. If $p < N/5$ and $N \gg 1$, (3.7) reduces to:

$$\tau_p = a^2 z \zeta_0 / 6\pi^2 p^2 kT$$

The prediction of the Rouse theory is clearly a set of discrete relaxation times, although in practice these merge together so closely that the discontinuities are not evident. The result may be expressed in a mathematically more sophisticated manner, using the Dirac delta† function δ, by giving the relaxation

† The Dirac delta function is defined by the equation:

$$\int_{-\infty}^{\infty} \delta[\ln(\tau/\tau_p)]\, d\ln\tau = 1$$

spectrum as:

$$H(\tau) = GkT \sum_{p=1}^{N} \delta\,[ln(\tau/\tau_p)] \tag{3.74}$$

where $H(\tau)$ gives the contribution to the modulus associated with relaxation times between $\ln\tau$ and $\ln\tau + \mathrm{d}\ln\tau$.

The physical equivalence of the Rouse theory with the Blizzard and Bueche theories given previously is not quite as clear as the physical equivalence of the Blizzard and Bueche theories. The same damping term is obviously introduced, and the direct calculation of entropy will give the same result as for a spring with a force constant given by an anlysis of the entropy of chain conformations. The remaining feature to be established is the equivalence of the three-dimensional analysis of the motion in Rouse's theory with the one-dimensional treatment of the other theories. This seems likely in view of the fact that the free energy of the chain is a function only of the values of the scalar distances $r_{i,\,i+1}$ between junction points i and $i + 1$ and the energy loss is a function only of the scalar velocity of the beads at the junction points. The equivalence can be more explicitly proved.

Even though Rouse's theory is based on three-dimensions, it loses this character when the scalar energies are calculated, and is not directly applicable to a full analysis of anisotropic viscoelasticity.

The Rouse treatment was generalised by Zimm to take account of hydrodynamic interaction between solvent and polymer molecule, and then solved for the two special cases of zero interaction, which is the Rouse theory, and infinite interaction, which is usually referred to as Zimm theory. A range of partial interactions was made possible in a more general analysis by Tschoegl, and agreement between experiment and this theory is shown in Fig. 3.22.

3.5 GELS AND MELTS

3.5.1 Gels

We can go on to consider two sorts of system in which there are greater interactions between chains. The first is a gel.

Suppose we have a fairly concentrated polymer solution, with a considerable intermingling of chains, and cross-links are formed between chain segments either by chemical reaction or physical entanglement. The complete movement of whole chains away from one another is then prevented. The elastic character will predominate over the viscous. However because of the large number of solvent molecules trapped within the system, the network will be very easily deformed and weak. It will also form by solidification of the complete solution without loss of solvent.

In the traditional formation of jelly, a polymer solution is made up, by dissolving gelatin in hot water. At the elevated temperature, the thermal vibrations will carry the molecules past one another, but, on cooling, the thermal vibrations cease to be strong enough to overcome the attractive forces at entanglements. The structure solidifies as a gel, which will be reversible: it will 'melt' on heating. If the gel is formed by chemical crosslinking, it will not be reversible.

A similar situation could be reached by allowing a cross-linked rubber to take up a liquid by swelling. Thus a swollen rubber is a form of gel, and the results of the theory of rubber elasticity, described in the next chapter, can be applied.

There will be viscoelastic effects in gels similar to those in concentrated polymer solutions, except that there will be no purely viscous flow of the system as a whole.

3.5.2 Viscosity of melts

Interactions can also be increased by removing solvent completely. If the temperature is high enough, and chemical degradation has not set in, a linear polymer material will be a melt. The viscosity of melts can be predicted by the methods already mentioned. The factors to be considered are the entanglement of chains with one another, and the ease with which neighbouring segments can slide past one another. Bueche has developed a treatment in terms of a parameter s which defines the slippage between successive chains. Thus if one chain has a velocity v, it will drag the next chain along with a velocity sv.

The easiest situation to imagine would be the viscous drag on a rod stirring the polymer melt. Suppose this rod collects n_0 chain molecules. Following (3.52), we see that, in the absence of entanglement, there would be a viscous drag $n_0 N_v \zeta_0 v$. But if each chain entangles with m others, the second group will be moving with a velocity sv, and so would give an additional viscous drag $mn_0 N_v \zeta_0 sv$. In general, this simple argument suggests that the ith group of entangled molecules would give a drag $(ms)^i n_0 N_v \zeta_0 v$. Consequently the total drag would be $\sum_0^\infty (ms)^i n_0 N_v \zeta_0 v$ or, introducing an effective value N^* in place of N_v in (3.52):

$$N^* = \left[\sum_0^\infty (ms)^i\right] N_v \tag{3.75}$$

If $ms \ll 1$, only the first term is important and we have $N^* \simeq N_v$: the viscous drag will be proportional to the molecular weight.

As ms approaches close to 1, there will be a rapid change in behaviour since the value of the sum begins to increase rapidly. In fact for $ms \geqslant 1$, the sum $\sum_0^\infty (ms)^i$ becomes infinite – the simple argument breaks down because it does not take account of the fact that some of the entanglements will be multiple entanglements between the same groups of chains, or even entanglements of a chain with itself. It is necessary to adopt a more exact, and complicated, analysis, which takes account of the chain conformations and so reduces the value of terms in the summation by eliminating the ineffective entanglements. This analysis has been carried out by Bueche, on the assumption that the chains take up Gaussian conformations. Bueche obtains a rather complicated general relation. However this reduces to simpler forms which are strictly valid for $m \ll 1$ and $\gg 1$, but, because the change near $m = 1$ is so sharp, can be used for $m < 1$ and $m > 1$. The equations are most conveniently expressed in terms of N_x, the number of units between entanglement points. It is obvious from Fig. 3.28 that $N_v/N_x = (m + 1)$. The change-over from the slightly entangled to the highly entangled systems at $m = 1$ will thus occur at $N_v = 2N_x$.

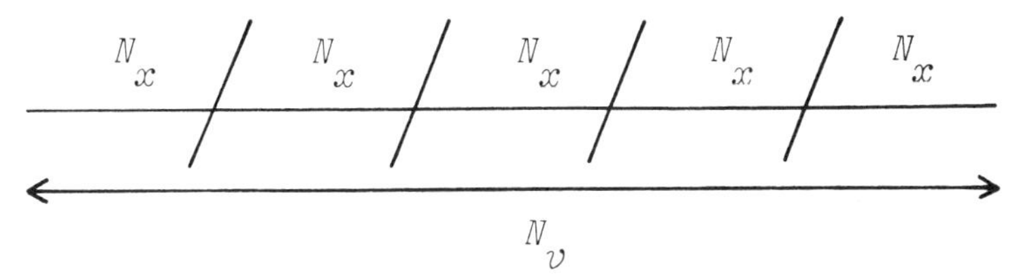

Fig. 3.28 – Schematic representation of chain with entanglement.

In giving the simplified form of Bueche's equations, it is convenient to identify the viscous resistance unit of the chain with the effective random link; in other words, we let $N_v = N_e$, and define N_x as the number of effective random links between entanglements. Bueche's first equation is the relation which can be predicted from the direct argument:

$$N^* = N_e \quad \text{for} \quad N_e < 2N_x \tag{3.76}$$

giving $\zeta = N_e\zeta_e$ where ζ_e is the frictional co-efficient for an effective random unit.

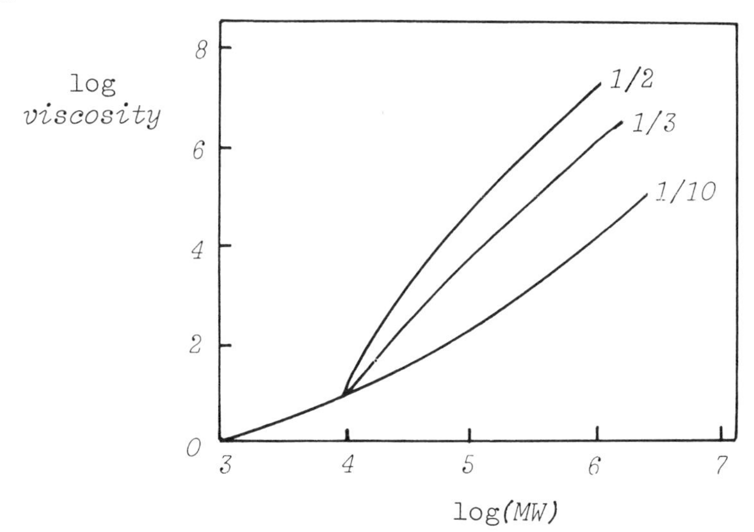

Fig. 3.29 – Prediction of the Bueche theory, for the values of s indicated.

Bueche's second equation can be written as:

$$N^* = N_e^{3.5}(cN_0a_e^3/48N_x^2M_e)\sum_{i=1}^{\infty} s^i(2i-1)^{3/2} \quad \text{for} \quad N_e > 2N_x \tag{3.77}$$

where c is the concentration of polymer in solution.

Alternatively we can write:

$$N^* = KN_e^{3.5} \tag{3.78}$$

where K is a constant for the particular system.

The value of the entanglement length N_x which is of course concentration dependent, can be obtained by putting $N_e = 2N_x$ and equating the two expressions.

It can be observed that there is a very sharp break at least on a logarithmic plot and so the use of two separate equations is justified except very close to $N_e = 2N_x$. Empirically, the predicted relations are a viscosity proportional to N below a given degree of polymerisation (molecular weight) and proportional to $N^{3.4}$ (very close to the theoretical 3.5) for higher degrees of polymerisation.

Fig. 3.30 which is typical of the behaviour of polymer melts, shows the sharp change from the first power dependence to the higher power dependence. For many experimental determinations, the viscosity fits most closely as proportional to $M^{3.4}$, close to the theoretical prediction of $M^{3.5}$.

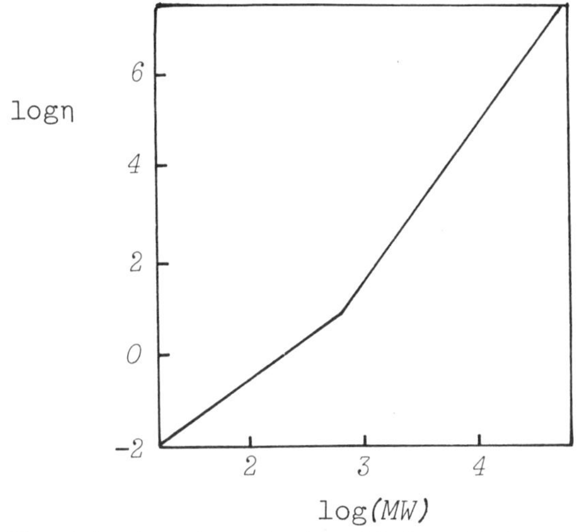

Fig. 3.30 – Plot of viscosity against molecular weight for polyisobutylene at 217°C.

3.5.3 Viscoelasticity of polymer melts.

The flow of a polymer melt, like a polymer solution, will lead to extended chain conformations, and in contrast to the flow of ordinary liquid, to interactions over the lengths of the chain molecules. Consequently an elastic component of force will develop and viscoelastic phenomena will be observed. In general terms the effects will be similar to those discussed in section 3.4 for concentrated solutions.

CHAPTER 4

Rubber Elasticity

> Even after the publication of Meyer's theory, many years elapsed before scientists generally were prepared to acknowledge its superiority over rival interpretations of rubber elasticity. This reluctance to accept what is now regarded as a major scientific advance may be attributed partly to a very proper scepticism with respect to a subject which had become overburdened with somewhat fanciful theories with little experimental backing, . . . and partly to the revolutionary nature of the kinetic theory itself, this last factor being closely related to the corresponding reluctance to accept the concept of a high polymer in any form whatever . . . .
>
> L. R. G. Treloar, *The Physics of Rubber Elasticity*

4.1 THE NATURE OF RUBBER

4.1.1 Common sense and history

The general physical characteristics of rubber are well known, and were known long before our present concepts of polymer structure. To the ordinary man, rubber is a substance which, without much effort, can be stretched several fold in length, which will bend and twist, which will give in localised compression, which will spring back to its original form when it is released, and which is tough and strong. More acute observation shows that it is changes of shape which are easy, while changes in volume are difficult and need large pressures. Strictly, these commonly accepted properties are not found in rubber in the natural state: under prolonged stress, raw rubber will flow like a viscous liquid – it has no permanence of shape, and can be moulded into new forms. It is the process of vulcanisation – a chemical reaction with sulphur – which gives to natural rubber its more perfect elasticity, and removes the tendency to flow.

More detailed scientific experiments on rubber also have a long history. John Gough, publishing in the proceedings of the Manchester 'Lit and Phil' in 1805 reported his observations that when rubber was stretched it became sensibly warmer, and cooled on contracting, and that rubber under tension contracts as its temperature is raised – in marked contrast to the usual thermal expansion of materials. Further experiments were carried out by Joule, following thermodynamic derivations by Lord Kelvin.

At that time, there was no molecular explanation of the properties. It was not unitl 1932, that the kinetic theory of rubber elasticity was proposed by

Meyer, von Susich and Valko. This theory recognised that flexible chains would be in a state of motion, always wriggling about and trying to take up a state of maximum disorder. In other words it is an entropy-dependent deformation, and not the energy-dependent deformation of the rigid molecular structures of crystals, that gives rubber its characteristic high elasticity.

4.1.2 The structure of rubbers

The chemical evidence shows that all natural and synthetic rubbers (using the word **rubbers** to mean materials having the physical properties described in the last section) are composed of linear polymer chains which are lightly cross-linked to form a network. Most commonly, each junction point joins four chain segments, as it will when cross-linking is caused by a chemical reaction like vulcanisation which joins separate chains: but three-chain points can occur due to the introduction of a trifunctional component during polymerisation, and more complex junction are theoretically possible. The number of free links between junction points is usually of the order of magnitude of a hundred. There will also be some free chain ends in a real network.

Structural studies, using methods like X-ray diffraction, give no evidence of any appreciable degree of order in the structure. At most, rubbers possess no more than the local ordering between closely neighbouring molecules, which is characteristic of liquids. Ordinarily, rubber is isotropic when it is free of stress. The chemical nature of the polymers which are rubber-like indicates that chain flexibility wiil be high, and that interactions between chains will be weak.

4.1.3 Quantitative observation of deformation

A typical load-elongation curve for a strip of rubber under tension is shown in Fig. 4.1. Its most important features are a very low initial modulus, extremely high breaking extension, and reasonably high strength. On release, the strip shows good recovery up to the highest extensions; even after rupture, there is very little permanent extension of the broken pieces. As shown by the figures in Table 4.1, the properties of a rubber are quite different from those of a typical hard solid; the bulk modulus is similar to that of a liquid; but, different from a liquid, there is a definite, though small, elastic shear modulus.

The results of a set of experiments carried out by Treloar on a particular sample of vulcanised rubber, deformed in various ways, are interesting. Fig. 4.2(a) shows behaviour in simple extension, and demonstrates the almost perfect reversibility up to 450% extension: at higher extensions, there is some hysteresis, though there is no permanent deformation at zero load. These results were obtained by observing the length of a strip on the successive addition of weights at intervals of one minute.

The second experiment was a biaxial deformation. The behaviour of rubber sheet when it is stretched in two perpendicular directions can be examined by studying the inflation of a balloon or of a clamped circular specimen: the strain

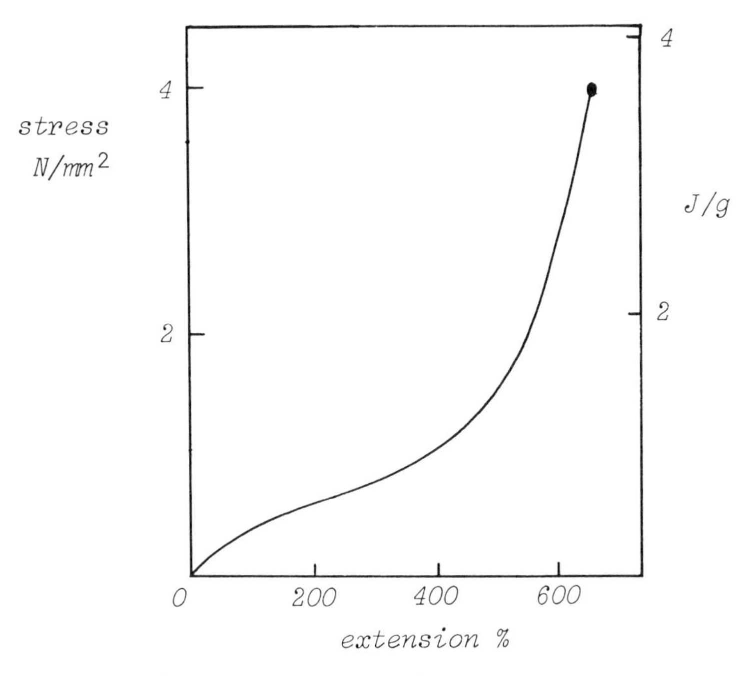

Fig. 4.1 – Load-elongation curve for a typical vulcanised rubber.

Table 4.1 – Comparison of the physical properties of rubber with those of other substances.

Property	Typical rubber	Typical metal	Typical glass
Density, g/cm^3	0.92	8	2.5
Young's modulus, N/m^2	10^6	10^{11}	5×10^{10}
Breaking extension %	500%	large plastic	3%
Elastic limit, %	500%	2%	3%
Tensile strength, N/m^2	10^7†	5×10^{10}	5×10^9
Shear modulus, N/m^2	3×10^5	5×10^{10}	2×10^{10}
Bulk modulus, N/m^2	3×10^9	10^{11}	5×10^{10}
Poisson's ratio	0.49	0.3	0.25
Specific heat	0.4	0.1	0.2
Coefficient of volume expansion per °C	7×10^{-5}	3×10^{-5}	2×10^{-5}

† $>10^9$ on area at break.

can be calculated from the area change on inflation, and the tensile stress can be calculated from the pressure and the curvature. Near the pole of the spheroidal balloon, the strain is reasonably uniform. Fig. 4.2(b) shows the stress-strain relation for such a biaxial deformation. From such a test it is possible to calculate the behaviour in compression: this is shown in Fig. 4.2(c).

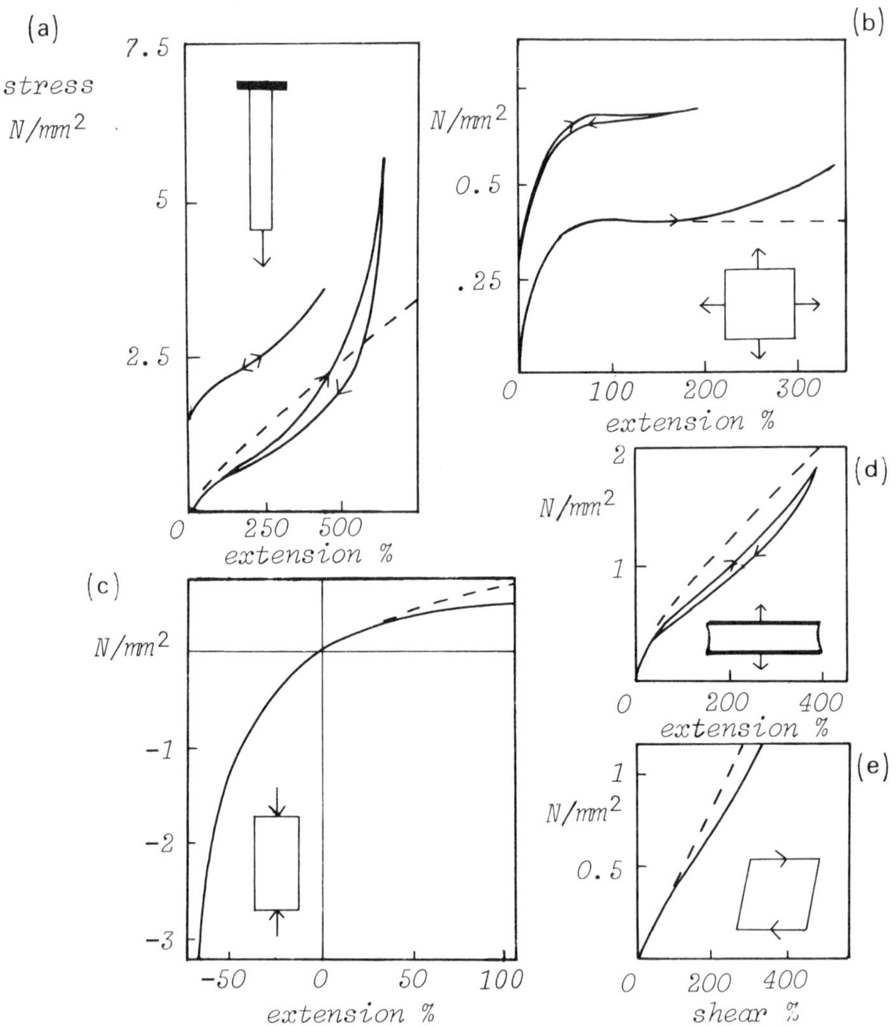

Fig. 4.2 – Treloar's experiments on various forms of deformation of a typical vulcanised rubber. The dotted lines are the theoretical predictions according to the Gaussian approximation. (a) Simple extension, with lateral contraction, up to 600% extension. The upper curve (displaced for clarity) shows the good recovery from 450% extension. (b) Equal biaxial extension – displaced curve shows recovery from 200% extension. (c) Calculated curve for uniaxial compression and extension. (d) Extension without lateral contraction. (e) Calculated curve for simple shear.

Another form of deformation, related to pure shear, is obtained by stretching a wide sheet held in clamps so that it cannot contract in width (except to a negligible extent near the edges): this gives the load-elongation curve shown in Fig. 4.2(d). Values for simple shear, calculated from the behaviour in this test, are shown in Fig. 4.2(e).

Since rubber is isotropic, its behaviour under small strains can be characterised by two elastic constants. We have noted that a typical value of Young's modulus E is about $1 N/\text{mm}^2$. The bulk modulus K is over $1000 N/\text{mm}^2$. For most purposes, we can clearly make approximations based on the relation $K \gg E$: this is equivalent to regarding the rubber as incompressible, and is analagous to the assumption of incompressibility made in hydrodynamics. The other elastic constants – Poisson's ration σ and shear modulus G – are given by the usual relations:

$$\sigma = \tfrac{1}{2}(1 - E/3K) \to 0.5 \quad \text{for } K \gg E \tag{4.1}$$

$$G = \tfrac{1}{2}E/(1 + \sigma) \to E/3 \quad \text{for } K \gg E \tag{4.2}$$

4.1.4 Thermal effects

The interaction of thermal and mechanical effects may be studied in various ways. A simple system consists of a piece of rubber which can be stretched in length under conditions of constant atmospheric pressure. We are concerned with four variables which can be controlled: tension F, length l, temperature T, and heat Q gained or lost from the atmosphere. In addition, we must remember that the lateral dimensions (or volume) can change in a way which is not capable of being controlled, unless the system under consideration is changed to allow the application of transverse force (pressure). The system, as it has been defined, has two degrees of freedom: for instance, in order to extend the specimen to a given length at a given temperature a certain tension will be needed and a certain heat interchange will occur. The state of the system is fully defined when any two variables are given.

There are twelve simple forms of variation of parameters. The most interesting are tabulated in Table 4.2 together with the corresponding commonly quoted properties, which have been normalised to take account of the specimen dimensions. Use of the equations relating partial differential coefficients – or more generally the transfer of data from one set of curves to another – will enable many interconversions to be made.

Fig. 4.3 shows the results of some measurements of the force-temperature relation at constant length. Ignoring the change below 213°K (−60°C) we see that as the temperature rises, the force increases: this is the opposite to the behaviour of most materials, where the tendency to expand on heating would lead to a reduction in tension.

If we neglect any work done by the atmospheric pressure – in other words,

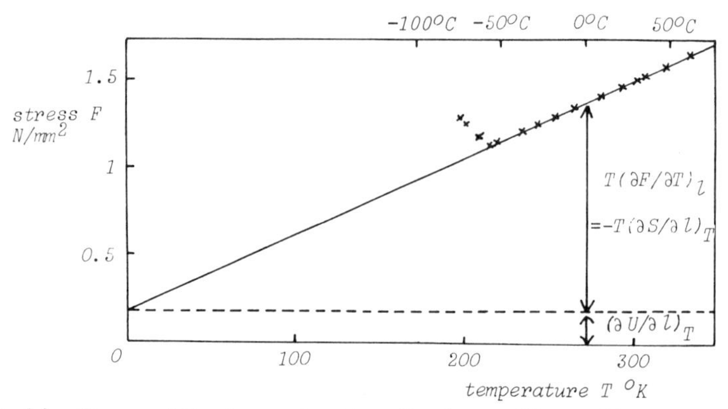

Fig. 4.3 – Meyer and Ferri's experiments on the change of force with temperature in rubber held at constant length.

Table 4.2 – Forms of variation with length and temperature, at constant pressure.

Variables studied	Constant quantity	Nature of variation	Differential coefficient	Normalised parameter
F,l	T	Isothermal load-elongation curve	$(\partial F/\partial l)_T$	Young's modulus $= \dfrac{l_0}{A}\left(\dfrac{\partial F}{\partial l}\right)_T$
F,l	Q	Adiabatic load-elongation curve	$(\partial F/\partial l)_Q$	Adiabatic modulus $= \dfrac{l_0}{A}\left(\dfrac{\partial F}{\partial l}\right)_Q$
T,l	Q	Heating due to adiabatic extension	$(\partial T/\partial l)_Q$	$l_0\left(\dfrac{\partial T}{\partial l}\right)_Q$
l,T	F	Contraction on heating under constant load	$(\partial l/\partial T)_F$	Coefficient of expansion $= \dfrac{1}{l_0}\left(\dfrac{\partial l}{\partial T}\right)_F$ (usually with $F = 0$)
F,T	l	Change in tension with temperature at constant length	$(\partial F/\partial T)_l$	Stress-temperature coefficient $= \dfrac{1}{A}\left(\dfrac{\partial F}{\partial T}\right)_l$
Q,T	F	Thermal capacity at constant load	$(\partial Q/\partial T)_F$	Specific heat $= \dfrac{1}{m}\left(\dfrac{\partial Q}{\partial T}\right)_F$ (usually with $F = 0$)
Q,T	l	Thermal capacity at constant length	$(\partial Q/\partial T)_l$	$\dfrac{1}{m}\left(\dfrac{\partial Q}{\partial T}\right)_l$

F = force; l = length; T = temperature; Q = heat; A = area of cross-section; l_0 = original length; m = mass.

Note: The other five forms are $(\partial F/\partial Q)_l$, $(\partial F/\partial Q)_T$, $(\partial F/\partial T)_Q$, $(\partial l/\partial Q)_F$, $(\partial l/\partial Q)_T$. In addition there are the reciprocals of the twelve forms.

we assume the conditions are effectively equivalent to zero pressure – we find from (A.3) and (A.9):

$$F = (\partial U/\partial l)_T - T(\partial S/\partial l)_T$$
$$= (\partial U/\partial l)_T + T(\partial F/\partial T)_l \quad (4.3)$$

The form of this equation is such that when F is plotted against T, the intercept at absolute zero gives $(\partial U/\partial l)_T$ and the part above the intercept gives $T(\partial F/\partial T)_l$ or $-T(\partial S/\partial l)_T$.

The results in Fig. 4.3 show that in vulcanised rubber the energy contribution is very small, and the entropy term is predominant, particularly at higher temperature. The linearity of the plot shows that the changes in energy and entropy with length are unaffected by the change in temperature over the range −60°C to +60°C.

For a first broad discussion of the behaviour of rubber, we can neglect the small energy contribution and assume that the material follows the idealised equation:

$$F = T(\partial F/\partial T)_l = -T(\partial S/\partial l)_T \quad (4.4)$$

As was pointed out in deriving (A.10) another consequence of an increase in order (decrease of entropy) on extension is that coefficients of expansion should be negative, in other words there should be a contraction in heating. For a system in which the change in entropy with length is the only factor concerned, we find by substitution from (4.4) in the first line of (A.10):

$$\text{coefficient of expansion} = -(1/l)(\partial l/\partial T)_F$$
$$= -(F/lT)(\partial l/\partial F)_T$$
$$= -(f/T)(\partial f/\partial \epsilon)_T \quad (4.5)$$

where $f = F/A$ = stress, and $\epsilon = \Delta l/l$ = strain.

Equation (4.5) shows that, subject to any variation in the modulus $(\partial f/\partial \epsilon)_T$, the coefficient of linear expansion, which is negative, should be proportional to the stress. At zero stress, the coefficient of linear expansion should be zero.

In 1859, Joule reported on measurements of the change in length with temperature for a strip of rubber with a weight hanging on it; his results, on which Fig. 4.4(a) is based, generally confirm the above predictions. There is one small, but important, difference: at zero stress there is a positive coefficient of linear expansion of 0.0022/°C. This corresponds to a coefficient of volume expansion of 0.0066/°C, and represents the superposition of another mechanism on the length-dependent effects, which we have been discussing. The other factor is the decrease of order and increase of entropy resulting from an increase of volume. Mechanistically, the volume expansion is due to the fact that the molecules will take up more space when they are vibrating more vigorously at a higher temperature. The volume expansion will remain as the predominant effect when the change in length predicted by (4.5) is small, as it will be at low tensile stresses.

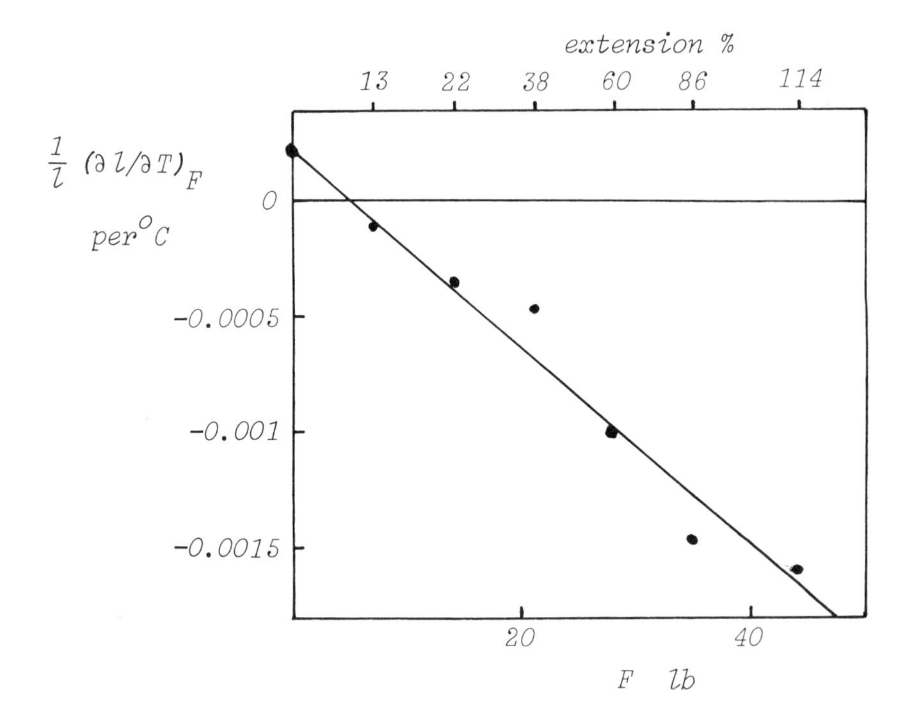

Fig. 4.4(a) – Joule's experiments in 1859 on thermal effects of extension of rubber. Variation of coefficient of linear expansion $(1/l)(\partial l/\partial T)_F$ with tension F.

Measurements of the force-temperature relations at a range of extensions also show up the consequences of volume expansion. At high extension, there is a rise in force with a rise in temperature; but at low extensions the reverse is true – the tendency to volume expansion leads to a decrease in force. Indeed, at zero extension, an increasing negative force would be needed to prevent the length increase associated with volume expansion. At a certain level of extension, known as the thermal inversion point, the consequence of the shape and volume changes with temperature just balance, and the force is independent of temperature.

Joule also measured the rise in temperature on adiabatic extension. Fig. 4.4(b) shows a plot of the values found, amounting to a rise of 0.14°C for a force of 40 lbs. By substitution from (4.4) in (A.12), remembering that constancy of entropy S is equivalent to constancy of heat Q, we get:

$$(\partial T/\partial F)_Q = (F/C_F)(\partial l/\partial F)_T \tag{4.6}$$

where C_F is the thermal capacity of the rubber.

If we can assume, approximately, that the thermal capacity and the modulus,

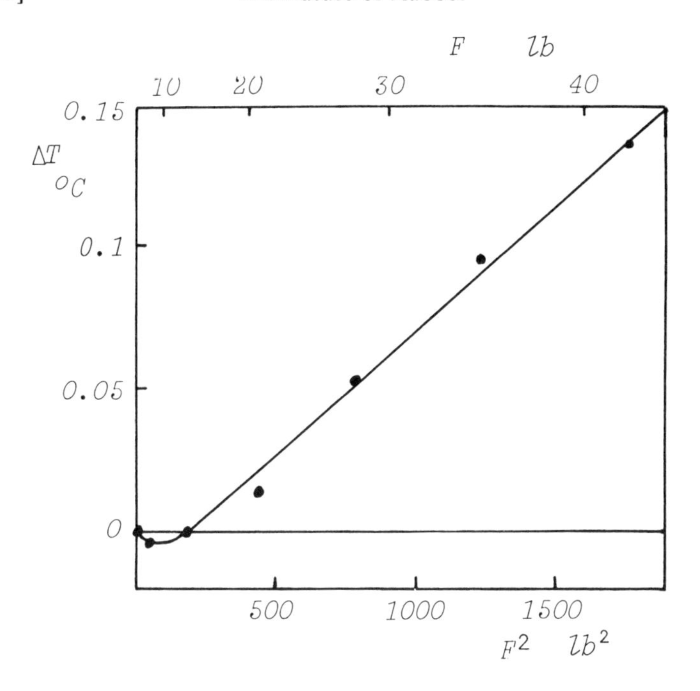

Fig. 4.4(b) – Joule's experiments in 1859 on thermal effects of extension of rubber. Rise in temperature ΔT with adiabatic application of force F. Experiments carried out by James and Guth in 1943 fall on same line when plotted against percentage extension.

which is inversely proportional to $(\partial l/\partial F)_T$, are constant, we can integrate this equation and find:

$$\text{rise in temperature} = \Delta T = \int_0^F \frac{F}{C_F}\left(\frac{\partial l}{\partial F}\right)_T \mathrm{d}F = \frac{F^2}{2C_F(\partial F/\partial l)_T} \tag{4.7}$$

Fig. 4.4(b) shows that, apart from some deviation at very low stresses, where the temperature falls due to the effect of the volume change, the theoretical prediction, based on the assumption that only entropy changes with length are involved, is confirmed. In passing, it may be noted that (4.7) could be derived directly from the fact that, in an entropy-dependent elastic deformation, all the work done is converted into heat, and in an adiabatic extension all this heat goes to raise the temperature of the sample.

Joule's two sets of measurements can be compared by the application of

(A.12) which may be rewritten in the form:

$$\frac{\Delta T}{(\partial l/\partial T)_F/l} = -\frac{Tl\Delta F}{C_F} \tag{4.8}$$

Except at small forces, where the temperature change is very small indeed, this ratio is found to be constant within reasonable limits of experimental error, and gives a value of specific heat in agreement with direct experimental measurements. It should be noted that (4.8) should apply to any material for measurements under any equilibrium conditions since it is derived directly from the general thermodynamic relations, without the use of the special form of (4.4).

The results of Joule's experiments have been confirmed by many later workers. For example, some results obtained by James and Guth fall on the same curve of temperature rise against extension. When larger forces are applied, with extensions up to 600%, a temperature rise of 10°C is observed. Summarising we can say that the various experiments on the interrelation between thermal and mechanical effects all go to show that, apart from some small secondary effects which may be particularly important at very small strains, the high elastic deformation of rubber is an entropy-dependent phenomenon. By analogy with the difference between ideal gases and real gases, an ideal rubber may be defined as one in which there is no change in internal energy with changes in length at constant temperature, with the additional proviso that there are no changes of volume as a result of changes in temperature or the application of stress.

4.1.5 When a rubber is not a rubber

Before leaving this opening discussion of the observed properties of rubber, we should note that some deviations from rubber-like behaviour are due to the rubber ceasing to be a true rubber.

The three main effects can best be noted by reference to natural rubber:

(a) below about −60°C, rubber changes to a glass, and the change of tension with temperature reverses in sign as shown in Fig. 4.3;

(b) below about 10°C, rubber crystallises and ceases to be an amorphous polymer – the crystallisation rate is usually slow except at very high extensions;

(c) if the rubber is insufficiently vulcanised, there will be appreciable viscous flow of molecules past one another.

For the present we shall be concerned with rubbers whether natural or synthetic while they are acting as elastic, cross-linked networks of flexible, disordered chains.

4.2 THE BASIC THEORY

4.2.1 Rubbery networks

In Chapter 2, we indicated what would be expected, on theoretical grounds, to be the molecular behaviour of a rubber – regarded as a condensed assembly of flexible polymer chains, in which the interactions are weak enough to allow easy movement of chains past one another. The chains would naturally take up coiled conformations, subject to the restriction that the condensed state of the system must be maintained. Deformation giving extended chain conformations would generate an elastic restoring force; but if the chains were independent, they would flow past one another and relieve – or relax – the force. The rubber would be visco-elastic, and in many circumstances the viscous behaviour would be dominant.

However, if the rubber is lightly cross-linked to form a network of chain segments the flow of chains past one another is prevented. The elastic behaviour predominates. This system, which, as we have seen, is available in commercial rubbers is an excellent one for the study of the fundamentals of rubber elasticity.

The elasticity will be kinetic in origin, resulting from the thermal fluctuations of the chain segments, which will naturally try to take up disordered conformations of maximum entroy. In a first treatment of the subject, changes in internal energy due to changes of shape at constant temperature are assumed to be negligible. Under stress, the rubber deforms rather easily as the chain segments are pulled into extended conformations. The chain segments can be stretched by a considerable amount before they are pulled straight and begin to break: consequently the network will show the high breaking extensions characteristic of rubbers.

In the next few sections, the behaviour of such a system will be analysed: the treatment is a simple one, and, in addition to explicit approximations, it contains some tacit assumptions.

4.2.2 Affine deformation of a network

Fig. 4.5 illustrates by analogy the form of a model network. For convenience, we assume that there are no free chain ends, so that each chain segment terminates in two junction points. When the network is deformed, the local strains may vary somewhat due to the local differences in arrangement. However, for simplicity, we assume that the deformation is affine: this means that the whole network of junction points deforms in a similar way. Mathematically, we can say that the components of the vector length of any chain segment change in the same proportions as the components of a corresponding vector drawn in the bulk rubber; or that if any two similar triangles are drawn, one connecting three junction points and one in the bulk rubber, then after deformation the triangles will still be similar to one another. If the specimen has a simple shape and is uniformly strained, then the large vectors or triangles may be drawn in the whole specimen; but if the deformation is non-uniform, then they must be drawn in a volume of rubber which is large on a molecular scale, so that it contains a great many chain segments, but small in relation to strain variations in the specimen.

It might seem simplest to restrict consideration initially to an analysis of the behaviour of a uniform strip in simple extension; but, in fact, the analysis is no more complicated when we consider any pure homogeneous strain, characterised by three principal strains taken along three mutally perpendicular directions. The increase both in understanding and generality is considerable. Any uniform state of strain of a body can, by a suitable choice of axes, be described in terms of a pure homogeneous strain; for an isotropic material, like rubber, the strain is related to a corresponding pure homogeneous stress. Thus all states of uniform stress and strain are included. Further, provided the scale of the pattern of strain variation does not encroach on the scale of the fine structure of the material, any non-uniform strain (such as torsion, bending, local compression, or deformation of a body of complicated shape) can be regarded as made up of a compatible set of pure homogeneous strains in an assembly of elements, and the problem analysed in macroscopic terms by the theory of elasticity.

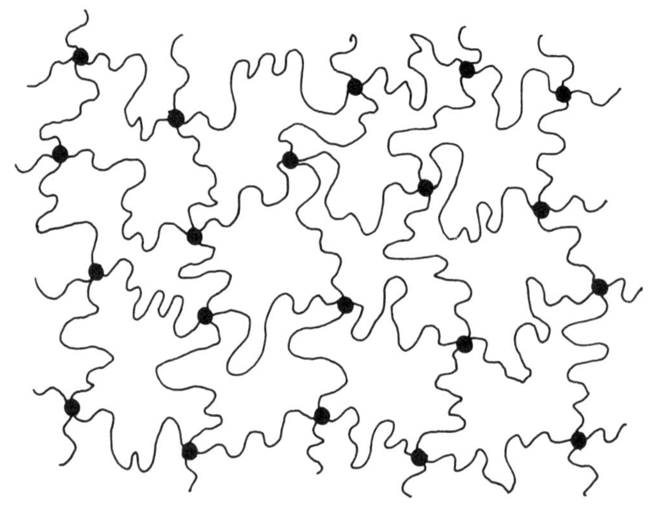

Fig. 4.5 – Schematic diagram of cross-linked network. Note that, apart from other simplification and artistic licence, a two-dimensional drawing can only be a rough analogy for a three-dimensional network.

4.2.3 Stored energy in a deformed network

We consider as a reference state a unit cube of the cross-linked rubbery network. This is uniformly deformed into a rectangular parallelepiped by a change in length of each side as shown in Fig. 4.6. This gives us the required most general form of pure homogeneous strain. The edge lengths in the deformed state are taken to be μ_1, μ_2 and μ_3.

For reasons, which will be apparent later, we choose the cube so that, when there are no stresses present, each side of the cube is of length λ_0. The reference state, when the sides of the cube are all unity, differs from the stress-free state by a small amount, which, for the moment, can be regarded as arbitrary in magnitude. Formally, we are thus assuming that $(\lambda_0 - 1) \ll 1$.

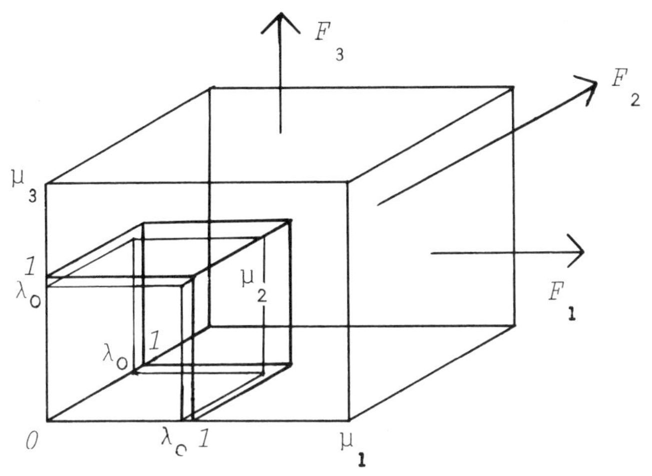

Fig. 4.6 – Deformation of a reference unit cube, 111, to a rectangular parallelepiped, μ_1, μ_2, μ_3. The cube, $\lambda_0\, \lambda_0\, \lambda_0$, is the state under zero stress.

We can define strains and extension ratios which are related to either the reference state or the stress-free state: the former is more convenient for our theoretical analysis, while the latter will be needed to relate the theory to reality. The expressions are:

	extension ratios	strain
related to reference state:	μ_1, μ_2, μ_3	$(\mu_1-1), (\mu_2-1), (\mu_3-1)$
related to stress-free state:	$\lambda_1 = \mu_1/\lambda_0$	(λ_1-1)
	$\lambda_2 = \mu_2/\lambda_0$	(λ_2-1)
	$\lambda_3 = \mu_3/\lambda_0$	(λ_3-1) (4.9)

Extension ratios are, of course, greater than one for extension and less than one for compression. Theoretically, it is easier to calculate the stored elastic energy in the system at given extension ratios than it is to try and calculate the forces directly. The stresses can be found from the energy function.

A single chain segment in the reference state is shown in Fig 4.7, with one end placed at the origin O of a set of Cartesian co-ordinates, OX, OY, OZ, parallel to the sides of the cube, and the other end at x, y, z. After deformation, if the axes are moved so that their origin remains at the first end, the other end will be at (x_e, y_e, z_e). As a result of the assumption of affine deformation, it follows that:

$$x_e = \mu_1 x\ ; \quad y_e = \mu_2 y\ ; \quad z_e = \mu_3 z\ . \tag{4.10}$$

The end-to-end lengths of the segment before and after deformation are given by:

$$l^2 = x^2 + y^2 + z^2 \tag{4.11}$$

$$l_e^2 = x_e^2 + y_e^2 + z_e^2 \tag{4.12}$$

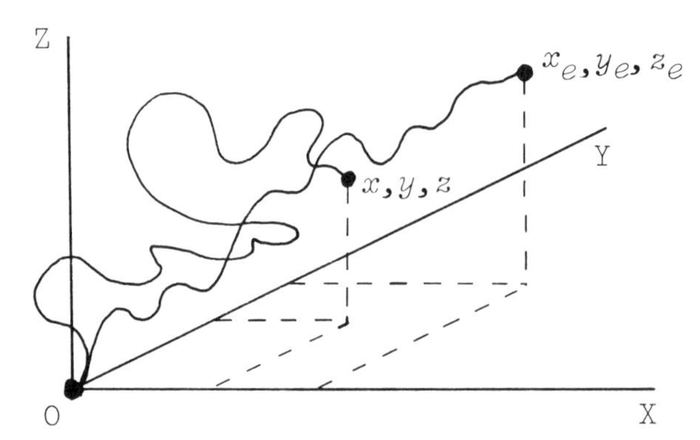

Fig. 4.7 A typical chain segment O (x, y, z) in reference network, and deformed to O (x_e, y_e, z_e).

If we now assume that a chain segment between junction points behaves like an isolated free chain containing the same number of links, its force-length relation will be given by (2.19):

$$F = (kT/a)\,[\text{arc}\,\mathcal{L}(l_e/L)] \tag{4.13}$$

where F is the tension in the chain segment, a is the length of a free link, L is the fully extended length of the chain segment, and arc $\mathcal{L}$ is the symbol for the inverse Langevin function.

The stored elastic energy A_s of the single chain segment will be given by:†

$$A_s = \int_0^{l_e} F\mathrm{d}l = \int_0^{l_e} (kT/a)\,\text{arc}\,\mathcal{L}\,(l_e/L)\mathrm{d}l_e \tag{4.14}$$

The individual energy terms must now be summed over all the n chain segments in the network, in order to give the total stored energy. These chain segments will differ because: (a) they will contain different numbers of links, giving different values of extended length L: (b) for any given extended length, there will be different values of (x, y, z) corresponding to different initial lengths and orientations of the segments. The averaging procedure must take account of these differences: this would not cause any difficulty in numerical computation, but further progress is algebraically difficult unless we approximate by taking only the first term in the series expansion of the inverse Langevin function, which is also the form given by the Gaussian approximation. Despite the loss of generality it is worth continuing with the Gaussian form of the analysis in order to illustrate the physical principles involved. With this form we have from (2.21) or (3.35):

$$F = 3(kT/a)(l_e/L) = 2kTb^2l_e \tag{4.15}$$

† Regarded as a macroscopic phenomenon. It is immaterial to the present argument, that the force is entropic in origin so that the energy is actually stored as heat in the environment, which is so large that its temperature change is negligible. The important point is that energy coming from the work done in extension is stored in a form which is recoverable.

where $b^2 = (3/2Na^2)$ and N is the number of links, equal to L/a. Hence

$$A_s = \int_0^{l_e} F \mathrm{d}l = b^2 kT l_e^2 \tag{4.16}$$

For all chains having the same value of b, that is with the same number of links, we can sum (4.16), substitute from (4.12), and write:

$$\begin{aligned} \underset{\text{same } b}{\Sigma A_s} &= b^2 kT \Sigma l_e^2 \\ &= b^2 kT[\mu_1^2 \Sigma x_b^2 + \mu_2^2 \Sigma y_b^2 + \mu_3^2 \Sigma z_b^2] \end{aligned} \tag{4.17}$$

where Σx_b^2 is the sum of squares of x-components of all chains with a particular value of b, and so on.

It follows from (4.11) that:

$$x_b^2 + y_b^2 + z_b^2 = l_b^2 \tag{4.18}$$

where l_b is the initial length of a segment with a particular value of b. Hence:

$$\Sigma x_b^2 + \Sigma y_b^2 + \Sigma z_b^2 = \Sigma l_b^2 \tag{4.19}$$

But if we assume the network to be isotropic in the reference state, it follows that for any statistically large group of segments, the values of Σx^2, Σy^2 and Σz^2 must be equal. Therefore:

$$\Sigma x_b^2 = \Sigma y_b^2 = \Sigma z_b^2 = \tfrac{1}{3} \Sigma l_b^2 = n_b \overline{l_b^2}/3 \tag{4.20}$$

where n_b is the number of segments with the given value of b, and $\overline{l_b^2}$ is the mean square end-to-end length for these segments.

The network is assumed to have been cross-linked when it was in a random state. It therefore seems reasonable to assume that the mean-square-length of unstrained chain segments equals the value for a corresponding free-chain, given by (2.29) or (3.27) as:

$$\overline{l_b^2} = 3/2b^2 \tag{4.21}$$

Substitution from (4.20) and (4.21) in (4.17) then gives:

$$\begin{aligned} \Sigma A_s &= b^2 kT(\mu_1^2 + \mu_2^2 + \mu_3^2) n_b \overline{l_b^2}/3 \\ &= \tfrac{1}{2} n_b kT(\mu_1^2 + \mu_2^2 + \mu_3^2) \end{aligned} \tag{4.22}$$

It is important to note that the form of this expression is independent of the value of b, and hence of the number of links in the chain segments. Consequently we can simply add up the numbers n_b in the corresponding sums for groups will all possible values of b, to get the total stored energy for all n segments in the cube:

$$\begin{aligned} A_c &= \tfrac{1}{2} nkT(\mu_1^2 + \mu_2^2 + \mu_3^2) \\ &= \tfrac{1}{2} G(\mu_1^2 + \mu_2^2 + \mu_3^2) \end{aligned} \tag{4.23}$$

Intuitively we can now see that the grouping needed to obtain large enough groups with a particular value of b was really only an artificial device, and that a sum over each individual element would have yielded the same equation (4.22) provided n was a large number.

The elastic stored energy A_c which we have calculated is the Helmholtz free energy for the system. Since we have assumed that there are no internal energy effects due to change of chain conformation, it is entirely entropic in origin and equals $(-TS)$. This means that the energy is in fact stored in the surrounding atmosphere, having been given out as heat. However, this is taken up again on recovery: there are no energy losses and the deformation is completely recoverable.

The number of chain segments n is related to the density of polymer molecules in the network and to the degree of cross-linking. If the number average molecular weight of chain segments, left between cross-links, is M_c, the density ρ of the unit cube must be given by:

$$\rho = nM_c/N_0 \tag{4.24}$$

where N_0 is Avogadro's number, the number of molecules in 1 gram-molecule. Hence:

$$G = nkT = \rho N_0 kT/M_c = \rho RT/M_c \tag{4.25}$$

We may also express G explicitly as a function of the degree of cross-linking. If the fraction of units in the polymer chain which are cross-linked is f_x, the number of junction points n_x will be $f_x \rho N_0/M_0$, where M_0 is the molecular weight of a chain unit (when the value of f_x is high, this expression must be corrected for the mass in the cross-links). Since each chain segment is attached at two junction points, it follows that, if X chains come from each junction point.

$$n = n_x X/2 = \tfrac{1}{2} X f_x \rho N_0/M_0 \tag{4.26}$$

$$G = \tfrac{1}{2} X f_x \rho (N_0/M_0) kT = \tfrac{1}{2} X f_x \rho RT/M_0 \tag{4.27}$$

Commonly, $X = 4$, giving:

$$G = 2 f_x \rho RT/M_0 \tag{4.28}$$

We note that the precise definition of a chain unit is not important. Although the derivation relates to an effective freely rotating link, the values of f_x and M_0 will change proportionately for any other definition of chain unit.

4.2.4 The restriction to a condensed system

The derivation in the last section looks plausible, until we note that the minimum value of A_0, which determines the equilibrium state of the specimen, appears to be at $\mu_1 = \mu_2 = \mu_3 = 0$ – in other words, when the cube has shrunk to zero volume! In terms of the method of analysis, it is easy to see how this has arisen. We have treated the problem as a sum of individual chains, and, as was shown in section 2.2.4, the stress-free state of an individual chain occurs when its ends are coincident, namely when the extension ratio λ is zero.

But, in reality, we are not dealing with a set of separate free chains, we are dealing with a condensed assembly of chains. Clearly some other restrictions must be imposed. The values of μ_1, μ_2 and μ_3 are not free to take up any set of values, and (4.23) cannot be used as it stands, since an energy expression is only useful if it is expressed in terms of independently variable parameters. The number of such parameters must equal the number of degrees of freedom of the system.

The added restriction is related to the fact that the system is a condensed one, and so the possible volume changes are limited. Very tight conformations are impossible because they would demand multiple occupation of the same space: very open conformations will be impossible because they will separate the chains in such a way that the material is no longer a condensed, rubbery solid.

Volume changes are strongly resisted in any condensed system, whether liquid or solid, and are associated with the bulk modulus of the material. The equilibrium volume is determined by the balance between attractive and repulsive internal energy forces between molecules, modified slightly by the entropy changes with volume. It is reasonable to regard the determination of this balance as separate from the effects of changes in chain conformations which are mainly associated with shape changes in the specimen.

The volume of the deformed cube is $(\mu_1\mu_2\mu_3)$ and since the initial volume of the reference state is one, the volume strain will be $(\mu_1\mu_2\mu_3 - 1)$. If we assume that volume changes are small, and that Hooke's law is followed, with a bulk modulus K, we get:

$$\text{energy due to volume change} = \tfrac{1}{2}K(\mu_1\mu_2\mu_3 - 1)^2 \tag{4.29}$$

The total elastic stored energy A_T is therefore given by:

$$A_T = \tfrac{1}{2}G(\mu_1^2 + \mu_2^2 + \mu_3)^2 + \tfrac{1}{2}K(\mu_1\mu_2\mu_3 - 1)^2 \tag{4.30}$$

The three parameters are now free to vary independently in taking up an equilibrium state. In ordinary rubbers we can assume that the modulus G associated with change of shape is much less than the bulk modulus K, and that the volume $\mu_1\mu_2\mu_3$ remains almost constant and equal to 1. In swollen rubbers, appreciable volume charges may occur.

4.2.5 General force relations: the stress-free and reference states

The relations between stress and strain can be determined by differentiation of (4.30). The most general situation is one in which the cube is in equilibrium with forces F_1, F_2, and F_3 applied to the three faces, as shown in Fig. 4.6. If there is a general change in dimensions, due to small changes in the applied forces, we must have by the law of conservation of energy:

$$\begin{aligned} \text{work done} &= F_1\,\mathrm{d}\mu_1 + F_2\,\mathrm{d}\mu_2 + F_3\,\mathrm{d}\mu_3 \\ &= \text{change in stored energy} = \mathrm{d}A_T \end{aligned} \tag{4.31}$$

Hence with the use of (4.30):

$$F_1 = \left(\frac{\partial A_T}{\partial \mu_1}\right)_{\mu_2\mu_3} = G\mu_1 + K\mu_2\mu_3(\mu_1\mu_2\mu_3 - 1) \tag{4.32a}$$

Similarly:

$$F_2 = G\mu_2 + K\mu_3\mu_1(\mu_1\mu_2\mu_3 - 1) \tag{4.32b}$$

$$F_3 = G\mu_3 + K\mu_1\mu_2(\mu_1\mu_2\mu_3 - 1) \tag{4.32c}$$

The system has three degrees of freedom (for example, the values of μ_1, μ_2, and μ_3 or of F_1, F_2, and F_3 can be independently specified, and by means of these three equations, the unknown (or dependent) forces and deformations can be calculated when values of any three are given.

Since the reference state is an isotropic cube, it is obvious on grounds of symmetry, and is also clear from the appearance of the equations, that whenever the three applied forces F_1, F_2, and F_3 are equal, then the lateral dimensions μ_1, μ_2, and μ_3 must also be equal. In the stress-free state, $F_1 = F_2 = F_3 = 0$, and we put $\mu_1 = \mu_2 = \mu_3 = \lambda_0$. Equations (4.32) then give:

$$G\lambda_0 + K(\lambda_0^3 - 1)\lambda_0^2 = 0 \tag{4.33}$$

Ignoring the incorrect solution, $\lambda_0 = 0$, we have:

$$\lambda_0^4 - \lambda_0 + G/K = 0 \tag{4.34}$$

Since, by assumption, we are looking for a solution where the stress-free state is very close to the reference state, in which $\mu_1 = \mu_2 = \mu_3 = 1$, it is convenient to substitute $\lambda_0 = (1 + \epsilon_0)$, giving:

$$\epsilon_0^4 + 4\epsilon_0^3 + 6\epsilon_0^2 + 3\epsilon_0 + G/K = 0 \tag{4.35}$$

With $\epsilon_0 \ll 1$, solution is:

$$\epsilon_0 = -G/3K \tag{4.36}$$

$$\lambda_0 = 1 - G/3K \tag{4.37}$$

$$\text{volume in stress free state} = \lambda_0^3 \cong 1 - G/K \tag{4.38}$$

Post facto, we see that it is reasonable to assume $\epsilon \ll 0$, when $G \ll K$.

The stress-free condition could alternatively have been obtained directly from the principle of minimum energy: at equilibrium, $(\partial A_T/\partial \mu_1)_{\mu_2\mu_3} = 0$. Consideration of the minimum energy principle also enables us to understand the physical significance of the reference state. It is the condition, $\mu_1 = \mu_2 = \mu_3 = 1$, when the second term in (4.32a) is zero, and thus when the second term in the expression for A_T in (4.30) is a minimum. This second term is associated solely with volume changes: so the reference state has the minimum free energy due to volume changes, analagous to those in liquids, without taking account of the contribution of changes in conformation of the chain molecules. The stress-free state is slightly more contracted because of the tension in all the cross-linked rubbery chain segments. If we wish to evaluate these two effects separately, we cannot easily use in the analysis a stress-free reference state determined by both together.

However, from a practical point of view, experimental measurements must be related to the stress-free state. It is therefore useful now to transform the final relations into this form. In the stress-free state the cube has sides of length λ_0, and extension ratios based on this state are given by (4.9). We also find:

conventional stresses, based on original dimensions in stress-free state

$$f_1 = F_1/\lambda_0^2 \; ; \quad f_2 = F_2/\lambda_0^2 \; ; \quad f_3 = F_3/\lambda_0^2 \tag{4.39}$$

true stresses based on defomed areas of cross-section

$$t_1 = F_1/\mu_2\mu_3 = F_1/\lambda_0^2\lambda_2\lambda_3 \; ; \quad t_2 = F_2/\lambda_0^2\lambda_3\lambda_1 \; ; \quad t_3 = F_3/\lambda_0^2\lambda_1\lambda_2 \tag{4.40}$$

free energy per unit volume in stress-free state

$$A = A_T/\lambda_0^3 \tag{4.41}$$

We also have $\lambda_0 \cong (1 - G/3K)$ and $\lambda_0^3 \cong (1 - G/K)$, and can make approximations on the assumption that $G \ll K$: usually this means that we can put $\lambda_0 = 1$, implying that there are no volume changes in the rubber, but we must be careful not to do this in any expressions of the order of $(\lambda_0 - 1)$, and hence of the order of G/K. We note particularly that the dimensional part of the second term in (4.32) is of this form, but is then multiplied by K, giving a total contribution of the order of G comparable to the shape term. Very small volume changes do play a significant part in some circumstances. Substitution in (4.30) and (4.32) gives the equations in Table 4.3.

Two points may be specifically noted. Firstly, there is the simple and useful form of the difference equations. Secondly, we note that the volume of the cube is $(\lambda_1\lambda_2\lambda_3)$ and, in many instances, this changes very little and can be put equal to 1, to give the forms shown by an arrow in Table 4.3. Care must however be taken in dropping terms, particularly when these are multiplied by the large parameter K.

4.2.6 An alternative method of derivation of the stress equations

In the orginal derivation of the theory of rubber elasticity an approach which differs from the one given above was adopted. As this appears in other books on the subject, it should be outlined briefly in order to avoid any confusion. Reference will be made in particular to the book by Treloar but similar derivations are given by Flory, Meares and others.

The first stage in the derivation is based on the Gaussian statistics for the conformations of chains, and leads to an expression for the entropy of the deformed network [Treloar's equation (4.8)], and hence to the free energy associated with changes of shape [Treloar's equation (4.9)]. The latter equation is similar to (4.22), except for a constant term corresponding to a different choice of the arbitrary origin of zero free energy. By contrast, in the analysis given in this book, the entropy calculation was done on isolated chains in section 2.2.5 and used to obtain a force relation for chain segments. The network was then analysed in terms of tensions and elastic energies. There are no differences in principle between the two methods: they merely confirm one another.

At the second stage, however, there is a more marked difference. In the present treatment by assuming that $G \ll K$, we have, in fact, assumed that volume changes in the rubber are very small. The volume term must however be included in the initial expression in order to avoid predicting that the cube

Table 4.3 – Equations of rubber elasticity, according to Gaussian approximation.

free energy

$$A = \tfrac{1}{2}G\lambda_0^{-1}(\lambda_1^2 + \lambda_2^2 + \lambda_3^2) + \tfrac{1}{2}K\lambda_0^{-3}(\lambda_0^3\lambda_1\lambda_2\lambda_3 - 1)^2$$
$$\cong \tfrac{1}{2}G(\lambda_1^2 + \lambda_2^2 + \lambda_3^2) + \tfrac{1}{2}K\,[(1 - G/K)\lambda_1\lambda_2\lambda_3 - 1]^2 \qquad (4.42)$$

conventional stress

$$f_1 = G\lambda_0^{-1}\lambda_1 + K\lambda_2\lambda_3(\lambda_0^3\lambda_1\lambda_2\lambda_3 - 1)$$
$$\cong G\lambda_1 + K\lambda_2\lambda_3[(1 - G/K)\lambda_1\lambda_2\lambda_3 - 1] \qquad (4.43)$$

rearranged as

$$f_1 = G[\lambda_1 - (\lambda_1\lambda_2\lambda_3)^2\lambda_1^{-1}] + K\lambda_2\lambda_3(\lambda_1\lambda_2\lambda_3 - 1) \qquad (4.44a)$$
$$f_2 = G[\lambda_2 - (\lambda_1\lambda_2\lambda_3)^2\lambda_2^{-1}] + K\lambda_3\lambda_1(\lambda_1\lambda_2\lambda_3 - 1) \qquad (4.44b)$$
$$f_3 = G[\lambda_3 - (\lambda_1\lambda_2\lambda_3)^2\lambda_3^{-1}] + K\lambda_1\lambda_2(\lambda_1\lambda_2\lambda_3 - 1) \qquad (4.44c)$$
$$\rightarrow f_1 = G[\lambda_1 - \lambda_1^{-1}] \text{ etc.} \qquad (4.44d)$$

difference equations

$$\lambda_1 f_1 - \lambda_2 f_2 = G(\lambda_1^2 - \lambda_2^2) \qquad (4.45)$$

true stress

$$t_1 = G\lambda_0^{-1}\lambda_1/\lambda_2\lambda_3 + K(\lambda_0^3\lambda_1\lambda_2\lambda_3 - 1)$$
$$\cong G(\lambda_1\lambda_2\lambda_3)^{-1}[\lambda_1^2 - (\lambda_1\lambda_2\lambda_3)^2] + K(\lambda_1\lambda_2\lambda_3 - 1) \qquad (4.46a)$$
$$t_2 = G(\lambda_1\lambda_2\lambda_3)^{-1}[(\lambda_2^2 - (\lambda_1\lambda_2\lambda_3)^2] + K(\lambda_1\lambda_2\lambda_3 - 1) \qquad (4.46b)$$
$$t_3 = G(\lambda_1\lambda_2\lambda_3)^{-1}[\lambda_3^2 - (\lambda_1\lambda_2\lambda_3)^2] + K(\lambda_1\lambda_2\lambda_3 - 1) \qquad (4.46c)$$
$$\rightarrow t_1 = G[\lambda_1^2 - 1] \text{ etc.}$$

difference equations

$$t_1 - t_2 = G(\lambda_1\lambda_2\lambda_3)^{-1}(\lambda_1^2 - \lambda_2^2)$$
$$\cong G(\lambda_1^2 - \lambda_2^2) \qquad (4.47)$$

collapses to zero volume, and to allow solutions to be obtained with three degrees of freedom, λ_1, λ_2, and λ_3.

The traditional method is to assume that rubber is incompressible, or in other words that K is infinite. Physically this is not true, but formally it is an acceptable procedure. It leads to the subsidiary condition:

$$\mu_1\mu_2\mu_3 = \lambda_1\lambda_2\lambda_3 = 1 \tag{4.48}$$

The number of degrees of freedom of the system is reduced to two and we can write:

$$A = \tfrac{1}{2}G(\mu_1^2 + \mu_2^2 + 1/\mu_1^2\mu_2^2) \tag{4.49}$$

From this equation, which is similar to Treloar's equation (4.11), the stresses can be obtained by differentiation. They are however always indeterminate in that any arbitrary uniform hydrostatic pressure can be added: this reflects the fact that the material is regarded as incompressible and so is unstrained by a uniform pressure. Only difference equations, similar to (4.45) and (4.47) can be obtained by this method.

The traditional method has not been used in this book because it is not physically realistic (rubber is not incompressible) and does not illustrate the fundamental behaviour so clearly; because it loses the mathematical symmetry of the relations in λ_1, λ_2, and λ_3; and because it cannot be applied to problems where small volume changes are playing a part. In going to the special case of a rubber strip in simple extension, the traditional method is slightly less cumbrous mathematically.

4.2.7 Special cases

Equations (4.44) can be used to calculate the stresses for a specimen under any form of pure homogenous strain. There are however a number of interesting special forms.

4.2.7.1 *Change of shape at constant volume*

For any deformation at constant volume, $\lambda_1\lambda_2\lambda_3 = 1$, and hence equations (4.44) and (4.46) become:

$$f_1 = G(\lambda_1 - \lambda_1^{-1}) \qquad t_1 = G(\lambda_1^2 - 1) \tag{4.50a}$$

$$f_2 = G(\lambda_2 - \lambda_2^{-1}) \qquad t_2 = G(\lambda_2^2 - 1) \tag{4.50b}$$

$$f_3 = G(\lambda_3 - \lambda_3^{-1}) \qquad t_3 = G(\lambda_3^2 - 1) \tag{4.50c}$$

Thus it is clear that G is an elastic constant associated with change of shape at constant volume. We note also that, at constant volume, the stress is a function only of the corresponding strain. This means that if a given value of λ_1, is imposed, the stress f_1, will be unaltered by any change in the relative values of λ_2 and λ_3, subject to the restriction that $\lambda_2\lambda_3 = \lambda_1^{-1}$ at constant volume.

4.2.7.2 *Uniaxial stress*

When stress is applied in only one direction, we can put $f_1 = f$, and $f_2 = f_3 = 0$. It is also obvious from symmetry (and can alternatively be proved by a detailed argument) that we can put $\lambda_1 = \lambda$, and $\lambda_2 = \lambda_3 = \lambda_\perp$. If we make the tacit assumption that volume changes in the rubber will be very small, we can put $\lambda_\perp^2 = \lambda^{-1}$. Substitution in the difference equations (4.45) and (4.47) then gives:

$$f = G(\lambda - \lambda^{-2}) \tag{4.51}$$

$$t = G(\lambda^2 - \lambda^{-1}) \tag{4.52}$$

In order to see more closely what happens, we can solve (4.44b) for f_2:

$$G_1(\lambda_\perp - \lambda^2\lambda_\perp^3) + K\lambda\lambda_\perp(\lambda\lambda_\perp^2 - 1) = 0 \tag{4.53}$$

Cancelling $\lambda_\perp$, this can be arranged to give:

$$\begin{aligned} V = \lambda\lambda_\perp^2 &= \frac{K\lambda - G}{(K - G)\lambda} \\ &= 1 + \frac{G(\lambda - 1)}{(K - G)\lambda} \\ &= 1 + (G/K)(1 - \lambda^{-1}) + 0(G/K)^2 \end{aligned} \tag{4.54}$$

To a first approximation this does give $\lambda\lambda_\perp^2 = 1$ as assumed above. To a second approximation, it shows that there is a small volume strain, equal to $(G/K)(1 - \lambda^{-1})$ or, if $\lambda = (1 + \epsilon)$ to $(G/K)\epsilon/(1 + \epsilon)$. Since G/K is typically of the order of 10^{-4}, the volume strain is very small. However if the volume strain is to be completely prevented, then as shown by (4.50) appreciable transverse stresses are needed and the axial stress is different.

Substitution of (4.54), including the term in G/K, in (4.44a) also yields (4.51). The second approximation must be used because the second term in (4.44) is multiplied by K.

4.2.7.3 *Uniform biaxial stress*

For uniform biaxial stress, we put $f_1 = f_2 = f$ and $f_3 = 0$; $\lambda_1 = \lambda_2 = \lambda$ and $\lambda_3 = \lambda^{-2}$ for constancy of volume. Equations (4.45) and (4.47) then give:

$$f = G(\lambda - \lambda^{-5}) \tag{4.55}$$

$$t = G(\lambda^2 - \lambda^{-4}) \tag{4.56}$$

4.2.7.4 *Uniform triaxial stress*

Under a hydrostatic pressure, we have $f_1 = f_2 = f_3 = f$, and $\lambda_1 = \lambda_2 = \lambda_3 = \lambda = 1 + \epsilon_v/3$, where ϵ_v is the volume strain. Substituting (4.44a) and assuming that $\epsilon \ll 1$, we find:

$$\begin{aligned} f &= G(\lambda - \lambda^5) + K\lambda^2(\lambda^3 - 1) \\ &\cong K\epsilon_v(1 + 2\epsilon_v/3) - G(4\epsilon_v/3) \\ &\to K\epsilon_v \quad \text{as} \quad \epsilon_v \to 0 \end{aligned} \tag{4.57}$$

Thus it is confirmed that K is the bulk modulus. Since it is not reasonable that Hooke's law would apply to large volume strains, as required in (5.30), it would not be right to assume that the above equation would be valid except when $\epsilon_v \ll 1$.

4.2.7.5 *Uniaxial extension at constant pressure*

The easiest experiment to carry out is to stretch a piece of rubber in the atmosphere. The atmospheric pressure p then acts in all directions on the sample, and if t and f are the additional measured stresses, we really have:

$$t_1 = t - p \; ; \quad t_2 = t_3 = -p$$

or approximately:

$$f_1 \cong f + p \; ; \quad f_2 = f_3 \cong p$$

Equation (4.53) should therefore be re-written:

$$G(\lambda_\perp - \lambda^2\lambda_\perp^3) + K\lambda\lambda_\perp(\lambda\lambda_\perp^2 - 1) = -p \tag{4.58}$$

This gives:

$$V = \lambda\lambda_\perp^2 = \frac{K\lambda - G - p/\lambda_\perp}{(K - G)\lambda} \tag{4.59}$$

But K, the bulk modulus, is of the order of 10^4 atmospheres: hence the term in $p/\lambda_\perp$ can be ignored, except at very high pressures, and constancy of volume assumed. Equation (4.47) then gives:

$$t_1 - t_2 = t = G(\lambda^2 - \lambda^{-1}) \tag{4.60}$$

This is the same as (4.52) and shows that the measured stress-strain relation is unaffected by the atmospheric pressure.

4.2.7.6 *Simple shear*

Simple shear in one direction as shown in Fig. 4.8(a) is, by definition, a constant volume deformation with an extension of one diagonal, a contraction of the

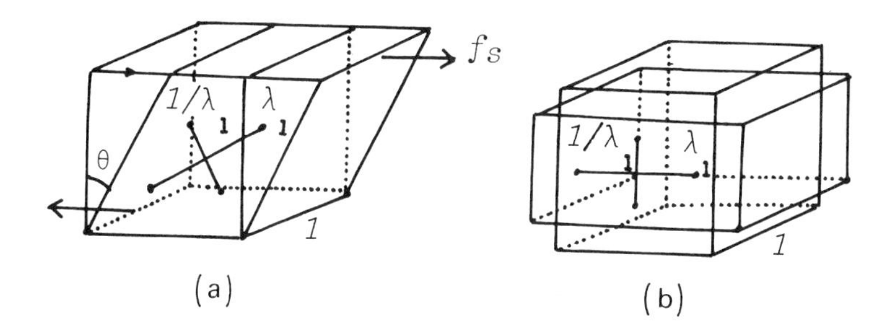

Fig. 4.8 – (a) Simple shear. (b) Pure shear.

perpendicular diagonal, and no strain in the third direction. We therefore have in the principal directions (which are not exactly coincident with the diagonals):

$$\lambda_1 = \lambda_1\,, \quad \lambda_2 = 1/\lambda_1\,, \quad \lambda_3 = 1 \qquad (4.61)$$

The shear strain γ may be shown to be:

$$\gamma = \tan\theta = \lambda_1 - 1/\lambda_1 \qquad (4.62)$$

Instead of transforming the stresses, it is easier to return to the energy equation (4.43), which becomes:

$$\begin{aligned} A &= \tfrac{1}{2}G(\lambda_1^2 + 1/\lambda_1^2 + 1) \\ &= \tfrac{1}{2}G(\gamma^2 + 3) \end{aligned} \qquad (4.63)$$

If the deformation energy is supplied solely through the shear stress f_s, we have:

$$f_s = \mathrm{d}A/\mathrm{d}\gamma = G\gamma \qquad (4.64)$$

This is an important result because it shows that if the strains are small enough for the first term in the inverse Langevin function series to be adequate, or in other words for the Gaussian approximation to be valid, then Hooke's law applies in simple shear, though not in extension, and the parameter G, which was introduced as nkT, is the shear modulus of the rubber.

4.2.7.7 *Pure shear*

The same state of strain in the material, namely $\lambda_1 = \lambda_1$, $\lambda_2 = 1/\lambda_1$, $\lambda_3 = 1$ can also be obtained as shown in Fig. 4.8(b). This is a shear deformation at constant volume but differs from simple shear in that there is no rotation of the axes: it is known as pure shear and can result from a tensile stress f_1 on one face and a compressive stress f_2 on another face.

Application of (4.50) shows that:

$$\begin{aligned} f_1 &= G(\lambda_1 - \lambda_1^{-1}) \\ f_2 &= -G(\lambda_1 - \lambda_1^{-1}) \\ f_3 &= 0 \end{aligned} \qquad (4.65)$$

4.2.8 Comparison with experiment

The dotted lines in Fig. 4.2 show the predictions of the theory. Provided the deformations are not excessively large in extension, there is good agreement between experiment and theory. It is not surprising that there is error at large strains because that is when the neglect of higher terms in the series expansion of the inverse Langevin function – the Gaussian approximation – will cease to be reasonable.

An alternative way of testing the theory is shown in Fig. 4.9. According to (4.47), the difference in stresses $(t_1 - t_2)$ should be proportional to $(\lambda_1^2 - \lambda_2^2)$.

Experimental results from a variety of test methods can thus be plotted together, and are found to be due to the single theoretical linear relation.

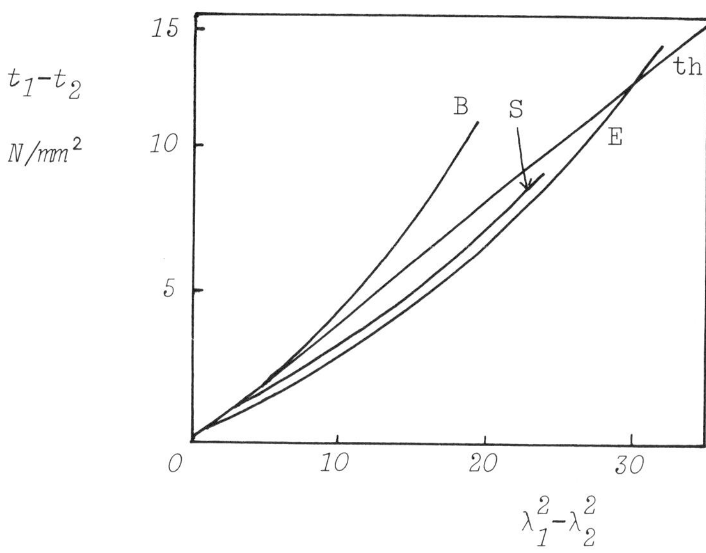

Fig. 4.9 – Plot (th) of the theoretical equation (4.47) with G=0.39 N/mm², compared with Treloar's results for (B) equal biaxial extension, (S) shear, and (E) uniaxial extension.

4.3 FURTHER EFFECTS

4.3.1 Behaviour at large strains

When rubber is highly strained, many segments will approach a fully extended state, and it is then clearly necessary to use an exact relation, such as the full inverse Langevin function form. Modern computational techniques would enable (4.14) to be evaluated to give the free energy of any chain segments in the unstrained rubber, subject to any imposed affine deformation, namely for any change from $l(x, y, z)$ to $l_e(x_e, y_e, z_e)$. Further computation would sum the values of A_s over any given distribution of chain segments, for example a network with a random distribution of junction points, in order to give the total free energy ΣA_s associated with chain conformations. A volume energy term could be added in as before to give:

$$A_T = \Sigma A_s + \tfrac{1}{2}K(\mu_1\mu_2\mu_3 - 1)^2 \tag{4.66}$$

As in (4.32a), the force could be computed from:

$$F_1 = (\partial A_T/\partial \mu_1)_{\mu_2\mu_3} \tag{4.67}$$

Other relations would follow as before. Thus if the network can be defined, there is no problem in evaluating the theoretical behaviour of the rubber.

The use of variational methods in the computation would, in principle, allow the assumption of affine deformation to be relaxed by allowing the co-ordinates of the junction points to adjust to a set of positions which minimised the values of A_T. In principle, one could also adopt a more correct molecular approach to the volume term, through derivation of the appropriate internal energy, though this is not likely to be significant in an ordinary rubber.

In a fundamental sense, the problem thus appears to be fully solved, though there may be operational problems in the computation, and physical problems related to the real nature of the network and the specification of the molecular properties.

4.3.2 Approximate analyses

Prior to the development of powerful computers, a number of other models were proposed in order to allow predictions to be made. These are still useful, both in giving physical insight and in providing relations which are adequate in their fit to be used in problems.

The simplest treatment is limited to uniaxial extension with the network contrained to deform at constant volume. The chain segments which run equally in all directions are replaced by three equal mutually prependicular sets of chain segments, undergoing affine deformation as illustrated in Fig. 4.10. If the

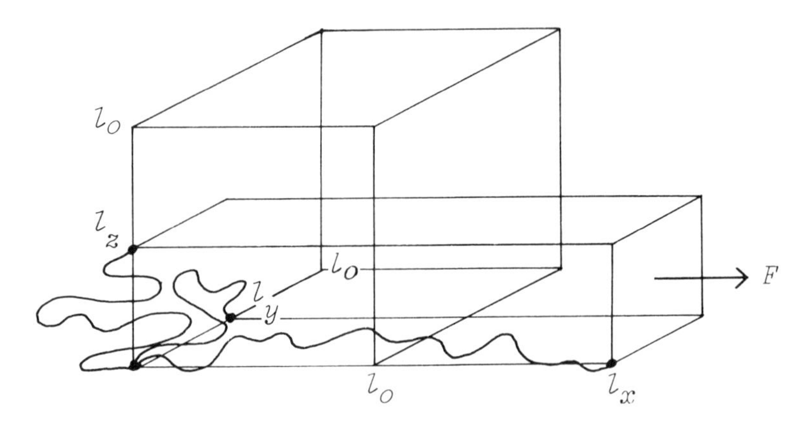

Fig. 4.10 – Deformation of the three-chain model.

original cube of length l_0 deforms to l_x, l_y, l_z under uniaxial extension with an extension ratio λ, then constancy of volume requires:

$$l_x = \lambda l_0\,, \quad \mathrm{d}l_x = \lambda \mathrm{d}l_0 \tag{4.68}$$

$$l_y = l_z = l_0/\lambda^{1/2} \tag{4.69}$$

$$\mathrm{d}l_y = \mathrm{d}l_2 = -\tfrac{1}{2}l_0\mathrm{d}\lambda/\lambda^{3/2} \tag{4.70}$$

Equation (2.19) will give expressions for the forces f_x, f_y, f_z in the chains:

$$F_x = (kT/a)\,[\mathrm{arc}\,\mathcal{L}(l_x/L)] \tag{4.71}$$

$$F_y = F_z = (kT/a)[\mathrm{arc}\,\mathcal{L}\,(l_y/L)] \tag{4.72}$$

If there are ν chains running in each direction, then the work $\mathrm{d}W$ done when the specimen is extended by a small amount $\mathrm{d}l_x$ under a force P must be given by:

$$\begin{aligned} \mathrm{d}W &= P\mathrm{d}l_x = Pl_0\,\mathrm{d}\lambda \\ &= \nu F_x \mathrm{d}l_x + \nu F_y \mathrm{d}l_y + \nu F_z \mathrm{d}l_z \\ &= \nu(kT/a)l_0\mathrm{d}\lambda[\mathrm{arc}\,\mathcal{L}(l_x/L) - 2\,\mathrm{arc}\,\mathcal{L}\,(l_y/L)/\lambda^{3/2}] \end{aligned} \tag{4.73}$$

If we assume that the chain segment length l_0 in the unstrained state is the root-mean square length of the free chain segment so that $L = N_s^{1/2} l_0$, where N_s is the number of free links per chain segment, then we have:

$$l_x/L = \lambda/N_s^{1/2}, \quad l_y/L = 1/\lambda^{1/2}N_s^{1/2} \tag{4.74}$$

Substitution for the relation between density of chains and the number of chains ν then leads to the expression for the stress f under a uniaxial extension ratio λ as:

$$f = 1/3nkTN^{1/2}[\mathrm{arc}\,\mathcal{L}\,(\lambda/N_s^{1/2}) - \lambda^{-3/2}\,\mathrm{arc}\,\mathcal{L}\,(1/\lambda^{1/2}N_s^{1/2})] \tag{4.75}$$

It will, of course, be noted that the three-chain model is not mechanically correct, since it is macroscopically postulated that force is applied in only direction but microscopically assumed that there are forces in all three directions: the link in the energy argument comes through the restriction to constant volume in a network with chains running in all directions.

A rather better treatment can be given by a four-chain model in which four segments linked at a common point terminate at the four corners of a tetrahedron as shown in Fig. 4.11. Treloar calculated the behaviour of such a model using the

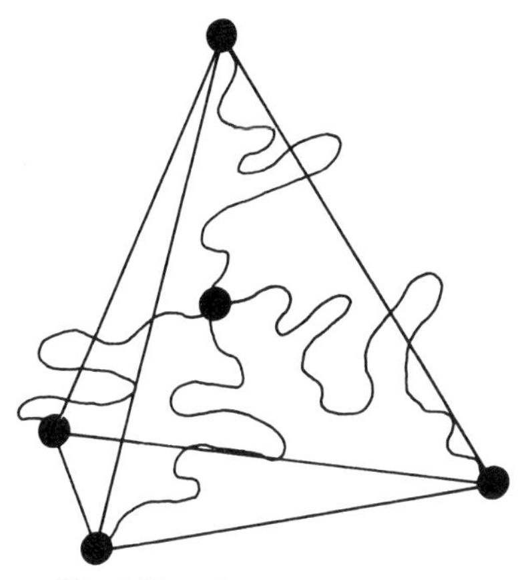

Fig. 4.11 – The tetrahedral model.

exact statistics of 25-link and 100-link chains. It is possible to make allowance for non-affine displacement of the mean position of the junction points.

Treloar has also shown that it is possible to obtain a series solution for the behaviour of a network, from the series expansion given in (2.21). This form is:

$$f = nkT\left(\lambda - \frac{1}{\lambda^2}\right)\left[1 + \frac{3}{25N_s}\left(3\lambda^2 + \frac{4}{\lambda}\right) + \frac{297}{6125N_s^2}\left(5\lambda^4 + 8\lambda + \frac{8}{\lambda^2}\right)\right.$$
$$\left. + \ldots\ldots\ldots \right] \qquad (4.76)$$

Treloar carries on the expression for 5 terms of the series. Alternatively, he carried out a graphical integration of the complete function for chain entropy.

A collection of these theoretical predictions is shown in Fig. 4.12. They all show a steep upward rise at high extensions in contrast to the Gaussian curve. The difference between the non-Gaussian predictions is relatively small and any of them can be adjusted to fit experimental results. Fig. 4.13(a) shows a comparison of experimental responses in uniaxial extension with the predictions of the three chain model, for a typical lightly cross-linked rubber. There is also

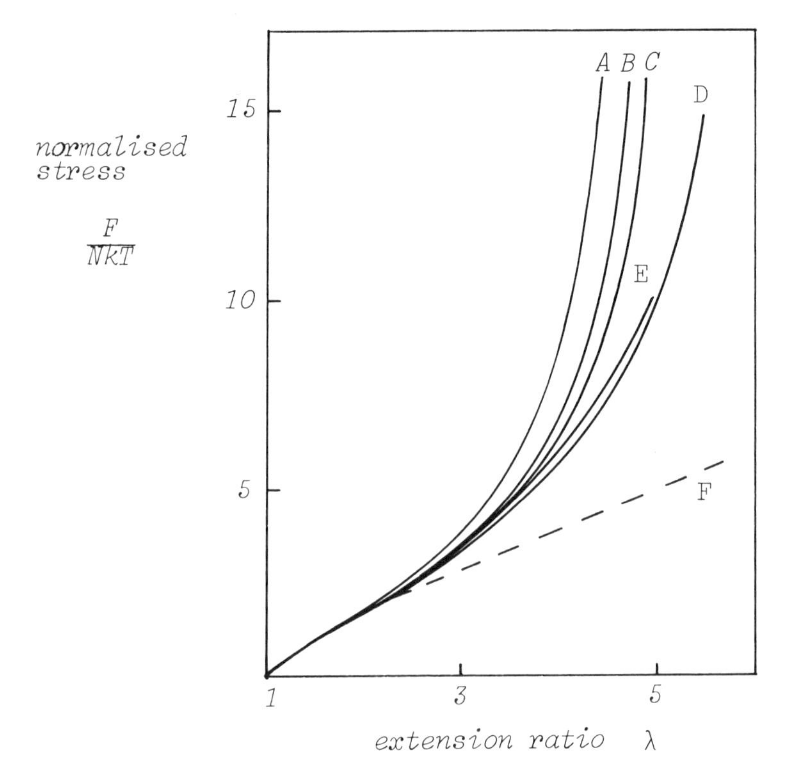

Fig. 4.12 – A variety of theoretical predictions of the rubber stress-strain curve in uniaxial extension.

agreement between theory and experiment, shown in Fig. 4.13(b) for the rather highly cross-linked protein matrix in wool, where the maximum theoretical extension is only 40%, except for a deviation at high stresses attributed to breakage of cross-links.

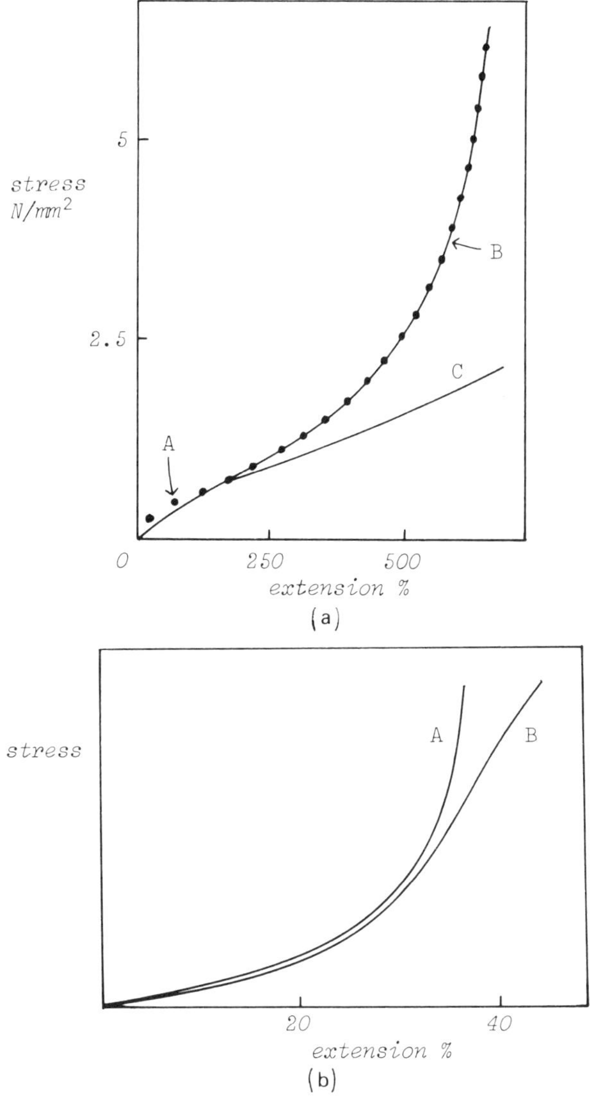

Fig. 4.13 – (a) Close fit between experimental results (dots A) and a typical rubber and the theory (B) from the three-chain model using inverse Langevin function form, (4.75), with N_s = 75 and nkT = 0.27 N/mm². The Gaussian form fails at high extensions. (b) Agreement of (A) the same (4.75) with N_s = 2 and (B) measured values of the stress-strain curve of the matrix in wool. After Chapman.

4.3.3 Phenomenonological representations

There is an alternative approach to the characterisation of rubber elasticity which depends on mathematical simplification instead of physical interpretation.

It can be shown (Appendix B.3.3) that the following quantities are independent of the choice of co-ordinate axes:

$$I_1 = \lambda_1^2 + \lambda_2^2 + \lambda_3^2 \tag{4.77}$$

$$I_2 = \lambda_1^2\lambda_2^2 + \lambda_2^2\lambda_3^2 + \lambda_3^2\lambda_1^2 \tag{4.78}$$

$$I_3 = \lambda_1^2\lambda_2^2\lambda_3^2 \tag{4.79}$$

They are thus referred to as **strain invariants**. Any function of the extension ratios λ_1, λ_2, λ_3 can be expressed in terms of these basic forms. It thus becomes convenient to adopt a general representation for the stain energy A:

$$A = f(I_1, I_2, I_3) \tag{4.80}$$

If we further adopt the assumption of constant volume, then:

$$I_3 = (\lambda_1\lambda_2\lambda_3)^2 = 1 \tag{4.81}$$

and the other two invariants can be written as:

$$I_1 = \lambda_1^2 + \lambda_2^2 + \lambda_3^2 = \lambda_1^2 + \lambda_2^2 + (\lambda_1\lambda_2)^{-2} \tag{4.82}$$

$$I_2 = \lambda_1^{-2} + \lambda_2^{-2} + \lambda_3^{-2} = \lambda_1^{-2} + \lambda_2^{-2} + (\lambda_1\lambda_2)^2 \tag{4.83}$$

A polynomial series for A, with a zero value of free energy at zero strain, namely $I_1 = I_2 = 3$ from $\lambda_1 = \lambda_2 = \lambda_3 = 1$, would then have the form suggested by Rivlin:

$$A_\text{T} = \sum_{i=0,\, j=0}^{\infty} C_{ij}(I_1 - 3)^i(I_2 - 3)^j \tag{4.84}$$

The first coefficient C_{00} for terms independent of strain can be made zero to give the zero value of A_T. The simplest form of (4.84) is then:

$$A_\text{T} = C_{10}(I_1 - 3) = C_{10}(\lambda_1^2 + \lambda_2^2 + \lambda_3^2 - 3) \tag{4.85}$$

This corresponds to the Gaussian approximation (4.43).

But the most general first-order relation is the form adopted by Mooney:

$$A_\text{T} = C_{10}(I_1 - 3) + C_{01}(I_2 - 3) \tag{4.86}$$

The Mooney equation does give a good fit to many experimental results, but it is unwise to read any great physical significance into this. The additional term is merely the most appropriate way of adding a small correction to the Gaussian form.

Various workers have examined expressions with additional correction terms. These, not surprisingly, can often be adapted to fit experimental results.

An alternative formulation by Ogden had the form:

$$A_T = \sum_r \frac{\mu_r}{\alpha_r} (\lambda_1^{\alpha_r} + \lambda_2^{\alpha_r} + \lambda_3^{\alpha_r} - 3) \tag{4.87}$$

in which a set of constants μ_r corresponds to the set of successive indices α_r.

The Gaussian and Mooney forms are special cases of this expression.

From the Ogden form, the stress-strain relations can be derived and a good fit obtained to experimental results natural rubber with three terms, and thus six adjustable constants, which is a convenient result – and in the end, the phenomenological representation is merely a question of operational convenience.

4.3.4 The status of the theory of rubber elasticity

Despite some complications and differences between experiment and prediction, the theory of rubber elasticity must rate as a major successful theory of the mechanics of the solid state, all the more impressive because it relates to an amorphous material whose structure must be characterised statistically. When the statistical mechanical argument for the change in energy with orientation of chain segments under load, which leads to the inverse Langevin function form derived in section 2.2.5, is combined with the mechanical analysis of a cross-linked network, there is reasonable agreement between experiment and theory over the whole range from zero stress to rupture.

There has been a good deal of effort devoted to attempts to improve the theory, but this work suffers from the problem that there is an interaction of difficulties in theoretical analysis with difficulties in the characterisation of real material networks, and this provides a great hindrance to meaningful or useful advance.

It is worth noting some of the points which contribute to the small differences between experiment and theory, although some of the special treatments take account of some of them.

(a) There are various mathematical approximations adopted in analysis.
(b) There are various deliberate physical approximations, such as assumptions of affine deformation, neglect of the entropy of fluctuations of junction points, the switch from average length of a chain at constant force to average force at constant length, and the use of the r.m.s. length of a free chain as the state of segments in the rubber at zero stress.
(c) There is the crude means of combining the conformational changes with the volume change.
(d) There are the problems of the network: the distribution of segment lengths, with some long segments contributing little, the presence of chain ends, with unlinked segments, entanglements, and so on.
(e) There are problems of molecular response, such as internal energy changes associated with chain conformation and the differences from freely rotating chain links.

(f) Finally, and remarkably, there is a 'confidence trick' at the very beginning of the treatment. The equations for the isolated chain segments assume that all conformations compatible with the co-ordinates of the ends of segments occur with an equal probability. But, in fact, in the solid condensed system, the great majority of these conformations will not be possible, and certainly their probability will be influenced by the conformations of the neighbouring chains. There is thus an implicit assumption that the available conformations have a distribution of forms which are geometrically similar to those of the total collection of conformations. This implicit assumption seems to be justified, but there has been no theoretical proof, and no knowledge of the magnitude of the error which may be involved.

4.3.5 Factors influencing the mechanical properties of rubbers

Fig. 4.14 is a representation of the stress-strain curve of rubber and shows up the significance of two features of resistance to extension: the initial stiffness, given by the modulus, and the maximum stretch at which large forces develop and breakage occurs.

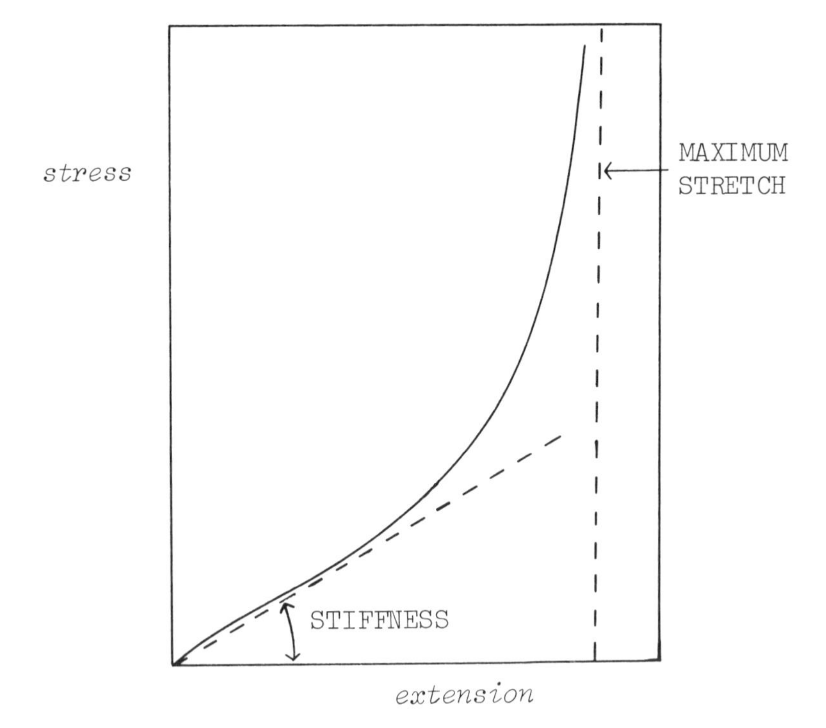

Fig. 4.14 – Schematic representation of stress strain curve of rubber in uniaxial extension.

Simple theory predicts that the initial modulus in extension would be $3G$, namely $3nkT$, where n is the number of chain segments in a unit cube. Clearly the modulus is proportional to the absolute temperature. The density of chain segments can be expressed in various ways, and another expression, (4.27) for the modulus is $3Xf_1\rho RT/2M_1$, where f_1 and M_1 relate to a monomer unit. The stiffness is thus directly proportional to the fraction of monomer units which are cross-linked, to the number of links emerging from each junction, to the density of the rubber, and inversely to the molecular weight of the monomer unit. Some considerations of the problem of the influence of the segment length distribution on the effect of chains in a rubbery network suggest that the modulus is about half the absolute value in the above expressions.

A typical example of the effect of cross-linking on the stiffness of a rubber is shown in Fig. 4.15 for a cross-linked natural rubber. If the molecular weight of the material is low then it is necessary to correct for the large number of loose

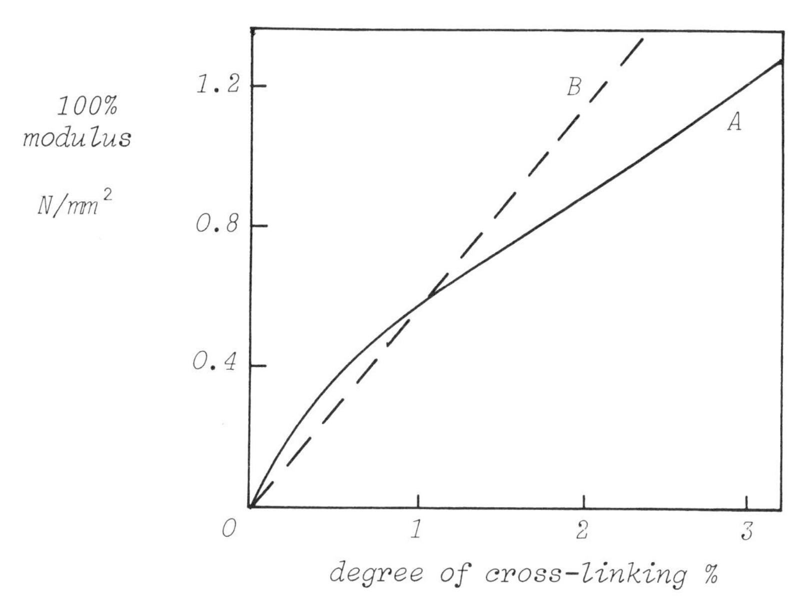

Fig. 4.15 – Modulus, as given by force at 100% extension, related to equivalent percentage of cross-linking agent applied to a natural rubber: (A) experiment; (B) theoretical prediction, namely $3\,nkT$. After Flory.

ends which do not contribute to the active network of linked chain segments. If N_T is the total number of monomer units in the chain molecule and f_1 is the fraction cross-linked, then each chain consists of f_1N_T segments of which the two at each end will be ineffective. The correcting factor by which the modulus is reduced is thus $(1\text{-}2f_1N_T)$, or, as often written, $(1\text{-}2M/M_c)$ where M is the mean chain molecular weight before cross-linking and M_c is the mean molecular weight between cross-links. The form of this correction is naturally similar to that discussed in section 2.5.3 for the effect of slip at chain ends. Fig. 4.16 shows results for the influence of this factor. Experiment and theory agree in form, though there was not an exact agreement with chemical estimates of M and M_c. It will be seen that there is a very rapid change from a low modulus to the value in which the correction is negligible over a narrow range of molecular weight, in this instance from 1×10^6 to 5×10^6.

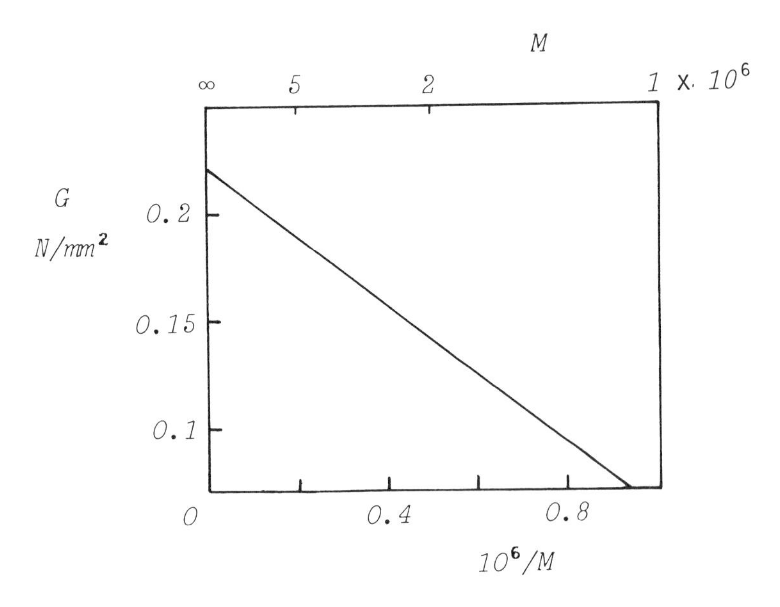

Fig. 4.16 – Effect of chain molecular weight (before cross-linking) on shear modulus of vulcanised rubber. After Flory.

The maximum extension ratio is (L/l_0). On the assumption that l_0 can be equated to the root mean square length of free chains, this means that the maximum extension ratio is $N_s^{1/2}$ where N_s is the number of free links between cross-links in the chain. The number of free links will be the reciprocal of the fraction f_x of free links which are cross-linked. The maximum extension ratio is thus also given by $1/f_x^{1/2}$ or by $(N_f/f_1)^{1/2}$, where N_f is the number of monomer units needed to form an effective free link and f_1 is the fraction of monomer units which are cross-linked. Fig. 4.13 illustrates rubbers with high and low degrees of cross-linking giving maximum extensions of about 40% and 600%.

4.3.6 The 'freezing' and 'melting' of rubber

The strict scientific definition of the terms freezing and melting apply to the change from a regular solid crystal to a mobile liquid melt, but analogous effects occur when an amorphous rubber is heated or cooled. Thus if a flexible rubber rod is immersed in liquid it becomes hard, rigid and glassy: the freedom of molecule motion which gives rise to the entropy-dependent rubber elasticity has been blocked. The transition to this glassy state will be considered in the next chapter.

Going up in temperature, a poorly cross-linked rubber, which appears to be a solid at room temperature albeit one which creeps under load, will turn into a mobile liquid on heating. The change will come when the thermal vibrations are strong enough to overcome the entanglement of long chains. At lower temperatures the entanglements will dominate, act as cross-links and cause an elastic resistance to rapid change of shape; but at high temperatures, or long times, the flow of molecules past one another will be dominant. The greater the molecular weight, the higher will be the temperature needed to allow easy liquid flow.

The effects of temperatures are illustrated in Fig. 4.17 for several forms of polystyrene. All the atactic amorphous samples show a transition from the high modulus of the glassy state to the lower modulus of the rubber. In the lightly cross-linked material, the rubber modulus is maintained as a plateau up to high temperatures, until chemical degradation occurs. But in the non-cross-linked

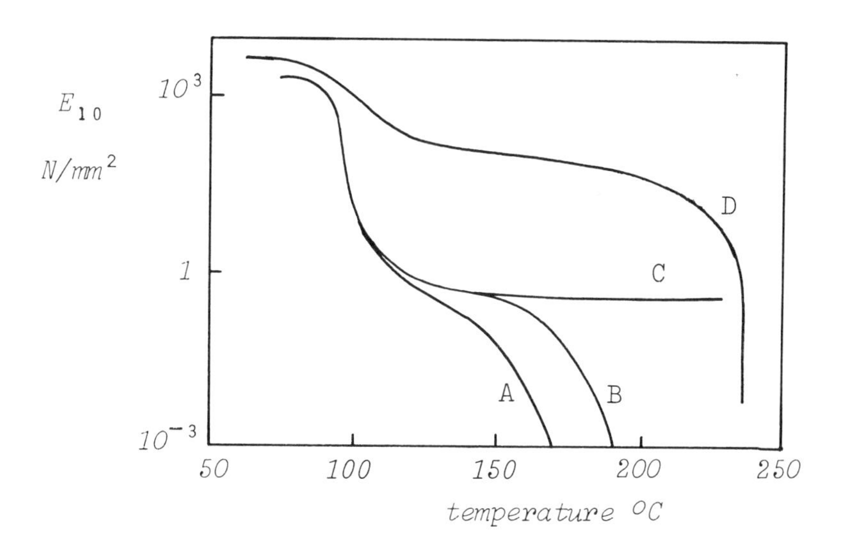

Fig. 4.17 – Relaxation modulus E_{10} of polystyrene (stress at 10 second ÷ strain). After Tobolsky (1960).

(A) : atactic with number-average molecular weight of 140,000
(B) : atactic with number-average molecular weight of 217,000
(C) : atactic, lightly cross-linked
(D) : isotactic, partially crystalline.

atactic material, the plateau is short, and above a certain temperature, the relaxation modulus falls away rapidly towards zero which would be characteristic of a perfect liquid: this occurs at a lower temperature with the lower molecular weight specimen. In passing, it may be noted that the partially crystalline isotactic polymer shows higher modulus values in both the glassy and rubbery regions, due to the stiffening effect of the crystals, and then truly melts at the crystalline melting point.

CHAPTER 5

Transition to the Glassy State

> Similar remarks apply to the famous glass-transition kink in specific-volume/temperature plots. For a time it was believed that this could be displaced down to absolute zero if only the rate of cooling could be reduced to 10^{-1000} deg C/millenium. According to this view a glass was essentially metastable and would flow like a liquid on sufficiently slow deformation (cf. the watch-glasses on Salvador Dali's famous painting 'The Persistence of Memory' (1931) in which time-pieces are seen flowing like treacle). But this view, avant-garde in 1931, may now confidently be rejected.
>
> M. Gordon in *Physics of Plastics* (ed. by P. D. Ritchie)

5.1 INTRODUCTION

In the discussion of rubbers in the last chapter we were concerned with the behaviour of polymer chains which were highly flexible and whose segments could easily slide past one another: in this chapter we examine the situation in which this freedom of movement is restricted. The basic ideas have already been introduced in Chapter 2 in terms of the behaviour of single molecules, but now the application to real assemblies of many polymer molecules must be taken up.

Two general comments may be made at the start. Firstly, we must remember that the restriction on freedom of movement may come either by a stiffening of individual chains or by a binding together of neighbouring chains. Secondly, we note that (a) for a given experimental procedure, carried out at a given rate and a given temperature, changes in chemical constitution can lead to observation of stiffening of the structure by increasing the energy barriers to movement; (b) for a given experimental procedure at a given rate on a given material, a lowering of temperature will lead to stiffening by reducing the magnitude of the thermal fluctuations available to overcome the energy barriers; and (c) for a given material and a given temperature, an increase in the rate of an experiment will lead to an apparent stiffening because less time will be available in which jumps over energy barriers may occur. There will thus be an interrelation between time, temperature and chemical constitution in the occurrence of a transition.

When the transition is complete and freedom of relative movement of neighbouring chain elements is prevented (at least under small forces) the material behaves as a hard glassy plastic. But, first, we look at the behaviour in the transition region, in which the freedom of movement is only partially restricted and the polymer material has a leathery character. The transition can be observed in many ways by following changes in density, specific heat, refractive index, electric properties, nuclear magnetic resonance and so on; but, for the present, we shall concentrate on mechanical behaviour.

5.2 THE TRANSITION IN MECHANICAL PROPERTIES

5.2.1 Stress-relaxation, creep, and stress-strain relations

The transition in mechanical behaviour can be observed in various ways. One can follow creep, stress relaxation, stress-strain curves or dynamic properties. In the transition region, the properties are very sensitive to changes in temperature. Typically a transition will span only 30°C in going through a change of 10^4 times from the low modulus ($\sim 10^5$ N/m^2) of a rubber to the high modulus ($\sim 10^9$ N/m^2) of a glassy plastic. The properties are much less sensitive to time changes: a similar change would have to span a range of a million times in rate; and to include also sufficient experimental data to establish the regions of constant modulus on either side would require a range of 10^{10} times (from a

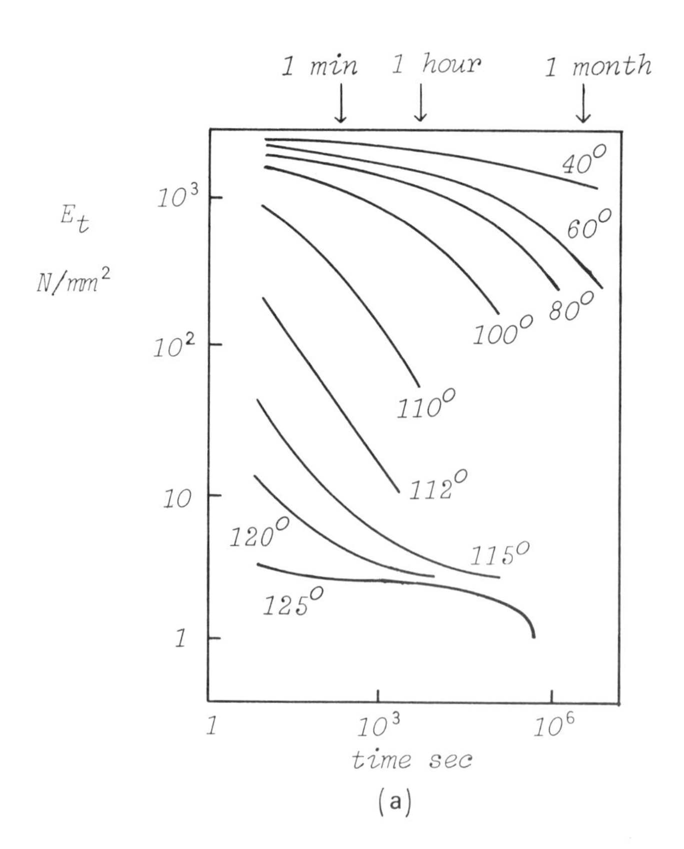

(a)

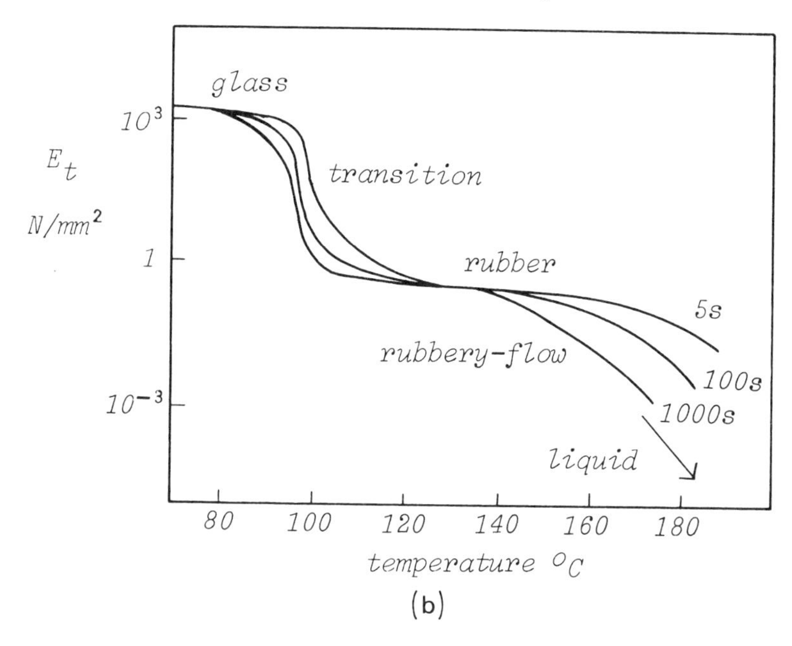

Fig. 5.1 – Variation of relaxation modulus E_t. (a) For polymethyl methyacrylate, plotted against time, for various temperatures in °C. After Tobolsky (1958). (b) For polystyrene, plotted against temperature for various times (same material as B in Fig. 4.17). After Tobolsky (1960).

microsecond to 10 days). No single experimental procedure can possibly cover such a wide range of times, and it is not even possible to do so by combining different techniques of mechanical measurement, although it can be covered in studies of electrical properties.

The general mechanical behaviour is well illustrated by the stress relaxation data shown in Fig. 5.1 for polymethyl methacrylate (Perspex) and polystyrene – two polymers which are typical glassy polymers at room temperature. Fig. 5.1(a) shows the relaxation modulus (that is, the stress after a given time divided by the constant applied strain) as observed at various temperatures over times ranging from 0.001 hours (3.6 secs) to over 1000 hours (40 days). Note the very rapid drop of nearly 1000 times between 100 and 120°C at ordinary times. There is an obvious suggestion that these results are portions of a family of curves of similar shape which are displaced to shorter times for higher temperatures. The same shape appears in the plot against temperature given in Fig. 5.1(b). The whole curve can be divided into five regions: at low temperatures, the high modulus glassy region, then the transition region, the rubbery-elastic region, the rubbery-flow region, and the molten liquid region. The last three of these regions have been discussed in previous chapters.

The sigmoidal response curve is also shown in creep, as illustrated for plasticised polyvinyl chloride in Fig. 5.2: there is little deformation in short times, but then an increasing rate of elongation with log (time) in the transition region, leading to a reduced rate at long times.

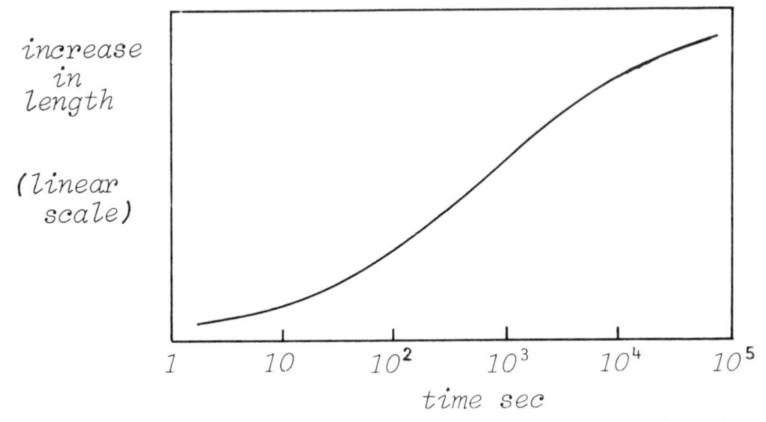

Fig. 5.2 – Recoverable creep of plasticised polyvinyl chloride at 20°C. After Alfrey (1948).

Stress-strain curves demonstrate a reduction in modulus as temperature increases, as shown by the change in initial slope of the curves for cellulose acetate in Fig. 5.3: but, a more marked effect is the earlier yielding at higher

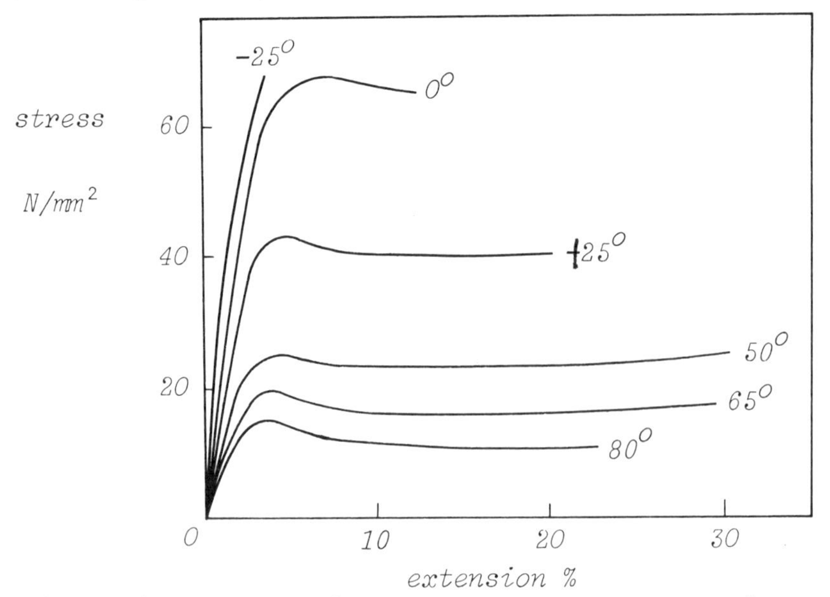

Fig. 5.3 – Stress-strain curves of cellulose acetate at various temperatures in °C. After Carswell and Nason (1944).

temperatures which suggests that an increase in stress will also take the material through the transition, although since this involves a major internal deformation and non-linearity of response it is a more complicated situation. The completion of the transition to the rubbery state can be regarded as occurring when the yield stress reaches zero. 'Stress-strain curves' can also be constructed artificially from creep or stress relation experiments. If a series of creep tests is carried out under different loads, (or relaxation under different extensions) then a set of associated stress and strain values for any particular time can be obtained. Fig. 5.4 shows some results for polyvinyl chloride obtained in this way. The change from high initial modulus at short times to low initial modulus at long times is clearly seen. There is also evidence of non-linearity in the stress-strain relaxation above 1.5% strain.

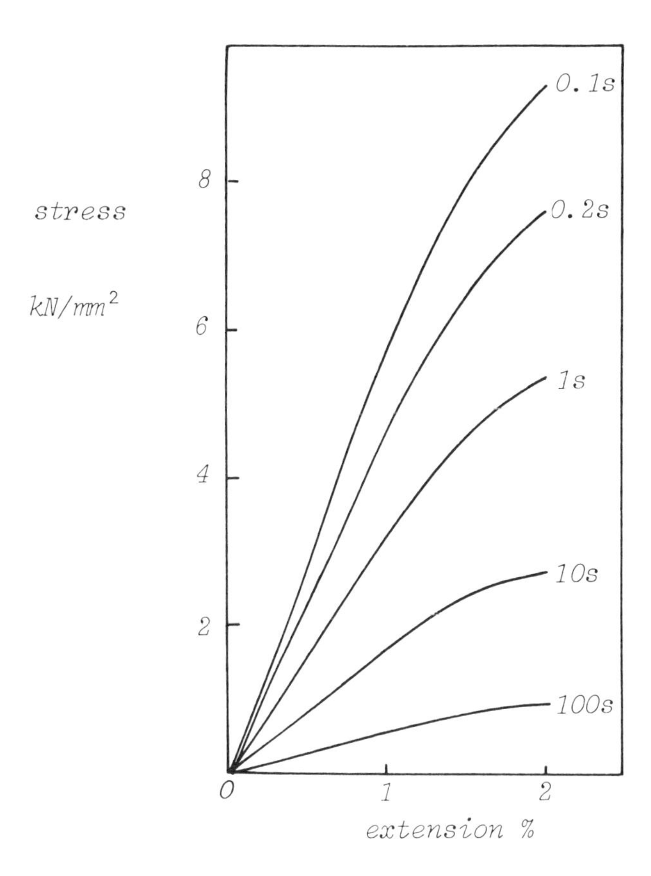

Fig. 5.4 – Stress-strain curves constructed from stress relaxation experiments, to correspond to different times, for polyvinyl chloride at 79°C. After Sommer (1959).

5.2.2 Energy loss in a transition

The increased energy loss, or reduction in resilience, in a transition region can be very directly demonstrated by dropping a steel ball bearing (which itself will deform elastically with little loss of energy) on to a sheet of the polymer material. The ratio of the rebound height to the drop height, which is the coefficient of restitution, shows the elasticity of the material. Alternatively, the result can be expressed as:

$$\text{fractional energy loss}$$
$$= \frac{\text{KE of ball before impact} - \text{KE of ball after impact}}{\text{KE of ball before impact}}$$
$$= \frac{\text{drop height} - \text{rebound height}}{\text{drop height}}$$

Fig. 5.5 shows a typical example. At room temperature the ball bounces back from a hard Perspex sheet to within 10% of its starting height; but as the temperature rises past 100°C the rebound height falls until at 160°C it is barely bouncing at all – it looks as though it is falling into putty. Then to the surprise of anyone watching the experiment, a further rise in temperature causes the ball to bounce right back up again with little energy loss, as the Perspex becomes an elastic rubber.

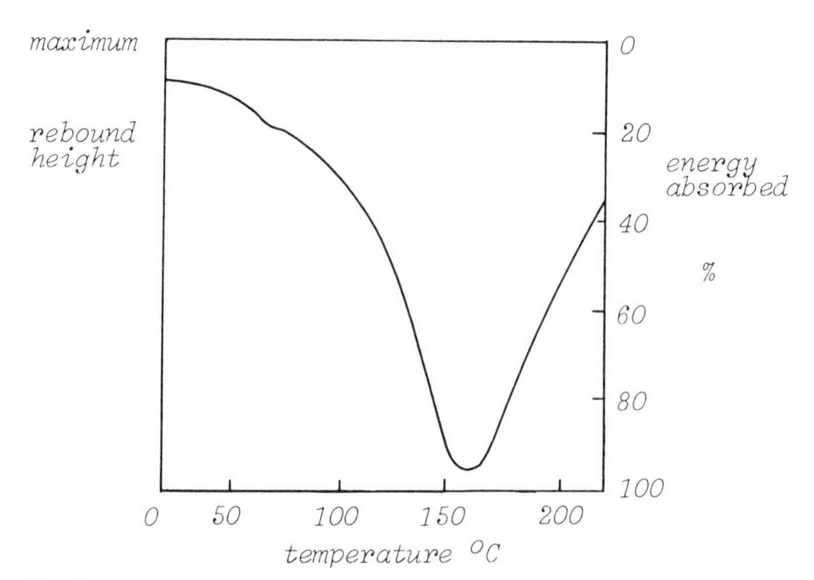

Fig. 5.5 – Change in rebound energy of a ball dropped on to a polymethyl methacrylate sheet at different temperatures. After Gordon and Grieveson (1958).

An impact test of this sort obviously corresponds to a fairly high rate of deformation, giving a rather high value for the transition temperature, but the time of deformation is not very well defined and may indeed change as the mechanical properties change. It therefore corresponds to an angled section through a three-dimensional plot of energy absorption against time and temperature.

5.2.3 Dynamic properties

Measurement of dynamic properties is probably the most useful way of studying the effects of time and temperature over a wide and well-defined range, and bringing out both real (stiffness) and imaginary (lossy) parts of the deformation.† Fig. 5.6 shows typical results obtained by shear testing of polystyrene and rubber at about 1 Hz. The general behaviour is clearly similar to the results shown in Fig. 5.1 for the relaxation modulus. In rubber the rapid decrease in modulus at the transition occurs below room temperature, whereas in polystyrene it is above room temperature. Coincident with the fall of the modulus (real part) there is a peak in the logarithmic decrement, which is related to the imaginary or loss modulus. Ths slight rise in modulus of the rubber above 0°C is a consequence of the kinetic theory of rubber elasticity already discussed; we can also note that in the absence of cross-linking flow starts just above the transition temperature.

A set of results where dynamic tests have been carried out over a wide range of frequencies (10^{-3} to 10^{2} c/s) is shown in Fig. 5.7(a). The characteristic sigmoidal variation of the real part of the modulus and peak in the imaginary part is clearly apparent. Another very full set of curves taken at different temperatures over a moderate frequency range is shown in Fig. 5.7(b) and these display the usual family relationship. It may be noted that similar effects are found in dielectric properties, as shown in Fig. 5.8.

The way in which a transition, as shown by a step in the real modulus or a peak in the loss modulus, varies in position with temperature and frequency can be shown by plotting collections of curves, Fig. 5.6 to Fig. 5.8, *or* a contour diagram Fig. 5.9, *or* a three-dimensional representation, Fig. 5.10.

5.2.4 Master curves: the WLF equation

The form of the sets of curves, such as Fig. 5.8, suggests that by suitable translations they might all be superimposed in order to give a single master curve. This has been recognised for a long time in both the dielectric and mechanical properties of many materials, and various particular procedures have been proposed.

† There are many different ways of representing the information. The energy loss in the oscillation can be expressed by several interrelated parameters: the imaginary part of the complex modulus, the loss factor, the loss angle δ, tan δ, the logarithmic decrement. Compliance (ease of deformation) may be used as an alternative to modulus (stiffness). See Appendix B.5.2.

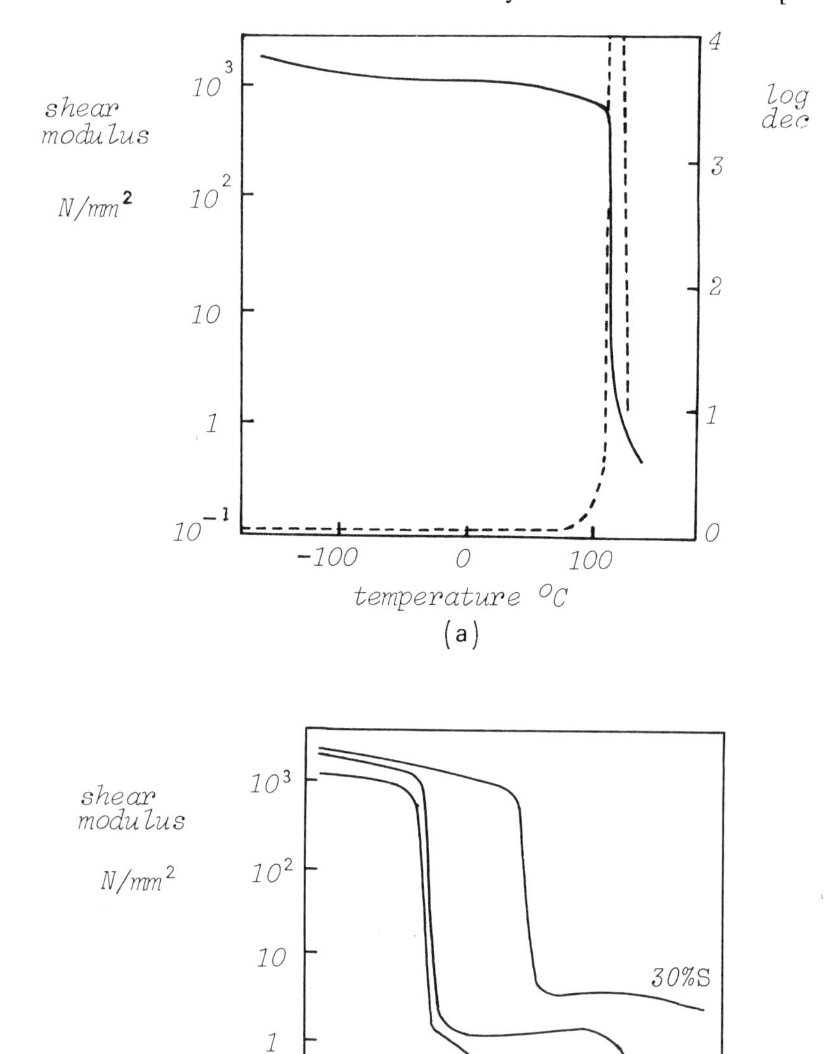

Fig. 5.6 – Variation in shear modulus in dynamic tests at about 1 Hz. After Schmieder and Wolf (1953). (a) For polystyrene, also showing the logarithmic decrement (dotted line) as a measure of energy loss. (b) For rubber, at different degrees of cross-linking as shown by sulphur content.

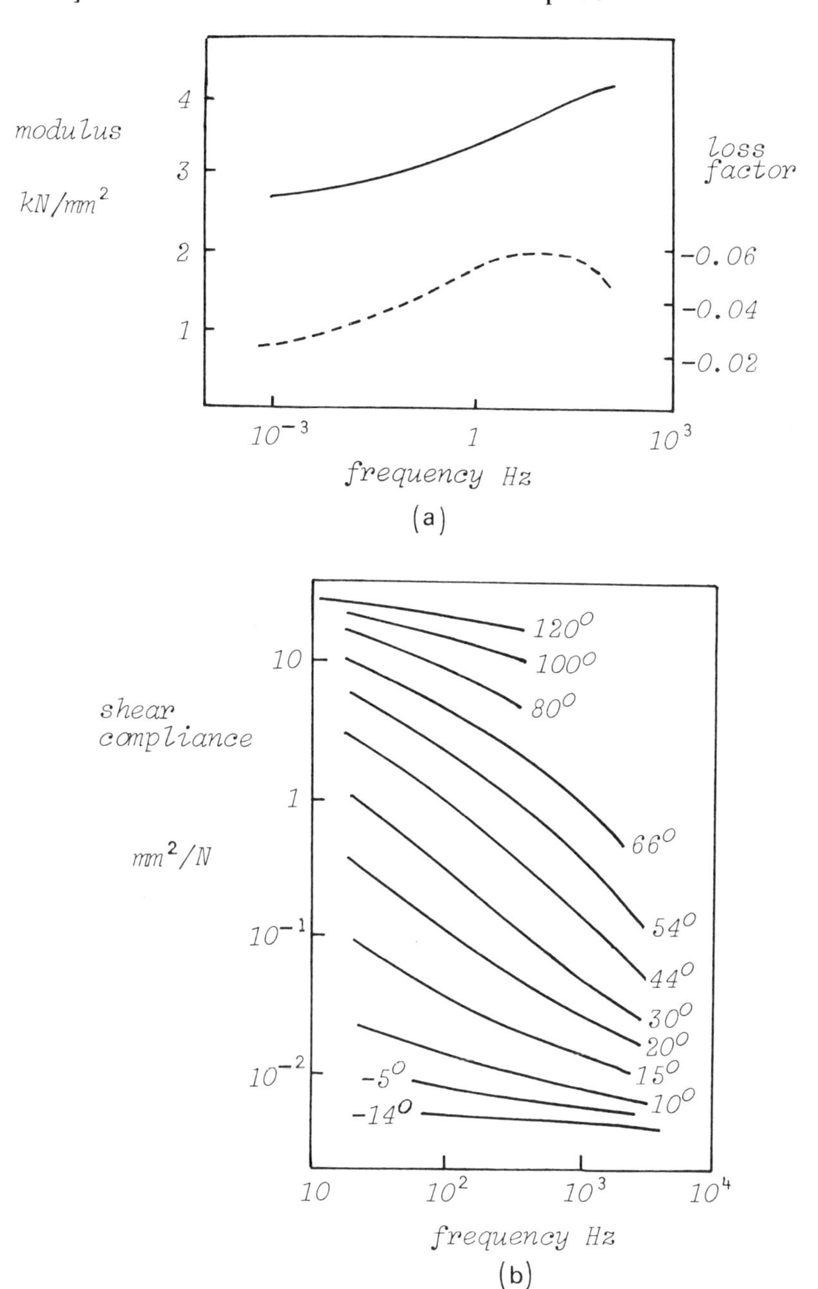

Fig. 5.7 – (a) Real modulus and loss factor for polymethyl methacrylate at 30°C, plotted against frequency of oscillation. After Maxwell (1956). (b) Shear compliance of poly-n-octyl methacrylate plotted against frequency, at 24 different temperatures. After Dannhauser, Child and Ferry (1958).

For amorphous polymers the most successful procedure is the time-temperature superposition suggested by Ferry. Although we shall see that this does have some theoretical basis, it is more conveniently introduced as an empirical procedure. There are two steps in the process of superposition. The curves of a log-log plot of compliance against frequency as in Fig. 5.7(b) are: (a) shifted upwards in compliance log J' by log $(\rho T/\rho_0 T_0)$ where ρ is the density at the experimental temperature T, and ρ_0 is the value at a standard temperature T_0; (b) shifted to the left in frequency log ω by an amount log a_T in order to superimpose the curves. This procedure is equivalent to log-log plotting of $J' T\rho/T_0\rho_0$

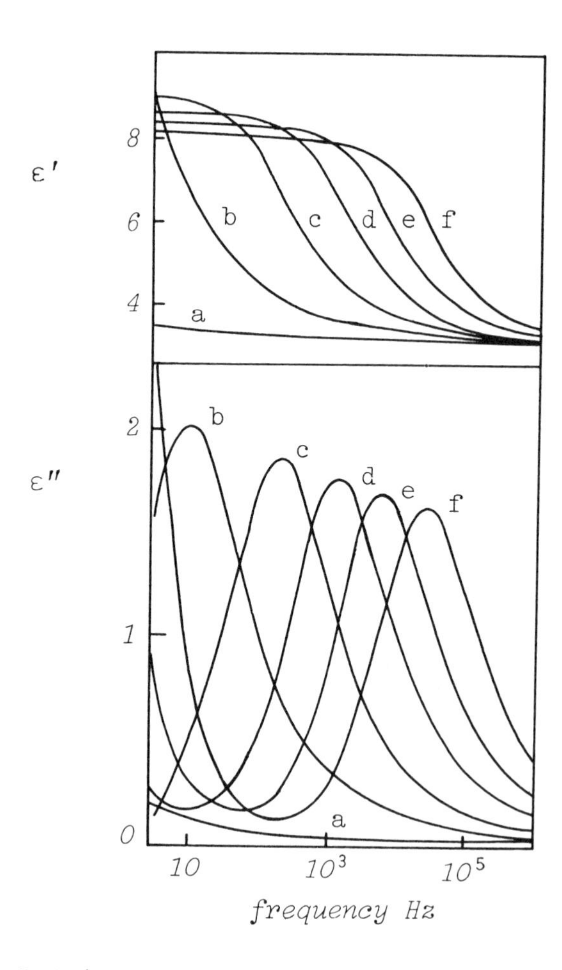

Fig. 5.8 – Real (ϵ') and imaginary (ϵ'') parts of dielectric constant of polyvinyl acetate. (a) 38°C; (b) 53.5°C; (c) 62.5°C; (d) 70°C; (e) 77°C; (f) 83.5°C. After Ishida, Matsuo, and Yamafuji (1962).

against t/a_T, where time t is the reciprocal frequency, since:

$$\log J'(T\rho/T_0\rho_0) = \log J'(t) + \log(T\rho/T_0\rho_0) \tag{5.1}$$

$$\log t/a_T = \log t - \log a_T \tag{5.2}$$

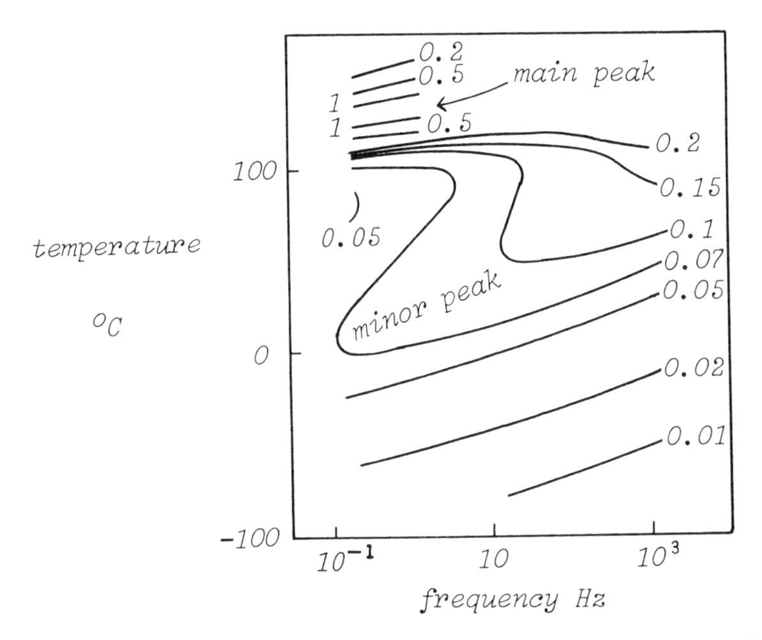

Fig. 5.9 – Contour plot of mechanical loss of polymethyl methacrylate. After Heijboer (1956).

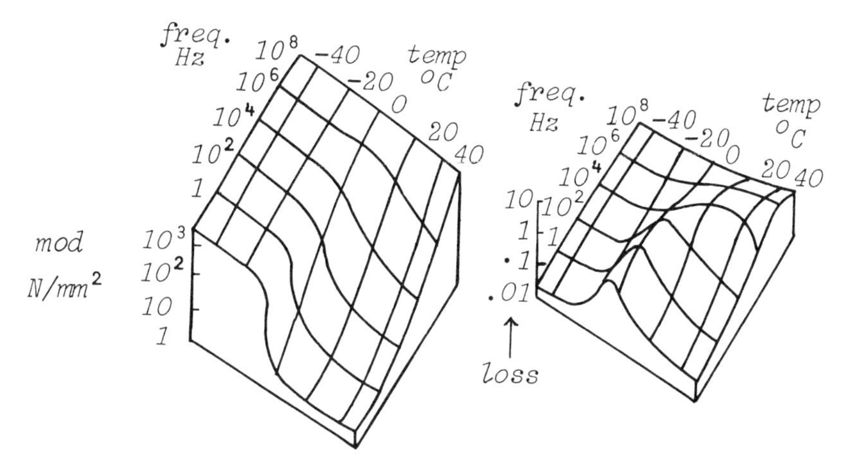

Fig. 5.10 – Three-dimensional plot of modulus and loss factor for a synthetic rubber, Buna N. After Nolle (1950).

Fig. 5.11 shows the data of Fig. 5.7(b) replotted in this way: the data clearly fall on a single master curve, indicating that all the curves have the same shape, although only a part is included in each experiment.

The time-temperature superposition principle has been introduced here in terms of measurements of compliance in a dynamic test. But equivalent procedures can be used for all the visco-elastic parameters: real and imaginary parts of dynamic modulus and compliance, creep compliance, and relaxation modulus.

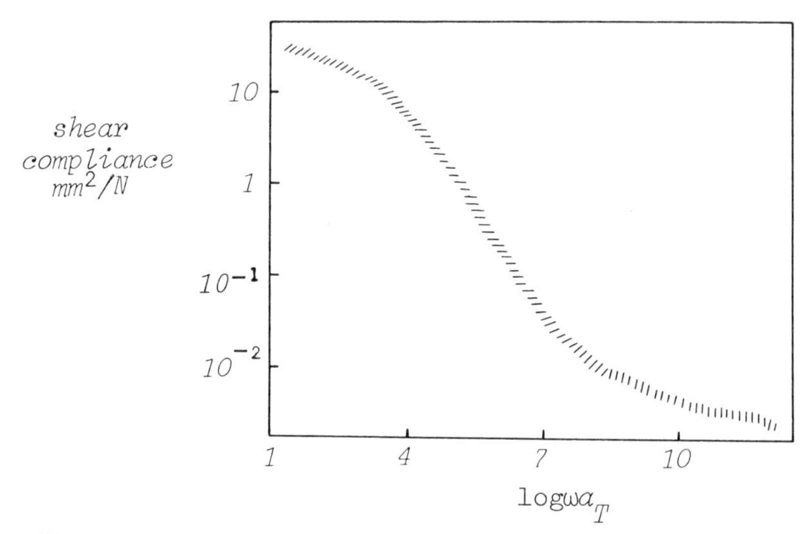

Fig. 5.11 – Master curve for the compliance of poly-n-octyl methacrylate, reduced to $T_0 = 100°C$. The breadth of the line represents the spread of results due either to experimental error or inexact superposition on the shift of the 22 curves of which a selection is included in Fig. 5.7(b). After Ferry (1970).

The tests of the applicability of the principle of time-temperature superposition are firstly the exact matching of the curves and secondly the use of the same values of a_T (which is a function of temperature) in order to superpose all the visco-elastic functions. The superposition of loss compliance, for the data from the same experiments as those used to give Fig. 5.7(b) and Fig. 5.11, is shown in Fig. 5.12: with the same values of a_T there is good agreement except at the highest frequencies.

Williams, Landel and Ferry found another great simplification. They found that experimentally determined values of a_T fitted an equation – the WLF equation – of the form:

$$\log a_T = -C_1'(T - T_0)/(C_2' + T - T_0) \tag{5.3}$$

The properties of the polymer over a wide range of times are thus determined solely by the temperature T, the density ρ and two arbitrary constants C_1' and C_2'.

As described so far, any temperature T_0 may be used for the superposition. However, a further simplification is often achieved by reducing the results to the glass transition temperature T_g, defined as the temperature at which there is an abrupt change of specific heat. We then have:

$$\log a_{T_g} = -C_1(T - T_g)/(C_2 + T - T_g) \tag{5.4}$$

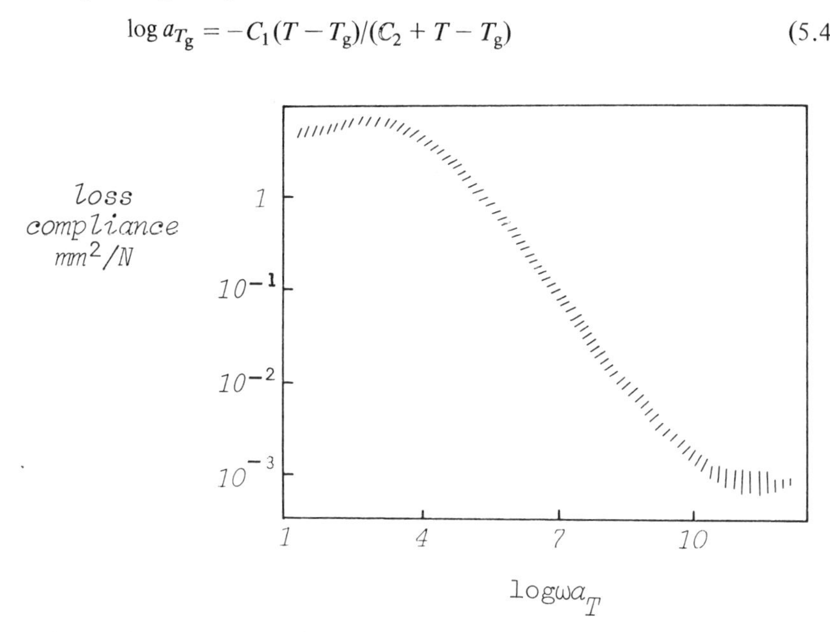

Fig. 5.12 – Master curve for the loss compliance of poly-n-octyl methacrylate, from the same experiments as used for Fig. 5.7(b) and 5.11, with the same value of a_T, reduced to $T_0 = 100°C$. After Ferry (1970).

The first results suggested that C_1 and C_2 were universal constants with the values 17.44 and 51.6°. In other words, when reduced to their glass transition temperature all polymers would show the same behaviour. In fact some differences are found – for instance polar polymers are rather different from non-polar polymers – and, for a variety of polymers quoted by Ferry, C_1 ranges from 15.6 to 26.1 and C_2 from 20 to 130°.

A rather wide collection of experiments has shown that time-temperature superposition and the WLF equation apply to amorphous polymers in the transition region from T_g to $(T_g + 100)°$. Somewhat similar procedures have been applied in other circumstances.

5.2.5 Theoretical basis of time-temperature superposition

We now approach the problem of time-temperature superposition in another way by seeing if we can define a procedure from theoretical arguments.

Firstly we must expect to be able to superpose the modulus (or compliance) values in the rubbery elastic plateau region, where there is no appreciable time lag in the response during the period of the experiment. The theory of rubber elasticity as given in the last chapter, shows that the shear modulus is nkT. Consequently the curves should have a vertical shift to take account of temperature and density of segments in order to bring them into register. This is the first part of the superposition procedure described in the last section: it is only a small correction, and, historically, it was only introduced because of the theoretical argument.

Secondly, we have to consider how the lag in response, as temperature falls or rates increase, can be introduced theoretically. The argument derives from the theories used to explain the viscoelasticity of polymer solutions; these were discussed in section 3.4 and derived, in turn, from the crude arguments of section 2.3.4.

Any particular molecule containing N random links is regarded as made up z sub-units each containing q random links (so $N = qz$); by a suitable choice, both q and z can be made reasonably large. Each sub-unit is then replaced by a bead (representing its mass and bulk) and an elastic string (representing its deformational character) as in Fig. 5.13(a). This is identical with the procedure used in discussing polymer solutions: the difference is that the viscous drag comes not from the solvent molecules, but from the difficulty of moving past neighbouring polymer molecules and from internal resistance to the bond rotation which gives chain flexibility. It is however still reasonably sensible to regard this viscous drag as represented by a friction coefficient $q\zeta_0$ where ζ_0 is the effective viscous drag which can be considered as acting on a single random link and slowing down its movement. The individual polymer molecule can thus be represented by a ladder model as shown in Fig. 5.13(b), where, as in polymer

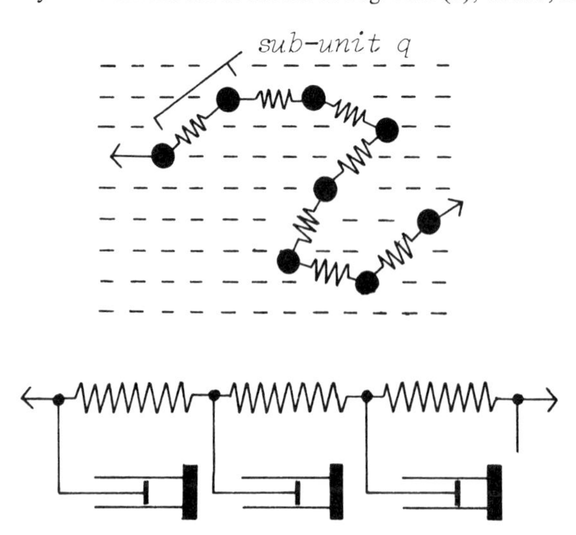

Fig. 5.13 – (a) Spring and bead model of one polymer chain surrounded by others. (b) Equivalent ladder network.

solutions, the springs correspond to the elastic restoring force in chain segments and the dashpots to the viscous drag on their motion. In describing the whole assembly of polymer molecules in this way, there is an element – doubtless justifiable – of 'sleight of hand' in that when any particular molecule comes under consideration it is regarded as a set of beads and springs, but when it is not under special consideration it joins all the other molecules as a diffuse source of viscous drag.

As discussed in sections 3.4.2 and 3.4.3 the ladder model is a complete representation in itself, though the same results can be obtained by analysis of the normal modes of vibration of the damped spring and bead model according to the theoretical treatments given by Rouse or Bueche. Whichever procedure is adopted the models give equations in terms of the relaxation times of the different modes, τ_p:

$$G(t) = NkT \, \Sigma \exp(-t/\tau_p) = \text{relaxation modulus} \qquad (5.5)(3.72)$$

$$G' = NkT \, \Sigma \, (\omega^2\tau_p^2/1 + \omega^2\tau_p^2) = \text{real part of modulus} \qquad (5.6)$$

$$\tau_p = zl^2q\zeta_0[24kT\sin^2\{p\pi/2(z+1)\}]^{-1} \qquad (5.7)(3.73)$$

The first term in (5.5) and (5.6) corresponds to the vertical shift due to the change in the modulus of the springs in the ladder model, on the basis of the theory of rubber elasticity. Otherwise temperature can only enter the equations through the values of τ_p.

The relaxation time of the first mode, $\tau_1 = zl^2q\zeta_0/24kT$, is clearly affected by temperature in several ways: (i) directly; (ii) perhaps slightly through variation in the length of a random link l, namely of the resistances to internal motion change; and (iii) certainly, very strongly, through change in the viscous drag given by the friction coefficient ζ_0. The really important feature to note is that all the relaxation times will be changed by the same factor due to change in temperature, since τ, is merely altered by multiplication by $[\sin^2\{p\pi/2(z+1)\}]^{-1}$ which is independent of temperature. Consequently the effect of change in temperature is to change the time-scale of the response by a constant factor, namely the factor a_T introduced in the empirical treatment on a logarithmic time-scale this is a lateral shift.

The use of a factor a_T in the procedure is thus justified. The value of a_T will be given by:

$$[\tau_p]_T/[\tau_p]_{T_0} = a_T = [l^2q\zeta_0]_T T_0/[l^2q\zeta_0]_{T_0} T \qquad (5.8)$$

If it is possible to carry out an experiment which is dominated by steady flow viscosity η of the polymer – for example a slow flow in a non-cross linked polymer at reasonably high temperature – then we would have:

$$a_T = \eta T_0 \rho_0/\eta_0 T \rho \qquad (5.9)$$

This gives a direct method of finding the value of a_T. Over a moderately narrow range of temperature, it is possible to neglect the effects of the factors T and ρ, and consider only the much larger change in η. This would give:

$$a_T \cong \eta/\eta_0 \tag{5.10}$$

As the material passes through its glass transition temperature, the onset of additional molecular vibration causes changes in internal structure which can be interpreted as an appreciable increase in free volume. Williams, Landel and Ferry therefore suggested using Doolittle's viscosity equation:

$$\log_e \eta = \log_e A + B(f^{-1} - 1) \tag{5.11}$$

where f = free volume as a fraction of total volume.

Thus:

$$\log_e a_T = \log_e \eta - \log_e \eta_0 = B(f^{-1} - f_0^{-1}) \tag{5.12}$$

It has furthermore been suggested by Fox and Flory that the fraction of free volume at the glass-transition temperature is a universal constant f_g. If we then assume a linear variation above T_g, we get:

$$f = f_g + \alpha_f(T - T_g)$$

$$\log \alpha_T = \frac{-B/2.303\, f_g(T - T_g)}{f_g/\alpha_f + T - T_g} \tag{5.13}$$

This is identical in form with the empirical WLF equation.

Arbitrarily, on the basis of Doolittle's results for liquids, Ferry puts $B = 1$, and his experiments on a number of polymers show that f_g typically has a value of about 0.025. The free volume is not a quantity which is well defined operationally – and indeed may not have much real physical significance – but the value of 0.025 (2.5%) at T_g seems plausible and is similar in magnitude to estimates by other methods.

The quantity α_f, which is a thermal expansion, shows more variation, although this may partly be due to neglect of the factor B. If it is assumed that the expansion above the glass transition temperature is partly due to the continued expansion resulting from the same mecahnism as in the glassy state and partly to the creation of new free volume determining α_f, as indicated in Fig. 5.14 we should have:

$$\alpha_r = \alpha_g + \alpha_f \quad \text{or} \quad \alpha_f = \alpha_g - \alpha_r = \Delta\alpha \tag{5.14}$$

where α_r is observed thermal expansion coefficient above the glass transition temperature and α_g is the value below it. Experiments show a reasonable agreement of the measured expansion with the mechanical data, as a consequence of change of temperature.

In summary, one can say that there is sound theoretical backing for the basic procedures of time-temperature superposition in the rubbery and transition regions. There are also plausible arguments in support of the exact form of the

WLF equation, though these results rest on the rather shaky foundation of the idea of free volume. What is probably justified is to say that the increased thermal vibration, due to rise in temperature, causes changes in both total volume and in viscosity and that it is reasonable to represent these by equations of the form shown: this demotes the 'free volume' to the status of an auxiliary variable of uncertain physical significance. It is worth noting that there are other ways of deriving the WLF equation by arguments which are different in form, but similar in effect.

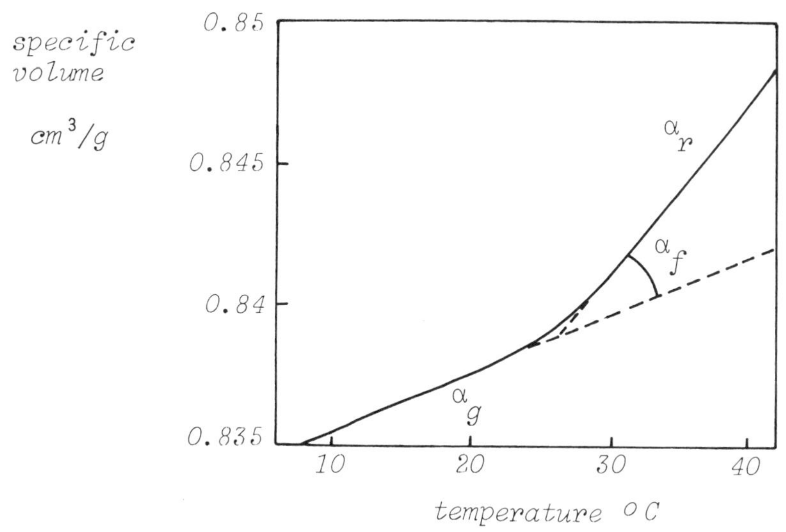

Fig. 5.14 – Change in volume of polyvinyl acetate. After Meares (1957). The slopes of the two lines are α_g and α_r, with a difference α_f.

The whole theoretical basis of time-temperature superposition as represented by Fig. 5.13(a) is, of course, based on the idea of a rubbery chain, in constant Brownian motion, with entropy dependent elasticity, subject to a viscous drag. It cannot therefore be expected to apply at temperatures below T_g, where the model would predict a complete absence of deformation due to an extremely high viscosity, which would tend to infinity and give a glass of infinite modulus. In reality, other modes of deformation, due to internal-energy dependent changes of bond angles and bond lengths become significant, and cut off the increase in modulus. The principles of time-temperature superposition discussed so far cannot be expected to apply in the lower temperature regions of the transition.

5.2.6 The glass-to-rubber transition more generally

The transition in properties from rubbery to glassy is of great technical interest, as well as fundamental scientific significance. The following are just some important aspects of its industrial importance.

(a) It gives the conditions (rate and temperature) at which the stiffness of the material changes dramatically; so that a body intended to be rigid fails (at high temperature, long times) by becoming soft, or a body intended to be flexible fails (at low temperature, high rates) by becoming stiff and often brittle.

(b) It indicates when there will be large energy losses and internal heating of a material during cyclic deformation. Sometimes, as in braking, this energy loss may be desirable, but it is usually a nuisance. As indicated in section 5.2.2, this is also a measure of the resilience of the material.

(c) By cooling through the transition temperature, a material can be set in new form. Sometimes this may be a nuisance, when a plastic gets overheated, deforms, and becomes set in the wrong shape on cooling; but often it can be deliberately used as a means of forming materials into required shapes. Since there is no change in the basic structure of the material, the set is essentially temporary; and there is the possibility of a return to the reference state by heating above the glass transition temperature and then cooling the material under zero stress. The forms of change are illustrated schematically in Fig. 5.15.

In addition to the influence of temperature and time, which have been explicitly dealt with in this discussion, and the influence of chemical constitution, which has been implicit, there are other factors such as pressure, addition of plasticisers, absorption of moisture, molecular weight, and so on which influence

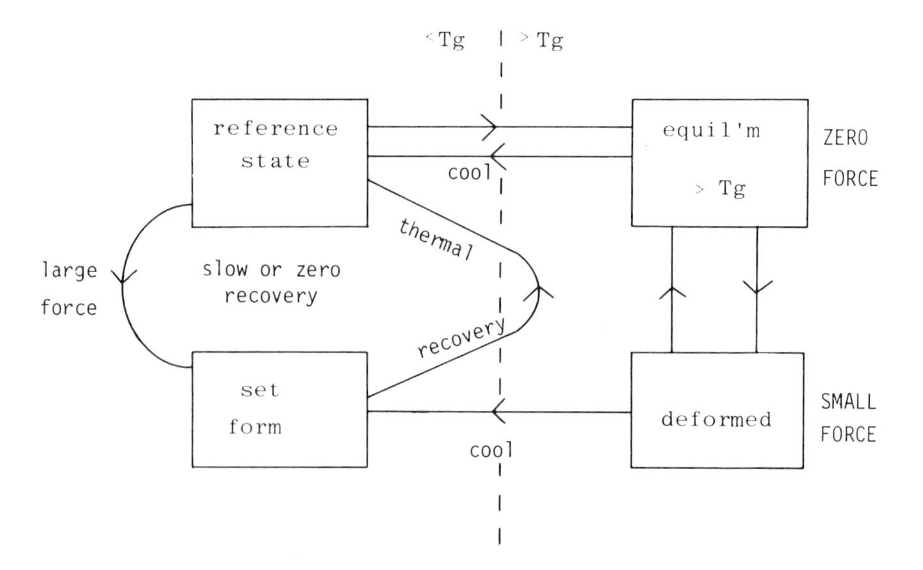

Fig. 5.15 – Schematic view of occurrence of temporary set at glass transition.

the transition. Furthermore the transition results not only in the changes in the various mechanical properties, but also in changes in the corresponding electrical properties, and in many other physical properties. To some extent, theoretical treatments, such as the WLF equation, can be applied to these other properties.

Fig. 5.16 shows examples of changes of thermal expansion and refractive

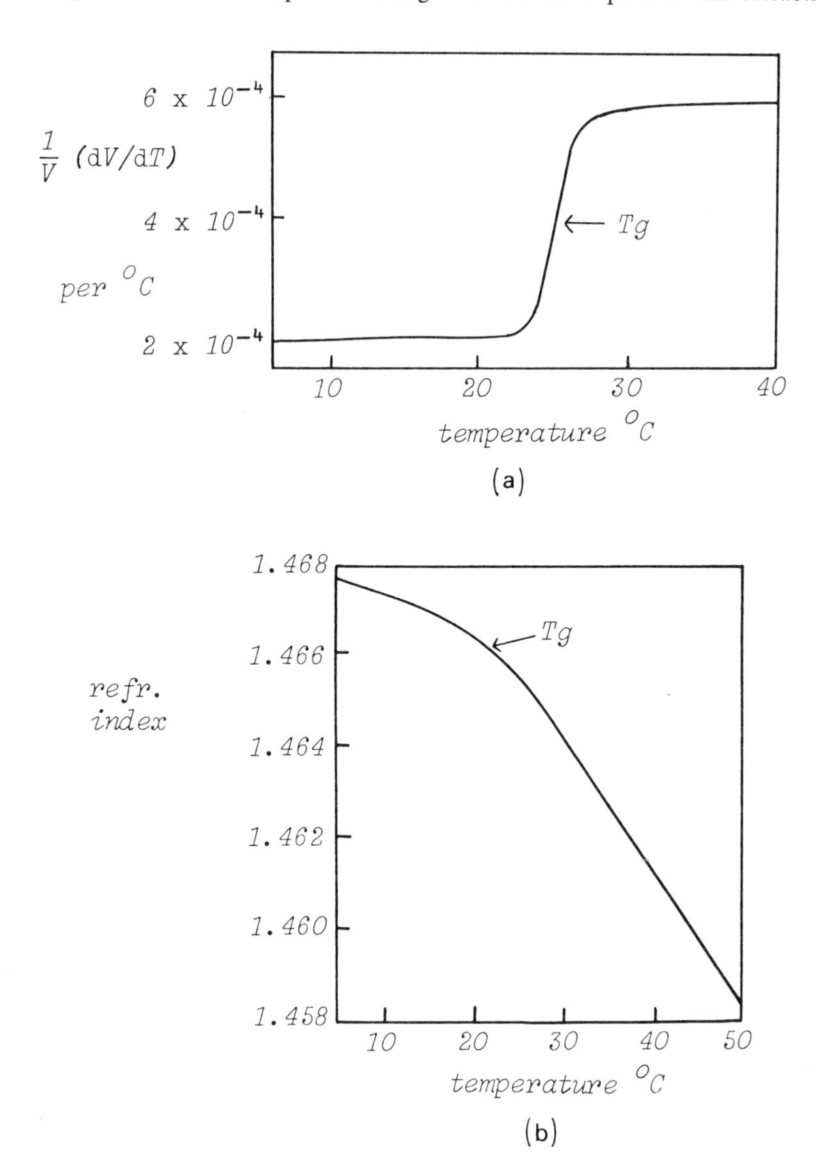

Fig. 5.16 – Changes in polyvinyl acetate at the glass transition. After Meares (1957). (a) Coefficient of volume expansion (b) Refractive index.

index of polyvinyl acetate at the transition from glass to rubber; and Fig. 5.17 shows the change in specific heat. Scientific interest centres on the thermodynamic nature of the transition, and in the structural interpretations. Usually there is no change in volume V or in heat content H (with a manifestation of latent heat) at the transition, but there is a change in the coefficient of expansion, dV/dT, and the specific heat, dH/dT. Thus the transition has the features of a second-order thermodynamic transition, and Ehrenfest's relation is often reasonably well obeyed. However, the transition is usually smeared out over a small, but definite, range of temperature, is influenced by the rate of change of temperature, and may show other deviations from the ideal pattern. More advanced discussion is difficult, because it is not easy to separate the influences of experimental error, of accidental inhomogeneities in the material structure, and of minor secondary effects, such as a thermodynamic development of structure in the glassy state, from a consideration of the validity of the basic theory.

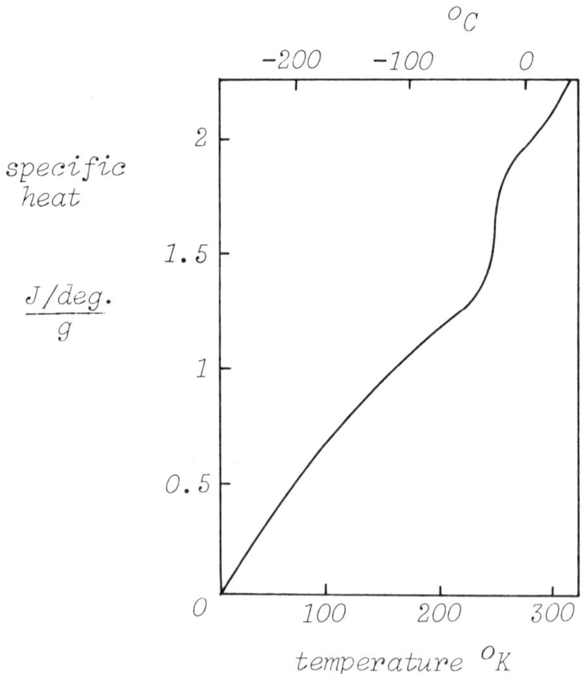

Fig. 5.17 – Change in specific heat of atactic polypropylene, with step at the glass transition. After O'Reilly and Karasz (1966).

Structurally, a second-order transition is one in which there is no discontinuous change of the pattern of form and packing of the molecules, such as the dissolution of a regular crystal lattice, but there is a sharp change in the response of the structure. This fits in with the simple model of the behaviour of polymers in which, at high temperatures, there is a flexibility of the chain which is 'frozen out' in the glassy state, but the unchanging picture of the amorphous glassy state is identical with an instantaneous picture of the amorphous rubbery state. At a deeper level of understanding – or ignorance – there may be subtle departures from this simple view for a variety of reasons.

5.3 MULTIPLE TRANSITIONS

5.3.1 A variety of mechanisms

In the last section, there has been a tacit assumption that a polymer shows a single transition from the rubbery to the glassy region. In reality, many materials show a multiplicity of transitions. Typical examples are given in Fig. 5.18: the shoulder on the low-temperature side of the rebound curve in Fig. 5.5 is another. In general, it is easy to see why there should be multiple transitions, although in particular instances there may be difficulties in deciding on the interpretation.

One cause is the occurrence of different groups along the polymer chain with different transition temperatures. If these groups are in separate molecules,

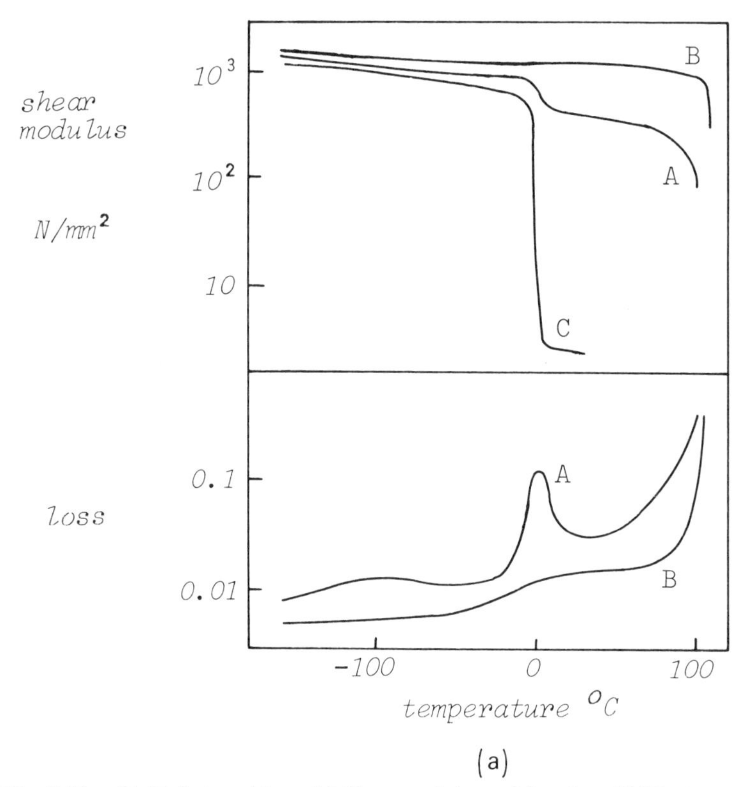

(a)

Fig. 5.18 – Multiple transitions. (a) Shear modulus and loss in a 50/50 styrene-butadiene block copolymer (A), with polystyrene (B) and butadiene (C) for comparison. After Angelo, Ikeda and Wallach (1965).

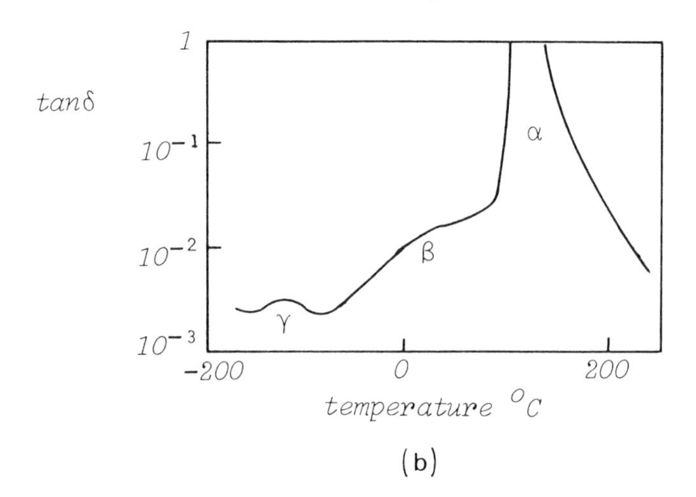

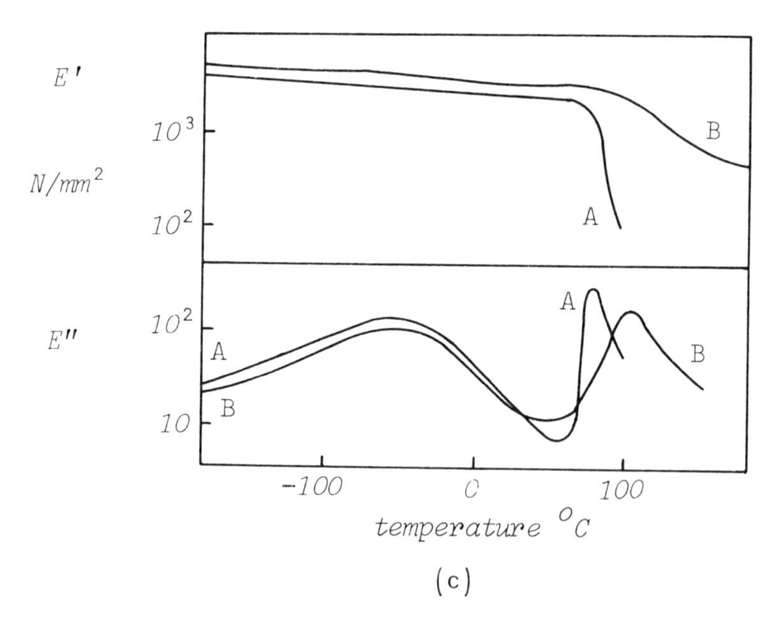

Fig. 5.18 – Multiple transitions. (b) Variation of tan δ at 0.5 Hz with temperature in a highly cross-linked atactic polystyrene. After Illers and Jenckel (1958). (c) Real (E') and imaginary (E'') moduli of polyethylene terephthalate: (A) very low (5%) crystallinity; (B) well developed (50%) crystalinity. After Takayanagi (1963).

or in substantial blocks in a copolymer then the material can be regarded merely as a mixture which shows up the transitions of each polymer, modified somewhat due to the difference in environment. This is the situation for the styrene-butadiene copolymers in Fig. 5.18(a).

Where the groups are more intimately mixed, for example in a homopolymer which contains different chemical types within a monomer unit, then the changes can be regarded as successive reductions in the length of a random link, or in other words all the transistions after the first are changes from one rubbery state to another. The situation is crudely illustrated in Fig. 5.19: at low

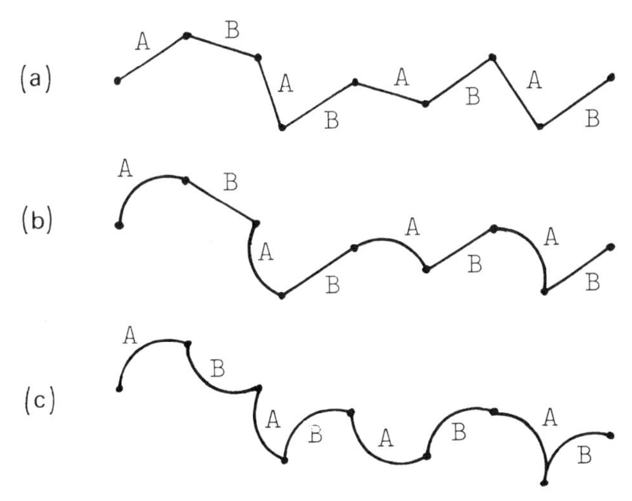

Fig. 5.19 – Schematic representation of a polymer molecule with different chemical units *A* and *B* in the chain: (a) both stiff; (b) *A* flexible; (c) both flexible.

temperature (a) both units are rigid and the chain is completely glassy; at a somewhat higher temperature (b) the A units become free and the material becomes a rubber with a random link length equal to some multiple of A + B (depending on the limits to the motion of A); at a higher temperature still (c), when B units become free, the length of a random link will be reduced. Alternatively, the two (or more transitions) may be due to successive changes in the nature of the motion allowed in a particular unit. For example, rotation over one energy barrier may give some freedom of movement at a certain temperature, but complete freedom of rotation may not occur until a higher energy barrier is passed at a higher temperature. In this instance, it would be the ratio of monomer length to length of an effective random link which would be changing in the minor transition but the observed effects would be similar.

The relative importance of the two transitions will clearly depend on the relative sizes and frequency of the different units. With a structure as in Fig. 5.19, there will be a very large drop from the glassy character to the rubbery in the first transition, and then a much smaller change of rubber modulus in the second transition. On the other hand, it must be remembered that, in a solid material, the chain fluctuations do have to occur in an environment of other chains. Consequently, with a structure as shown in Fig. 5.20 a freeing of the unit A will have only a small effect due to the overwhelming influence of the long stiff B units, which lock the structure into a glassy state. In this situation, the first transition would correspond merely to slight easing of the means of deformation of a glassy polymer, and the big transition would be the second one.

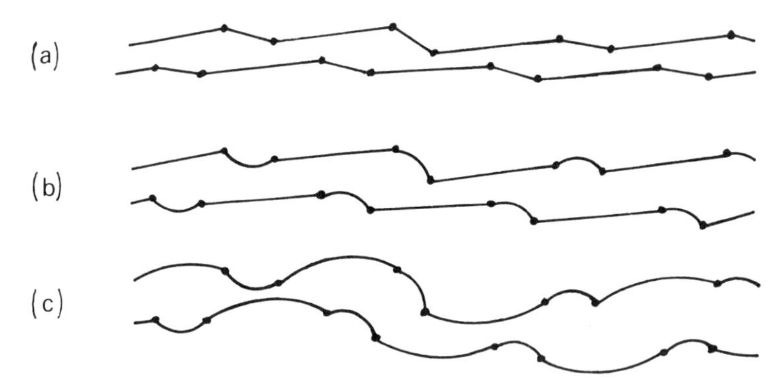

Fig. 5.20 – Schematic representation of polymer with short units A and long units B: (a) both stiff; (b) A flexible; (c) both flexible.

The freeing of groups in side chains will have similar effects. The problem is schematically illustrated in Fig. 5.21. Suppose we concentrate attention on

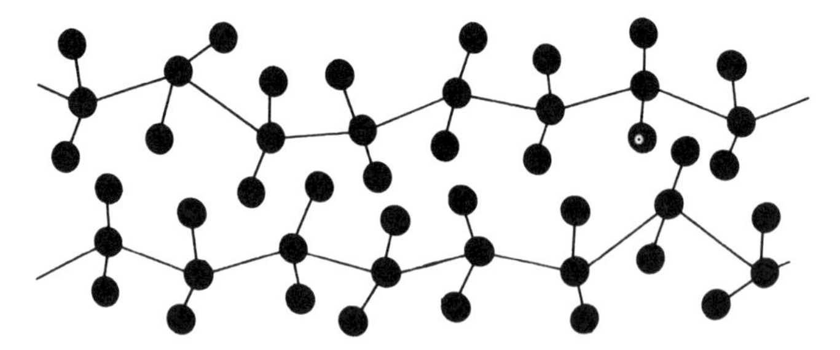

Fig. 5.21 – Schematic representation of polymer chain with side-groups.

a particular region of a glassy polymer material at very low temperature. Under stress, the main chains will be elastically deformed, but in addition there will be an elastic deformation of the side-groups lying between the chains – for example, as two neighbouring chain segments straighten, they will tend to compress the side groups in between, or vice versa. We can therefore write for the stress σ due to local strain ϵ:

$$\sigma = \frac{\partial U_{\mathrm{mc}}}{\partial \epsilon} + \frac{\partial U_{\mathrm{sc}}}{\partial \epsilon} \tag{5.15}$$

where U_{mc} is the elastic energy associated with main chain deformation and U_{sc} is the elastic energy associated with side-chain deformation.

At the low temperature, both terms in (5.15) will be large, and the actual values of U_{mc} and U_{sc} will depend on what detailed local deformation gives the minimum total elastic energy. But if the temperature rises, and the side-chains become mobile, then the effect of the second term will become negligible, and the stress corresponding to given strain (that is, the real part of the modulus) will fall to the value given by the first term alone (which may itself fall somewhat, since it can adjust to its own minimum, not the combined minimum of U_{mc} and U_{sc}). The matrix of side-groups between the main chains can be regarded as changing from a high modulus to a low modulus material. The overall effect of the change will be small due to the dominant effect of the main chain, so this will be a minor transition. In the transition region, the movement of the side-chains will be sluggish and all the usual transition effects (shown by the imaginary part of the modulus, or the energy losses) will be apparent.

The converse effect will come when the main chains become free at a lower temperature than the side-chains. This will lead to a minor transition in the rubbery region. As long as the side-chains are rigid, the freedom of rearrangement of the main chains will be more restricted than when the side-chains are free.

Another mechanism of multiple transitions is in successive breakage of cross-links. Suppose we have a polymer (for example, a polyamide) which at intervals along the chain contains groups capable of forming moderately strong cross-links such as hydrogen bonds. The structure is illustrated in Fig. 5.22. At low temperatures, the whole system would be glassy. But at some temperature, the chain segments will become free and the material will act like a rather highly cross-linked, and so rather stiff, rubber. Indeed if the cross-linking

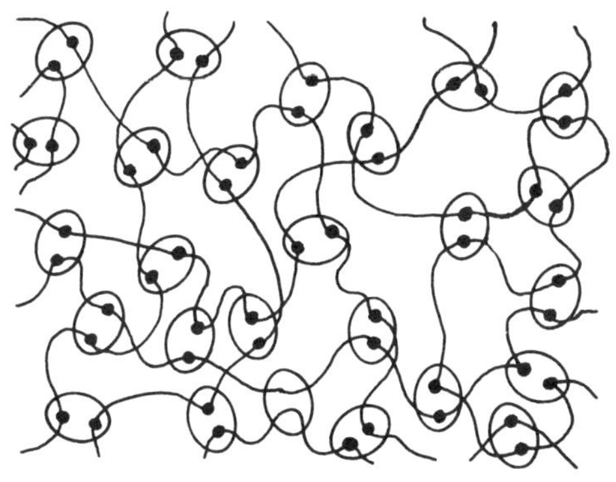

Fig. 5.22 – Schematic representation of an amorphous structure with temporary cross-links at intervals along the chain molecules.

is dense enough, the freeing of short connecting chains will have a very small effect, and the change will appear as a minor transition within the glassy region. On the other hand, if the connecting links are long, this may be a major transition. At a higher temperature still, the thermal fluctuations become large enough to cause an effective dissociation (or at least a rapid dynamic breaking and reforming) of the cross-links: the material thus goes through a transition from a stiff highly cross-linked rubber to a more flexible assembly of free rubbery chains.

The mechanism described in the last paragraph is probably the cause of the effects in polyethylene terephthalate, shown in Fig. 5.18(c). At about −60°C, the aliphatic sequences -CO-O-CH_2-CH_2-O-CO- become mobile due to bond rotation, but the benzene rings, which alternate with them, remain associated up to about 100°C. Many of the polymers of this sort which have been made and studied are, in fact, partially crystalline polymers and thus have a more complicated behaviour, as also indicated in Fig. 5.18(c).

While it is easier to discuss this last mechanism in terms of one type of chain segments linked by groups which form cross-links, similar arguments will apply when there is a more uniform interaction along the chain, but an interaction which becomes less effective at some distinct temperature different from other transitions. Thus polyacrylonitrile has one transition at around 100°C associated with the relative slippage of chains, as the dipole interactions of the -C≡N groups are overcome by thermal vibration, and another at a higher temperature when flexibility develops in the chain.

There are thus many possible mecahnisms, and many detailed effects to observe. Apart from the need to understand the individual behaviour of all the chemical groups concerned, there is the general theoretical problem of the interaction between the different modes of deformation, which may range from one extreme of completely independent effects, through some modification of responses, to an intimate coupling which combines two effects into a single transition.

5.3.2 The significance of co-operation in transitions

In discussing multiple transitions, one considers separately all the different ways in which a structure can develop greater freedom of internal movement. Whether or not they appear as separate transitions – and also the sharpness of each transition – is a more complicated matter. In general, one can write for any particular mode of deformation occurring in a possible transition of type A:

$$\begin{matrix}\text{total energy} & & \text{specific} & & \text{energy barrier} \\ \text{barrier due to} & = & \text{local energy} & + & \text{due to associated} \\ \text{transition A} & & \text{barrier A} & & \text{changes in environment}\end{matrix} \qquad (5.16)$$

By the first term, we mean the barrier which would be present in isolated chains in the polymer gas: the particular barrier to rotation or to the dissociation of a particular cross-link. But in a solid polymer freeing the unit in this way is not enough; other neighbouring molecules have to move out of the way before the change in conformation can occur. It is this which gives rise to the second term.

If only a single mode of transition is involved, then (5.16) becomes:

total energy barrier due to transition A	=	specific local energy barrier A	+	energy barrier due to transition to neighbouring units
			+	energy due to remaining deformation of environment

$= n \times$ (energy barrier A) + (energy due to remaining deformation of environment)

where n is the number of units which have to move together.

The value of n will depend on the detailed geometry of the structure, since it is determined by the minimum number of units needed to give realisable internal movement of the structure. (In reality, the changes probably occur due to a continuous sequence of coupled motion, but it is much easier to think of the problem as separate 'batches' of movement.) If n is large, the degree of co-operation will be high and the transition will be sharper and will tend to occur at a high temperature.

The polyamides are a good example. With the structure shown in Fig. 5.22, each hydrogen bond can break independently, with an energy barrier being due mainly to the bond energy with some addition due to elastic deformation of the environment of -CH_2- groups: the transition, which centres on about 50°C, is therefore a broad one. The whole situation is very much like the dissociation of a diatomic gas, in which at any temperature it is reasonable to talk about a certain proportion of monatomic molecules being present. In the same way, the dissociation of the hydrogen bonds can be regarded as gradually increasing as the temperature rises through a long transition range, thus gradually decreasing the degree of cross linking of the network.†

But when the degree of orientation increases, as illustrated in Fig. 5.23, it

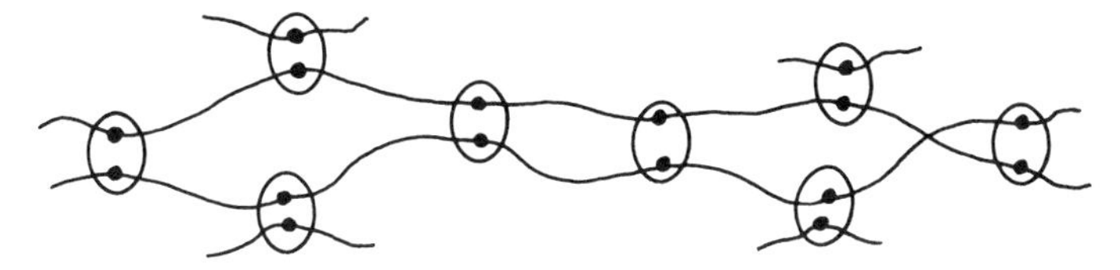

Fig. 5.23 – Small portion of temporarily cross-linked structure in an oriented state.

† It should be noted that this type of mechanism is rather different from the mechanism discussed in section 5.2.5, as leading to the WLF equation. Deviation from the WLF equation can be expected when other mechanisms are involved.

becomes more difficult to envisage a single cross link breaking: several neighbouring links have to go, and so the transition temperature rises and becomes sharper. In the limit, with the perfect assembly of the crystal, the degree of co-operation is so high than the transition occurs sharply at a single temperature, the melting point. When we consider two possible transition mechanisms, then equation (5.16) becomes:

$$\begin{array}{lclclcl} \text{total energy} & & n \times (\text{energy} & & m \times (\text{energy} & & \text{energy due to} \\ \text{barrier to} & = & \text{barrier A)} & + & \text{barrier due} & + & \text{remaining deformation} \\ \text{transition A} & & & & \text{to B)} & & \text{of environment} \end{array}$$

If the second term in this expression is relatively large, then the two transition mechanisms will merge into one, and we shall really have:

$$\begin{array}{lclcl} \text{total energy} & & \text{energy barrier} & & \text{energy barrier due} \\ \text{barrier to} & = & n \text{ A units} & + & \text{to remaining} \\ \text{transition AB} & & m \text{ B units} & & \text{environment} \end{array}$$

One must remember that the quantities in these equations are not absolutely determined: there can be a relative adjustment in order to give the minimum total energy barrier. Thus if the local barrier to transition B is very high, then it will be preferable to deform the environment by means other than deformations of type B; in other words $m \rightarrow 0$, and B remains as a quite separate transition.

Factors which will tend to cause two possible transitions to merge into one are:

(a) structurally, the absence of any easy way in which deformation of type A can occur without B also occurring;

(b) a similarity in the heights of the energy barriers so that both mechanisms can combine without an undue increase in the total energy: this is equivalent to saying that two mechanism which are inherently close together in transition temperature are more likely to merge than two which are far apart.

Whether or not transitions remain separate and identifiable depends on the balance of these two factors. The general situation is summarised in Table 5.1.

Table 5.1 – Combination of two transitions.

'Inherent' transition temperature	Strong structural coupling between motion *A* and *B*	Weak structural coupling between motion *A* and *B*
A = B	Transitions indistinguishable	Transitions indistinguishable
A < B	Merge to single broader transition	Appears as double overlapping transition
$A \ll B$	Minor transition at A Major transition at B	Major transition at A Lesser transition at B

One particular example of a strongly coupled pair of mechanisms is the internal rotation in a chain and the movement of neighbouring units past one another. These must occur together since the first is impossible without the second: in a single polymer, they therefore merge to give the single major glass transition, and it may be impossible to separate the two mechanisms.

The general argument as given above for two mechanisms can of course, be applied without any change of principle to polymers in which there are more possible mechanisms. Some of these may merge, while others remain separate.

CHAPTER 6

Hard Amorphous Polymers

> Little can be said with certainty about the structure of glassy plastics. Current (1972) opinion still ranges from a belief that the glass is completely without organisation to a belief that the glass has a subcrystalline order.
>
> R. P. Kambour and R. E. Robertson in *Polymer Science* (ed. by A. D. Jenkins)

6.1 THE STRUCTURE OF GLASSY AMORPHOUS POLYMERS

6.1.1 Order and disorder

The characteristics at low temperature of a hypothetical isolated polymer molecule, which has not crystallised, were discussed in section 2. At each bond, the geometry will have dropped into a conformation of minimum energy, with thermal vibrations allowing only minute fluctuations (tending to zero at absolute zero temperature) from this geometry. But since there are typically three minimum energy positions for bond rotation (see Figs. 1.18, 1.19 and 1.21) the chain has an irregular shape, resulting from a random selection along the chain of these three geometrical possibilities.

One view of the structure of a solid amorphous polymer might be that it is the result of the condensation of such irregular glassy molecules. But stiff rods of this sort could only pack together with extreme inefficiency, comparable to a stack of brushwood, to give a very low density solid. This is clearly not in the least like the high density structure formed by cooling a rubber through the glass transition temperature as described in the last Chapter. The length of polymer molecules thus leads to another difference from small molecules, where there is little structural difference between solids formed by condensation from the vapour or cooling of the liquid.

The simplest view of glassy amorphous polymers is that they are a 'frozen' form of rubber. The molecular mobility, due to jumps over energy barriers, is blocked – and we might regard the structure of the glass as being identical with an instantaneous picture of the structure of the rubber. This is not an easy answer because it leaves open the question of the extent to which there is structure in the rubbery melt, due to some necessary correlation between the conformation of neighbouring chains.

There are further difficulties. The instantaneous stopping of the rubbery structure is not a valid representation, because this would imply that the system had been frozen with the bond geometries at energy levels over the whole range, as indicated schematically in Fig. 6.1. In reality each bond would have to drop back to a minimum level (though not exactly at the minimum in an isolated chain, because of the interference of neighbouring segments). There will be some compacting of the material, although the evidence suggests that this occurs progressively, and not sharply at a temperature with a discontinuity in volume, which would imply a first order transition.

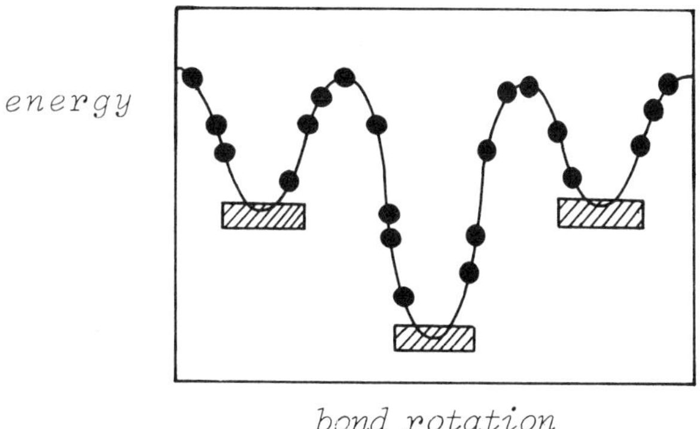

Fig. 6.1 – Schematic indication of instantaneous locations of bonds along a rubbery chain at high temperature (•) and in a glassy chain at low temperature (▬), in relation to energy level of position.

The next complication is that the 'frozen' structure is inherently metastable. There are many other more-or-less equivalent random arrangements, and some of these are likely to be of lower energy. A succession of small localised movements may therefore lead to a progressive change of structure, as it compacts further from the first 'frozen' form to lower energy forms. These movements may be due to several causes. Firstly, there may be very infrequent jumps over energy barriers due to the residual thermal vibrations. Even though, in macroscopic terms and on an ordinary time-scale, the material appears to be below the glass transition temperature, it may still for long times or in localised regions be within the transition region. Secondly, there are the molecular motions associated with the minor transitions at lower temperatures. These will allow some shaking down of the structure into preferred forms of packing. These two effects are possible causes of the **ageing** of glassy polymers: changes in properties over a period of time, as a consequence of structural changes, occurring more rapidly at higher temperature, when it may be termed annealing. Thirdly, molecular movement induced by yielding under stress may cause neighbouring segments to drop into more favourable conformations. Thus, on both a molecular and a macroscopic level, there may be a difference in structure between equally oriented samples depending on whether they are formed by cooling or stretching.

Perhaps the best crude analogy to the situation is a bag of stones. If this is roughly filled a certain volume will be occupied. But when the bag is shaken, or deliberately compacted, the packing can be improved, though remaining irregular, and the volume reduced.

There is another complication associated with the long-chain nature of polymer molecules. It is very likely that a preferred low-energy state will be one in which there is a relatively close packing in small localised regions, separated by zones where the packing is less effective. This would give a nodular structure, for which there is some experimental support. One speculative view of the arrangement is shown in Fig. 6.2.

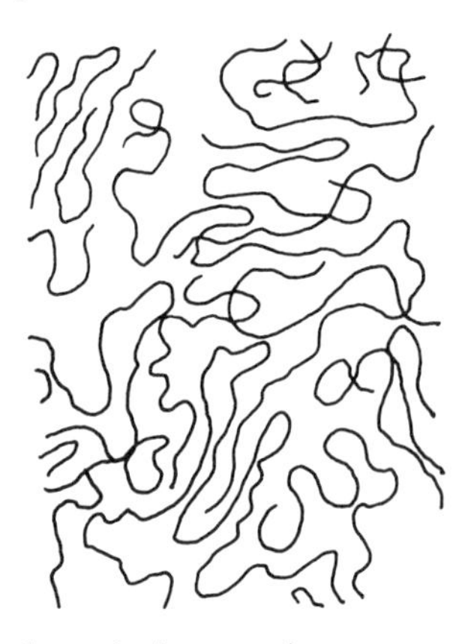

Fig. 6.2 – Model of a polymer in the amorphous state, as suggested by Yeh, with varying density of packing.

As a first approximation, the concept of a glassy polymer as a random assembly of irregular rigid chains is a sensible model; and this will be the view adopted in the rest of this chapter, since although there are theoretical ideas and experimental results which suggest other structures, neither are so far sufficiently conclusive or developed to be of clear value in explaining properties.

6.1.2 Orientation and molecular extent

A glassy polymer can certainly possess structure in terms of chain orientation. If an ideal rubber under zero stress, which is then isotropic by definition, is cooled below the transition temperature, it will give an isotropic glass. But, if the rubber is held under stress and then cooled, the orientation of the chain segments will be present in the glass, and will remain when the stress is released. The glass will be anisotropic, and the distribution of chain segments may correspond to an affine deformation of several hundred per cent from the uniform state.

The drawing of a solid polymer beyond the yield point will also lead to the development of orientation, which will remain set in the glassy plastic.

Rubber elasticity is concerned with changes in the orientation of individual segments between effective linkages, without any other change in the pattern of the path followed by a chain molecule. This is illustrated in the change from Fig. 6.3(a) to 6.3(b). However if the material is subject to appreciable viscous flow, the molecule as a whole may be pulled out into a more extended form as shown in Fig. 6.3(c). In an extreme case, there may be a relaxation over short lengths, which removes any preferential orientation of individual segments without allowing the general extended pattern to change, as indicated in Fig. 6.3(d).

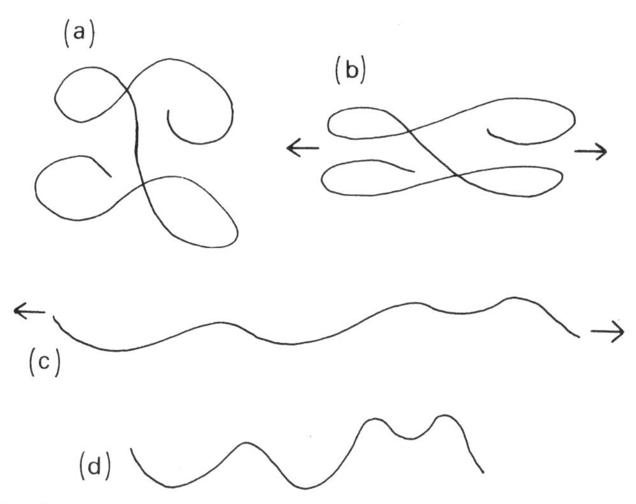

Fig. 6.3 – Schematic representation of a rubbery chain. (a) Equilibrium conformation under zero stress. (b) With preferential orientation of segments, but no change in basic pattern. (c) With preferential orientation of segments and increase in molecular extent due to the long-range path. (d) With changed pattern, but no preferred orientation.

We can thus see that the end-to-end lengths of the chains, or the molecular extent, may have different values, depending on the history of the material. This is likely to cause appreciable differences in some mechanical properties (see section 2.5.2).

In rubbers, the end-to-end length and orientation of segments are correlated as different manifestations of the same effect, since the segments are free to take up the equilibrium distribution of orientations subject to the constrained positions of the ends. In the glassy state, this is not so, and the orientation distribution of segments and the molecular extent are separate and different frozen-in features – or, rather, the two extremes of a spectrum of distributions related to the length considered.

6.2 THE MECHANICAL PROPERTIES OF GLASSY POLYMERS

6.2.1 Elastic behaviour of glassy thermoplastics

In the glassy state, amorphous polymers deform elastically under small stresses by the displacement of bond lengths and angles (including rotation) from their equilibrium position of minimum energy, and the associated separation of chains. These are the same mechanisms as operate in the deformation of a crystalline lattice. Consequently, we can expect the modulus of glassy polymers to be related to the modulus of the corresponding perfect crystals, although the situation is much more complicated because of the disorder in packing together of the chains. In a crystal, it is possible to define a unit cell, and use its deformation as representative of the whole structure. This is not possible in the amorphous polymer where the local environment varies from place to place.

The formal position is that, in the absence of stress, the atoms within the structure will take up the positions and orientations giving the minimum total internal energy U (assuming that entropy changes can be ignored because the number of conformations does not alter appreciably in deformation). These positions will depart slightly from the individual minimum values due to the need to accommodate the whole structure. The details of the state of the system could in principle (but not in practice!) be found by solving the enormous set of minimum energy equations containing the whole set of co-ordinates x_i, y_i, z_i, of each unit in the chain:

$$\frac{\partial U}{\partial x_1} = 0\,; \quad \frac{\partial U}{\partial x_2} = 0 \ \ldots \ \frac{\partial U}{\partial x_i} = 0 \ \ldots \ \frac{\partial U}{\partial y_i} = 0 \ \ldots \ \frac{\partial U}{\partial z_i} = 0 \ \ldots \tag{6.1}$$

If the specimen is deformed, the material will take up a new conformation of minimum energy U_{m} (subject to the external restraints, which hold some of the set x_i, y_i, z_i, at particular values). By definition this minimum energy is greater than the minimum energy achieved without restraints. The force F_j needed to maintain an external dimension l_j is then given by:

$$F_j = \partial U_{\text{m}} / \partial l_j \tag{6.2}$$

No progress has been made in any detailed solutions along these lines: the problem is an extremely complicated one.

As an approximate treatment, the simplest argument might be to say that the detailed changes are the same as those in a crystalline lattice of the same polymer, so that the modulus E_{a} of an amorphous polymer would be given as:

$$E_{\text{a}} \sim \bar{E}_{\text{c}} \tag{6.3}$$

where $\bar{E}_{\text{c}}$ is the crystalline modulus averaged over all directions.

However values of the modulus of the polyethylene crystal have been estimated to be of the order 10^{12} dyn/cm^2, whereas the modulus of a typical amorphous polymer, such as polystyrene, is not much more than 10^{10} dyn/cm^2.

This indicates that, in the more open and irregular structure of a glassy polymer, deformation may be one or two orders of magnitude easier.

The crude argument given in section 2 suggested that the modulus of a chain of N links might be about E_c/N^2. For a solid amorphous polymer, we might put:

$$E_a = p\bar{E}_c/n^2 \tag{6.4}$$

where p is a geometrical parameter related to the distribution of stress and strain in the structure, and n is the number of links which can deform as an independent unit.

An illustration of the structure with n of the right order of magnitude (about 5) is given in Fig. 6.4. It looks plausible.

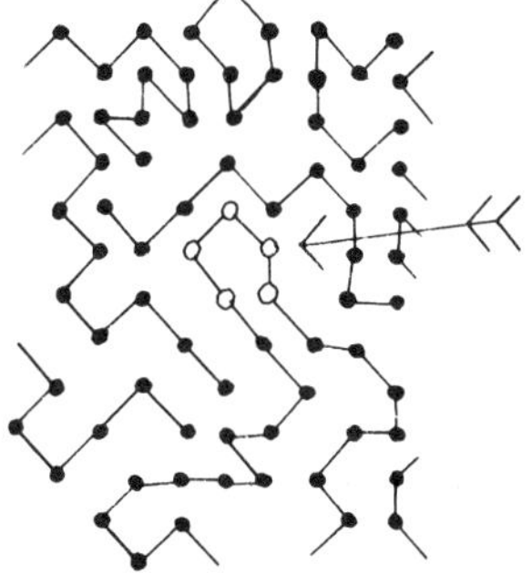

Fig. 6.4 – Schematic representation of an amorphous structure, illustrating a typical group of 5 chain units (○) able to deform independently.

An alternative viewpoint, though not necessarily a fundamental physical difference, would place more emphasis on p than n, and suggest that the modulus is lower because most of the deformation occurs through increased chain separation rather than chain extension. In other words $\bar{E}_c$ is an average which should be highly biassed towards the lower values of crystal modulus.

If the glassy polymer is oriented, then it will show an anisotropy in modulus values due partly to the fact that E_c is dependent on direction, intra-chain or inter-chain deformation being easier is some directions than others, and partly due to the fact that n may vary due to different effects in different directions.

On the simple theory outlined above, the modulus of a glassy polymer should be independent of temperature. In fact the modulus of common glassy polymers usually decreases by a factor of $\frac{1}{2}$ to $\frac{2}{3}$ between absolute zero and room temperature, which is still at least 70°C below the glass transition temperature and so well clear of the onset of the change to the rubbery state. This drop in modulus can be attributed to the minor transitions occurring in side-groups and elsewhere at low temperatures.

At low strains, up to 1 or 2% extension, and at temperatures appreciably below the glass transition temperature, glassy plastics are almost perfectly elastic. There may, however, be some slight viscoelasticity, manifested by creep and in other ways, due to the effects of minor transitions, as mentioned in the last paragraph, and to the effects of ageing, which will be biassed under stress, as mentioned in section 6.1.1. Close to the yield stress, the viscoelastic effects will become more pronounced as the effects of the transition to the soft rubbery state come into play.

6.2.2 Behaviour at high stresses

If the stress applied to a glassy plastic is increased above the level needed to deform the material by 1 or 2%, there are two alternative possibilities. The material will either fracture by crack propagation or yield and suffer irreversible deformation. At low temperatures or for rapid loading, there is more likely to be brittle fracture, whereas at higher temperatures or lower speeds the material will yield, as illustrated in Fig. 6.5. Yield in compression is illustrated in Fig. 6.6.

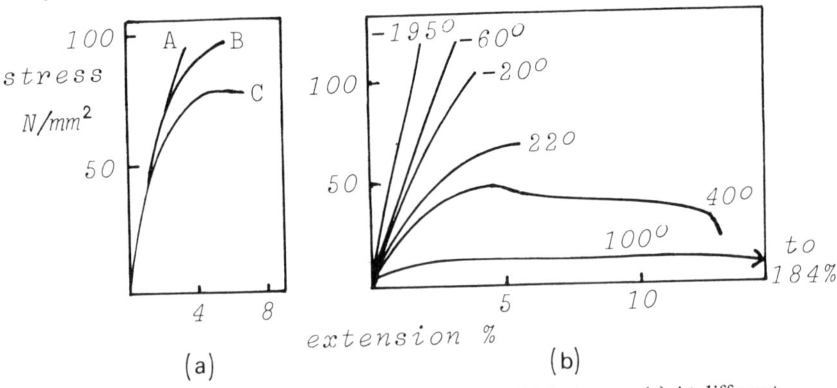

Fig. 6.5 – Behaviour of polymethyl methacrylate at high stresses. (a) At different strain rates at room temperature: (A) 128% per minute; (B) 32% per minute; (C) 2% per minute. After Knowles and Dietz (1955). (b) At different temperatures in °C. After Rabinowitz and Beardmore (1972).

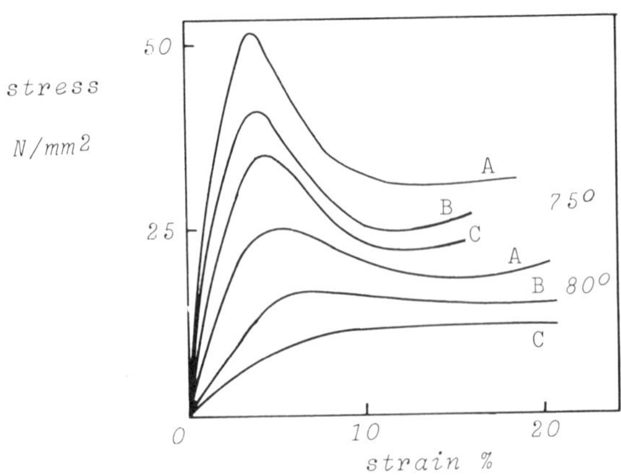

Fig. 6.6 – Polystyrene in plane strain compression at 75°C and 80°C. Strain rates are (A) 1.5×10^{-3}, (B) 3.6×10^{-4}, (C) 1.9×10^{-4} per sec. After Bowden and Raha (1970).

It is tempting to equate the brittle-ductile transition with the glass transition but this is not the case. Well below the glass transition temperature, glassy polymers may be either brittle or ductile. The real criterion is the competition between the fracture stress and the yield stress as indicated schematically in Fig. 6.7. There may also be additional competition among different modes of fracture and different modes of yield. This can lead to fracture by crack propagation after extensive yield.

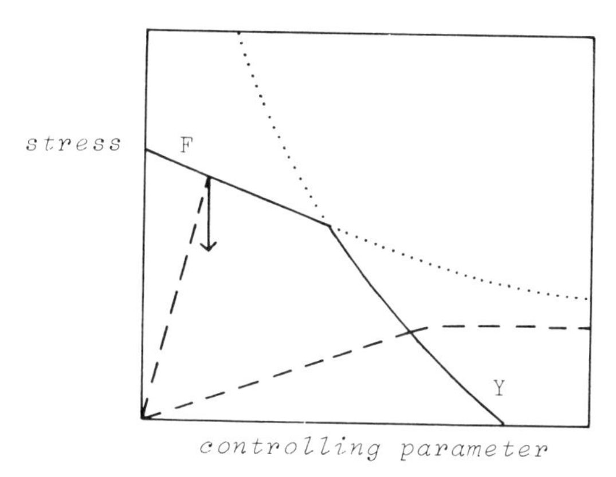

Fig. 6.7 – Schematic view of the brittle-ductile transition: (F) fracture limit and (Y) yield varying with a parameter such as temperature or time. The dashed lines show the shape of the stress-strain curves, though not on the same strain scale.

The following sections will show the problems of predicting either fracture or yield. In general, it can be said that yield is more likely to occur:

(a) at higher temperatures
(b) at slower rates of loading
(c) when there is a large shear component in the applied stress
(d) in compression rather than in tension
(e) in a pre-oriented, anisotropic glassy polymer
(f) in the absence of critical defects
(g) when there is a substantial low-temperature loss peak.

6.2.3 Fracture

In principle, the occurrence of fracture, whether brittle fracture at low strain or ductile fracture at high strain, is governed by the simple relations between energy change U_e and crack size c:

$$\frac{\partial U_e}{\partial c} < 0 \quad \text{for crack growth} \tag{6.5}$$

$$\frac{\partial^2 U_e}{\partial c^2} < 0 \quad \text{for unstable catastropic crack growth} \tag{6.6}$$

For brittle fracture, this is dealt with by the well-developed theories of linear elastic fracture mechanics, with a failure stress dependent on the size of flaws on the surface or inside the material. Strength is thus very much a result of the previous treatment of the specimen.

For the ductile fracture, the situation is not yet well understood.

6.2.4 Yield

In a simple way, as discussed in Appendix A4.4, it can be said that yield will occur when the deformation reaches the level of the inflection in the energy-deformation curve. The problem is to relate the macroscopic stress to the fine structural modes of deformation.

Beyond the yield point there are a number of possibilities, indicated in Fig. 6.8. The simplest would be continuing yield under constant stress, which, apart from the short curved region, would be an ideal elastic-plastic response.

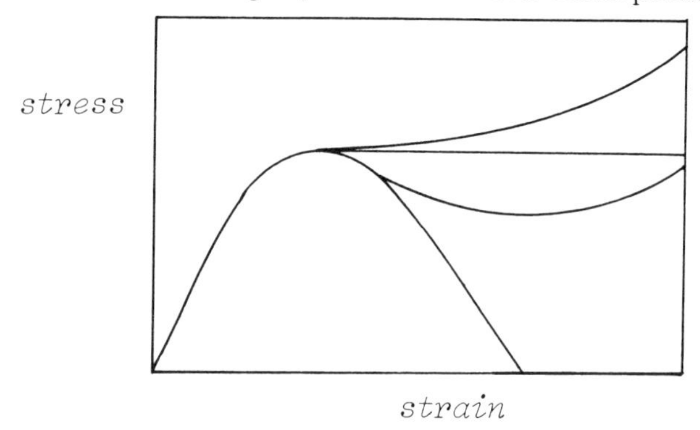

Fig. 6.8 – Schematic representation of different forms of yield behaviour.

The most common response is a reduction in load followed by an increase, but either the decrease or the increase may be absent.

A falling load implies instability and will lead to the formation of a localised neck, since the part which first starts to yield will continue to extend more easily. If the load decrease continues to zero, the neck will thin down until it separates, and this will be a mode of failure.

However, in the commoner case the yield force will increase again, so that further localised yielding will cease and the neck will grow. This is the common form of cold-drawing at a neck. When the whole specimen has yielded in this way (or from the start of yielding when there is no decrease), the material will continue to draw uniformly under increasing load.

There is also an inherent time and temperature dependence in the situation, since below the stress level of the inflection in the energy curve the small remaining energy barrier may be surmounted in time by a thermal fluctuation. Thus the yield stress will decrease with increasing temperature and decreasing rate of extension, as illustrated for example in Fig. 5.3, where the same subject was discussed as an example of the promotion of the glass to rubber transition by stress. Another set of data is shown in Fig. 6.9, which also illustrates the influence of the nature of the applied stress.

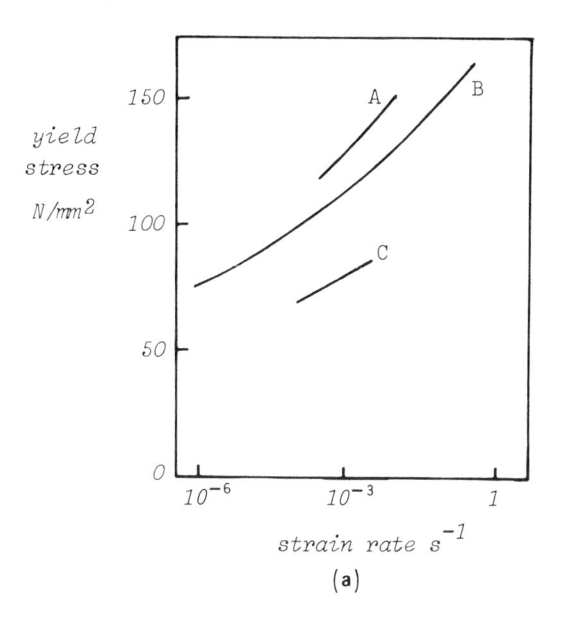

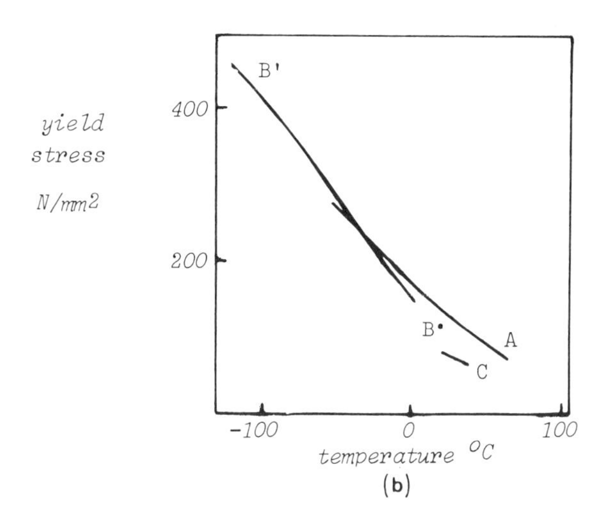

Fig. 6.9 – Yield stress of polymethyl methacrylate: (A) plane strain compression; (B, B′) uniaxial compression; (C) uniaxial extension. After Bowden and Raha (1970) Duckett, Rabinowitz and Ward (1970), and Crowet and Homes (1964). (a) As a function of strain rate at 22°C. (b) As a function of temperature at a strain rate of 10^{-3} per second.

Although the general arguments of this section are widely applicable there are many factors and possibilities involved, so that any detailed examination of the subject requires attention to the particular polymers concerned, to their particular state, to the particular environmental conditions of loading.

Apart from the inherent change in material response with rate of loading, there is an indirect consequence since at slow speeds deformation will be isothermal, with thermal equilibrium established with the surroundings, but at high speeds the deformation will be adiabatic, because there will be no time for heat loss. This will intensify the reduction in yield force by the shift down to the higher temperature curve.

The influence of orientation on yield is shown in Fig. 6.10 which shows the extension behaviour of amorphous polycarbonate, parallel and perpendicular to the direction of orientation. The behaviour in the perpendicular direction is almost the same as that of unoriented material, but in the orientation direction there is a higher modulus, higher yield stress and greater stiffening after yield.

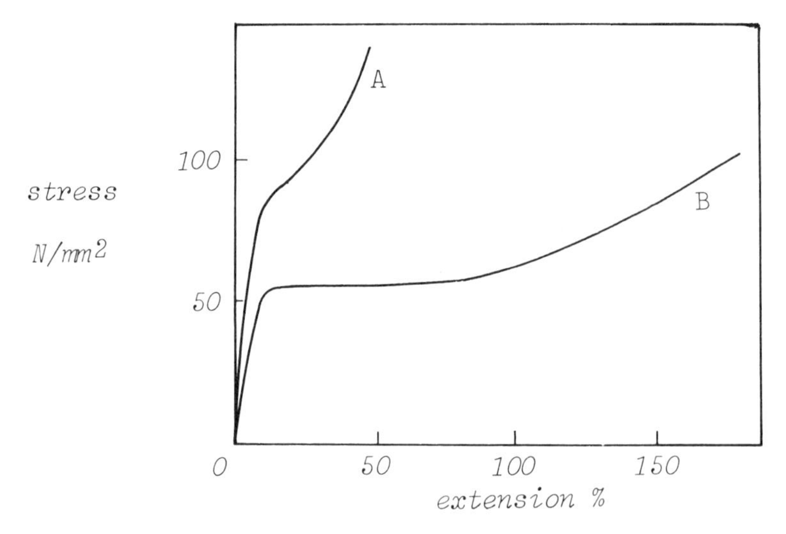

Fig. 6.10 – Stress-strain curve in extension of uniaxially oriented amorphous polycarbonate: (A) in orientation direction; (B) perpendicular. After Schnell (1964).

6.2.5 Mechanisms of yield and fracture: kink-bands and crazes

Yield could be due to the sum of a set of localised changes in chain conformation distributed fairly uniformly through the material. Fracture could be due to a simple cleavage of the material through a combination of chain breakage and chain separation. In practice other effects also occur.

Yielding frequently occurs through the development of large kink-bands in the material, as illustrated schematically in Fig. 6.11. The existence of kink-bands

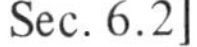

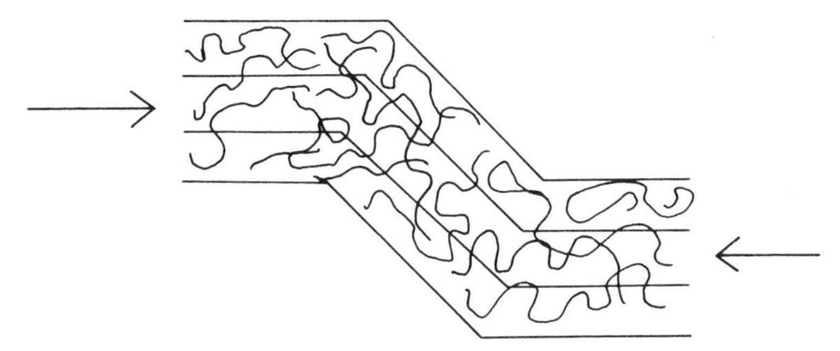

Fig. 6.11 – Schematic illustration of a kink-bend.

can be shown either by their gross appearance on the surface, or by their appearance in section, or by differential contrast in polarised light. An example is shown in Fig. 6.12. Kink-bands are a form of major local plastic deformation under shear, and may also be called shear bands. They occur most easily when a material is subject solely to shear, but will also occur along the lines of a shear component of a tensile stress. Although similar effects can be observed in other materials, the long chain continuity in polymers tends to encourage their formation.

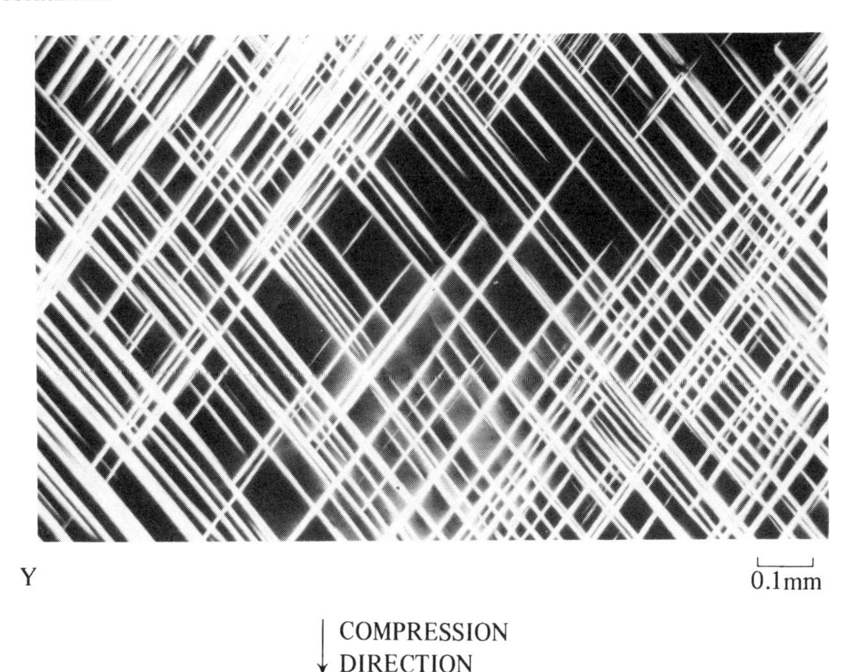

Fig. 6.12 – Kink-bands in polystyrene in plane strain compression test. (Oxborough, Ph.D. thesis, Cambridge University).

Fracture in glassy plastics is frequently preceded by crazing. Instead of a crack with a complete separation of the surfaces, as in Fig. 6.13(a) the continuity of long chain molecules can lead to fine fibrillar bridges across the crack: this is known as a craze, and is shown Fig. 6.13(b). The craze is not a complete failure of the material; but further loading may cause the bridges to break, so that a crack forms. The typical behaviour in fracture is thus a crack leading into a craze as shown in Fig. 6.13(c).

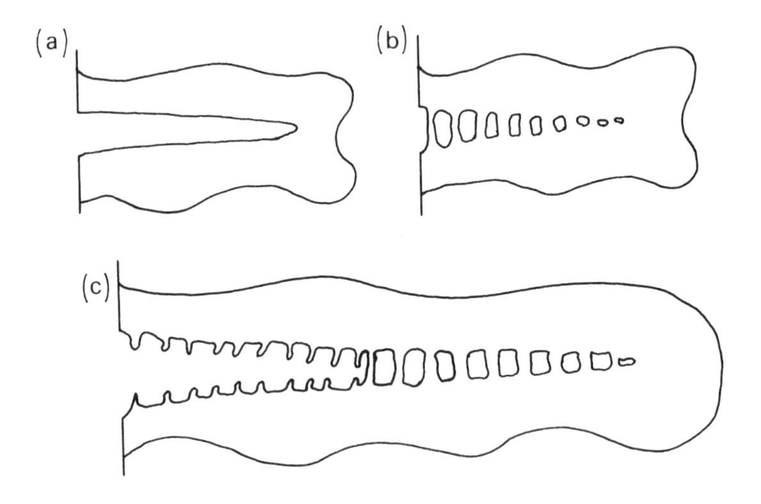

Fig. 6.13 – (a) Crack. (b) Craze. (c) Craze leading to crack.

Crazes may occur either on the surface or internally. If they are widespread, they cause a general whitening of a transparent plastic. Individually, they show up under illumination and may be examined in detail at high magnification. Some examples are shown in Fig. 6.14. The growth of crazes is considerably influenced by the chemical environment.

The failure of plastics by general yield, localised yeilding, crazing and cracking in various directions is a complicated subject with a diversity of experimental data, a number of theoretical treatments, and a lack of full understanding. It will be taken up in more detail in Volume 2.

6.3 THERMOSETTING POLYMERS

Although there is a great deal of practical technological experience, there has been little publication of detailed experimental or theoretical work on highly cross-linked thermosettting polymers.

Struçturally these will be dense amorphous networks. The pattern of order or disorder will depend on the prior state of the precursor and the circumstances in which the cross-linking reaction takes place. In the limit, the density of cross-links would be such that there is no freedom of rotation around bonds, so that any residual entropy-dependent rubber elasticity would be completely eliminated. However in typical products, such as epoxy resins, the cross-linking may be less: at room temperature, there is no freedom of rotation in the polymer links and the material is stiff with a modulus greater than 1000 N/mm^2, but when the temperature is raised freedom develops at the glass-rubber transition and the modulus drops to around 10 N/mm^2.

In the rigid state, deformation will depend on elastic deformation of bond angles and bond lengths. Generally, the material would be expected to be isotropic. The same general arguments for the prediction of modulus will apply as were discussed for glassy plastics in section 6.2.1, except that, because the network is three-dimensional there will be no easy mode of deformation by chain separation, and so the modulus will generally be higher.

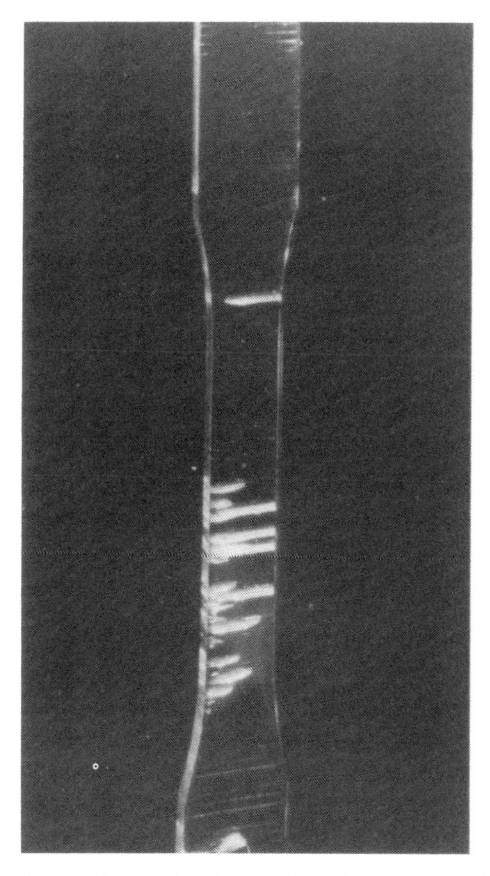

Fig. 6.14 – Crazes. (a) In polycarbonate in tension in alcohol. (Kambour, private communication.)

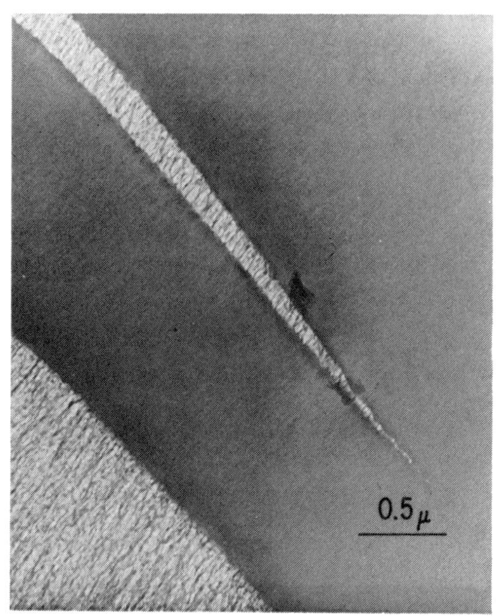

Fig. 6.14 – Crazes. (b), (c) In poly-(2,6-dimethyl-1,4 phenylene oxide). (Kambour, private communication.)

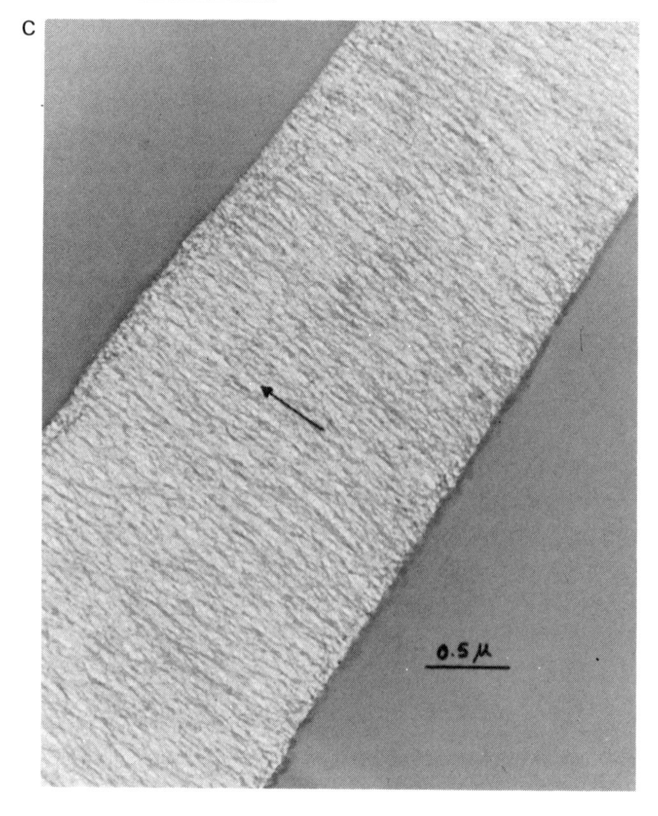

Furthermore, there will be no available mechanism for yielding. The continuity of the molecular network means that physical separation involves chemical degradation. When the stress gets too large, chain segments must break and this will lead to brittle failure.

In the ideal state, thermosetting polymers will thus be stiff, hard, inextensible, brittle materials showing good elasticity with no anisotropy or time dependence. They will also be insoluble and incapable of melting, so that there will only be minor changes in properties with environmental changes unless chemical degradation occurs.

CHAPTER 7

Polymer Crystals

> We see that an external dimension of the crystal is associated with a molecular feature, the fold length; hence the unique situation arises where the measurement of a crystal dimension can give quantitative information on a molecular property.
>
> A. Keller, *Polymer Crystals, Reports on Progress in Physics*, 1968

7.1 THE CRYSTAL LATTICE

7.1.1 The discovery of crystallinity

The subject of polymer crystallisation can be approached, as it was in Chapter 2, through a fundamantal discussion of the way in which polymer molecules may be expected to behave. But this approach owes much to the hindsight of experimental observation, and it is worth looking briefly at the way in which the subject developed historically.

In terms of the classical definition of a crystal, ordinary polymer samples seemed to be the very antithesis. Rubbers, fibres and plastics do not possess the regular faces or the brilliance of crystalline minerals, ice-crystals, or the crystals of inorganic salts. Nor could granular crystals be easily grown and observed in the optical microscope, as they were for metals. The evidence for crystallinity in polymers came only when the structure was probed more finely.

As the atomic theory developed, the old macroscopic criteria of crystallinity were replaced by a recognition that the essential feature of a crystalline material was a packing together of molecules in a regular three-dimensional lattice; and in 1912, von Laue showed that, because of the repetitive character of their lattices, crystals would scatter a beam of X-rays into regular patterns which are analogous to optical diffraction patterns. Fifteen years later, it was shown that natural polymeric fibres gave a pattern of regular reflections when exposed to a beam of X-rays. Subsequent work confirmed that this crystallinity at the level of molecular packing was shown by many polymer materials: cellulose and protein fibres; stretched rubber; polyethylene, polyamides (for example, nylon fibres), isotactic polystyrene and many other synthetic polymers. But besides the X-ray diffraction and other evidence of crystallinity, there was also evidence of appreciable disorder: this was shown by the macroscopic amorphous character, by the low density, by the diffuse scattering of X-rays superimposed on the regular pattern, and by other special evidence. A debate began, and still continues, about the interpretation of this evidence.

The subject of polymer crystallisation was transformed when, in 1957, Fischer in Germany, Keller in England, and Till in America showed that regular diamond-shaped crystals could be grown from dilute solutions of polyethylene.† Before this, there had been work on the kinetics of crystallisation and on the observation of spherulitic patterns in bulk polymers. The new discovery led to a great increase in research on polymer crystallisation.

In this chapter, we shall look at the results of fundamental laboratory experiments on polymer crystallisation and at the further theoretical development of the subject. In the following chapter, we consider the complicated morphologies which can easily develop, and so occur in commercial plastics and real fibres.

However, in studying this whole subject area, it is important to remember, as Keller wrote in 1976, "that the crystallization of polymers has been and still is in a state of flux and that any attempt to achieve a definitive formalization is premature".

Finally, we should note by way of introduction that polymer crystallisation requires a chemical and geometrical regularity of structure, as a prerequisite to the possibility of forming a crystal lattice, and that crystallisation is helped by a chemical constitution which gives a molecular shape suitable for close packing of chains, especially if the molecule is rigidly fixed in a favourable shape, and which gives strong attractive forces between neighbouring chains. If these features are inadequately present, the system will 'freeze' as a glassy polymer, without crystallisation having occurred.

7.1.2 Simple crystal lattices

What forms of regular lattice can polymer chains take up? In answering these questions we can get rid of complications by considering an infinitely large crystal made up of infinitely long molecules. With molecules of uniform finite length, a super-lattice may also be present; with molecules of irregular length, some disorder will be introduced. Folding of long chains would also introduce disorder.

By far the simplest form of lattice would have all the chains lined up parallel to one another. The structure would then be characterised by three features; the conformation of the chain around the linear axis; the pattern of packing of

† There had, in fact, been earlier reports of crystals in gutta-percha by Storcks (1938) and in a rather low molecular weight polyethylene by Jaccodine (1955), but these were not widely noticed or appreciated.

chains in a plane perpendicular to their length; and the displacement, parallel to the axis, of each chain relative to its neighbour. Certain restrictions follow.

For example, if each chain is to have identical neighbours, then only a limited number of ways of packing chains together is possible. More important are the chain conformations which are possible. In a simple lattice, it is necessary that the chain should show continuity from one unit cell to the next: this is a restriction which is not present in the unit cell of ordinary low molecular weight substances.

In the simplest type of unit cell, a whole number p of chain repeat units must make a whole number q of turns in order to match with the next unit cell. The possible chain conformation can therefore be categorised as helices of type p-q†. The simplest possibility is the 1-1 conformation, namely one repeat making one complete turn, which is not really a helix at all; it is the planar zig-zag conformation of polyethylene shown in Fig. 7.1(a) and alternatively described as 2-1. Another 2-1 planar zig-zag with a longer chemical repeat is shown in Fig. 7.1(b). Where there are side groups which interfere or interact this conformation is no longer possible, and a common form is the 3-1 helix shown, for example, by isotactic polypropylene, Fig. 7.1(c). But other forms are also found, and some examples, such as 7-2 and 4-1 are shown in Fig. 7.1. The systematic consideration of possible forms is a useful way of rationalising this aspect of the subject. Furthermore, certain rules concerning the diffraction of regular helices enable the conformation to be found.

The simplest way in which chains will pack side-by-side is in a hexagonal array. If all chains are identically aligned this leads to a unit cell with effectively one chain, as in the left-hand side of Fig. 7.2, but if there are two different alignments a unit cell of rectangular cross-section is present, as shown on the right-hand side of Fig. 7.2: the latter cell contains the equivalent of two chains (one in the centre, four quarters at each corner). With a greater range of alignments, the unit cell will be larger still, and in a more complicated array, which is not hexagonal.

The simplest side-by-side arrangement would have all chains in register. This would then make the whole unit cell a rectangular parallelepiped, and is,

† The full standard notation, applied to chains with a -C-C- backbone, has three figures n*p-q, where n is the number of backbone atoms in the chemical repeat unit, and p is the number of units in the q turns which make up the complete repeat. It follows that $2\pi q/np$ gives the mean angle between the direction of successive C-C bonds, projected on a plane perpendicular to the chain axis. The shortest chemical repeat is usually chosen, but polyethylene, Fig. 7.1(a), can be described *either* as 1*2-1 based on a $-CH_2-$ unit *or* as 2*1-1 based on a $-CH_2-CH_2-$ unit (or even as ∞-∞-0, if the repeat is regarded as a zero turn instead of a full turn through 2π). In defining the repeat unit, stereochemical differences must be taken into account. Thus isotactic polypropylene can be built up from $-CH_2-CH(CH_3)-$ and the conformation is 2*3-1, Fig. 7.1(c); but syndiotactic polypropylene has four backbone atoms in its stereochemical repeat $-CH_2-CH(CH_3)-CH_2-CH(CH_3)-$ and its conformation is 4*2-1, Fig. 7.1(b).

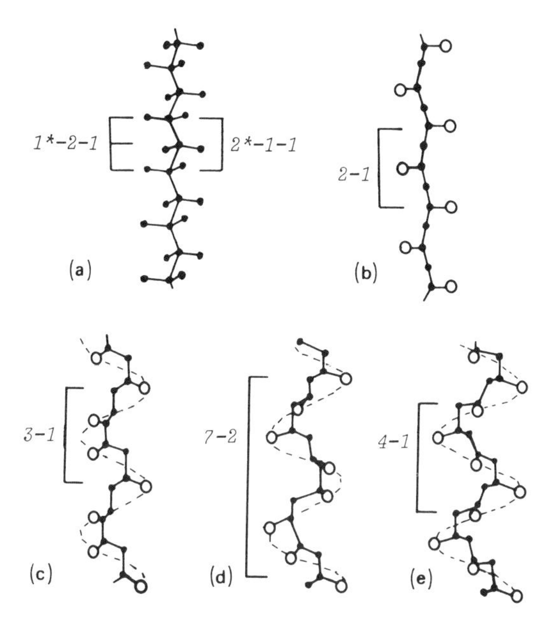

Fig. 7.1 – Various chain conformations. (a) Planar zig-zag of polyethylene 1-1 or 2-1, depending on whether repeat is taken as $(-CH_2-)_2$ or $-CH_2-$. (b) Planar zig-zag of a syndiotactic polymer, 2-1. (c) 3-1 helix of isotactic polymer. (d) 7-2 helix of isotactic polymer. (e) 4-1 helix of isotactic polymer. Except in (a), only the main chain carbon atoms (●) and the large side group (○) are shown.

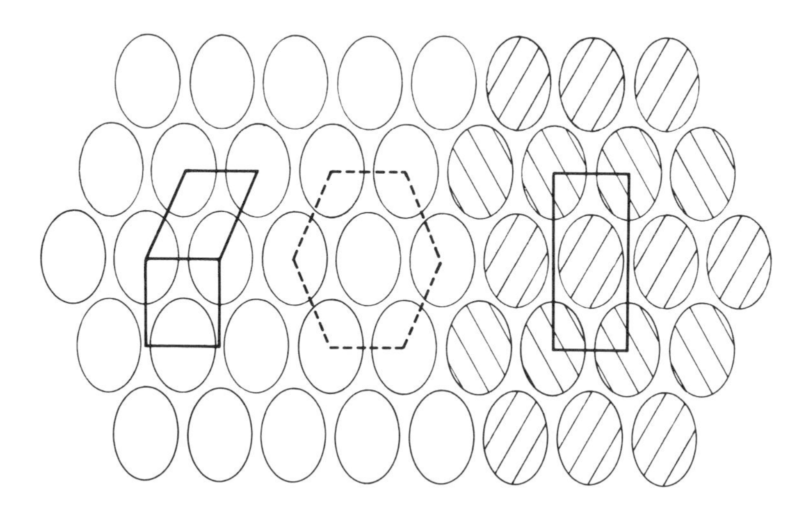

Fig. 7.2 – Hexagonal close-packing of polymer chains, with unit cells when chains are all identically aligned (left-hand side), or alternatively lined up in different directions (right-hand side).

in fact, the form taken up by polyethylene, Fig. 7.3. In other polymers, there may be some staggering of chains in order to provide a better fit or particular

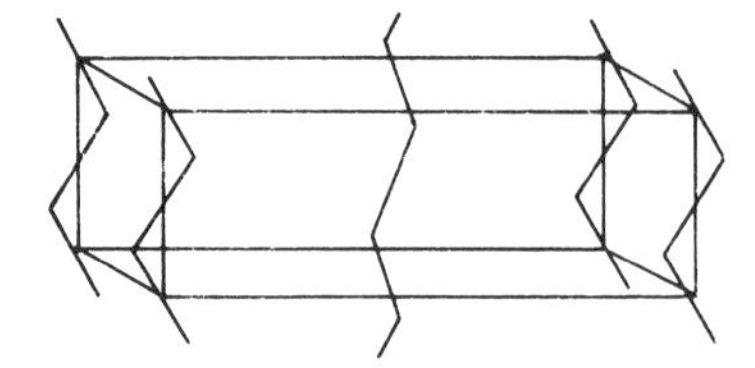

Fig. 7.3 – Regular side-by-side packing in polyethylene, giving a rectangular parallelpiped as unit cell.

inter-chain bonding. For example in nylon 66, Fig. 7.4, the hydrogen bonding leads to a displacement of one chain relative to the next.

The subject of the way in which polymer chains will pack into regular crystal lattices is an interesting study involving interpretation of X-ray diffraction diagrams, model building, and theoretical calculation; but consideration of the

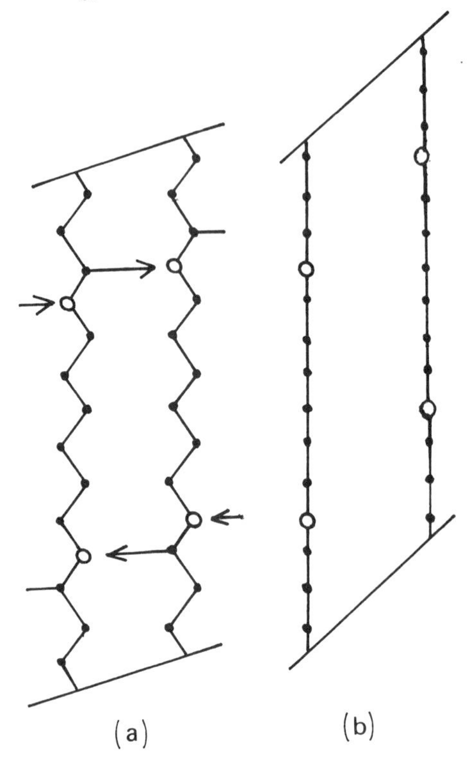

Fig. 7.4 – Staggered side-by-side packing in nylon 66 with hydrogen bonding between >C=O and >NH groups, ⟜→: (a) in hydrogen-bonded sheets; (b) between sheets.

wealth of detailed treatments of particular regular polymer crystal lattices would go beyond the scope of this book.

7.1.3 Other lattices

Although most polymers do fit into the simple lattices described in the last section, this is not universally so. More complicated patterns can exist, with repeats involving more units and more complex forms. The choice of lattice depends on two factors: first, there will be a preference for the lattice with the lowest free energy, which will often be the densest lattice and will usually be a simple repeat pattern; second, even if a complicated lattice does give a lower equilibrium free energy, the probability of the molecules getting into this form may be very small. In other words very complicated forms will be separated from the other forms by free energy barriers which are too high to be passed during crystallisation, unless, as may happen in biological systems, the molecules are presented appropriately as they are synthesised.

When it seems impossible to solve the structure of a crystal lattice the problem should be examined to see if it is being treated in too restrictive a manner. For example, in 1951, Pauling solved a problem which had been worrying X ray crystallographers for 15 years, when he postulated a helical conformation with no simple relation between the geometric repeat and the chemical repeat as the basis of the crystal structure of some polypeptides. This α-helix has 3.6 monomer (amino-acid) residues per turn, and is bonded between turns by hydrogen bonds as indicated in Fig. 7.5. The helices then pack side-by-side as cylinders.

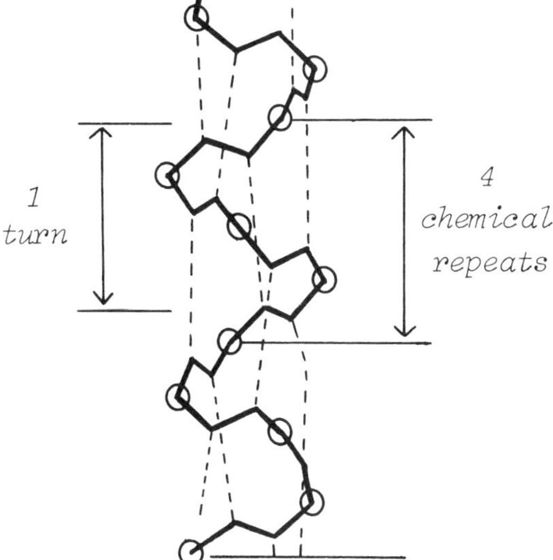

Fig. 7.5 – Simplified views of the α-helix polypeptides, $(\text{-CO-CH(R)-NH-})_n$, showing the helix of the main chain. The circles are the centre of each repeat unit, and the dotted lines show hydrogen bonding between turns.Note the progression of position of units along the helix.

When one considers the matter, there is no inherent reason why chemical and geometric repeats should coincide. All that is necessary is that a given point on a unit in one turn should be lined up with a particular point on the unit in the next turn – there is no reason for the two points to be identical. The difference between a helix where there is a simple number of monomer units per turn and one where there is not corresponds to the difference between an extended chain lattice with units in register and one with units staggered: this is indicated in Fig. 7.6.

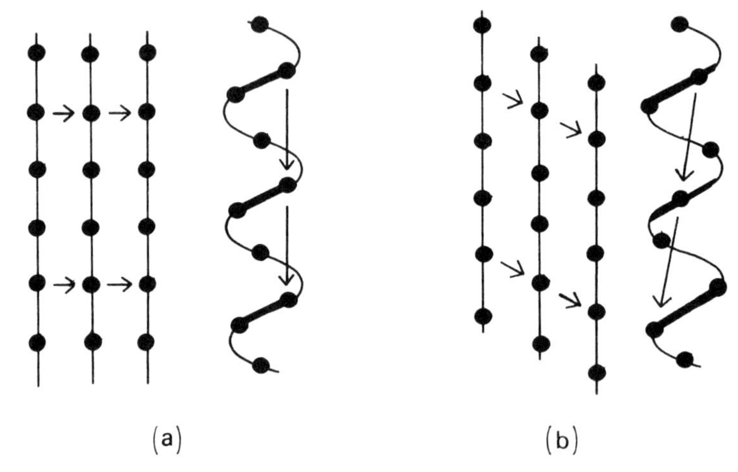

Fig. 7.6 – Comparison between (a) extended chains in register and a simple helix (b) staggered extended chains and helices without a simple relation between chemical and geometrical repeat.

The real importance of the α-helix is that it occurs, in somewhat modified form, in the crystals of many natural proteins. For example, in keratin – the protein in hair – the axial repeat is 5.1Å instead of the 5.5Å of the simple model which fits the synthetic polypeptides. It has been suggested that the helically wound chains are twisted together in pairs or triplets. The well-known double helix of DNA is another example of a complex crystal form. A bigger assembly is the helical array of polymer molecules in the fibres of collagen, which are ordered in a very complicated way. This degree of complexity is only possible with polymer chains with a very specific sequence of groups – the twenty building units of proteins along their length. It would not be possible with small molecules.

Other protein molecules such as haemoglobins and myoglobin, Fig. 1.13, coil up in a regular way to give globular particles which then pack together to make up the crystal.

As an example of a simpler synthetic polymer, which has an unusual form, we can cite polytetrafluorethylene in which the difficulty of packing the fluorine atoms into a planar zig-zag chain like polyethylene is overcome by a slight twisting of the chain which gives 13 -CF_2- units in one turn of a helix of length 33.8 Å.

There are other variants which could occur in polymer structures. For examples some lattices might be based on curved sheets rather than planar sheets it was not until this was realised that the crystal structure of asbestor was solved. In general, we should not allow the fact that so many polymers pack into very simples lattices to blind us to the possibility that much more complicated but still repetitive, crystal structures are possible.

7.1.4 Pseudo-lattices

In the proceding discussion, we have tacitly assumed that polymer crystal lattices are absolutely regular and repetitive. This is not necessarily so. In the proteins for example, there is a varied sequence of chemical side-groups along the chains, which can nevertheless pack in a basically regular fashion. In other polymer crystals, some atoms may have a moderate degree of freedom of movement although the backbone of the chain is fixed within the normal limits of crystal lattice vibrations. In some copolymers, it is possible to substitute units without so distorting the lattice that crystallisation is not possible, though in general irregular copolymers are amorphous systems.

In some other polymers, there may be even less regularity in the lattice. For example, it is generally thought that in polyacrylontrile, which shows a very poor, but still pseudo-crystalline, X-ray diffraction pattern, the helically would chains act as cylinders which pack into a regular hexagonal array, but do not show regularity in their packing in the axial direction. This is illustrated in Fig. 7.7. The chains themselves may, in fact, be atactic and irregular, so that irregularity in packing is inevitable.

Other forms of pseudo-lattice may also exist, as a consequence of the polymeric form which allows for partial regularity of packing as a possible form.

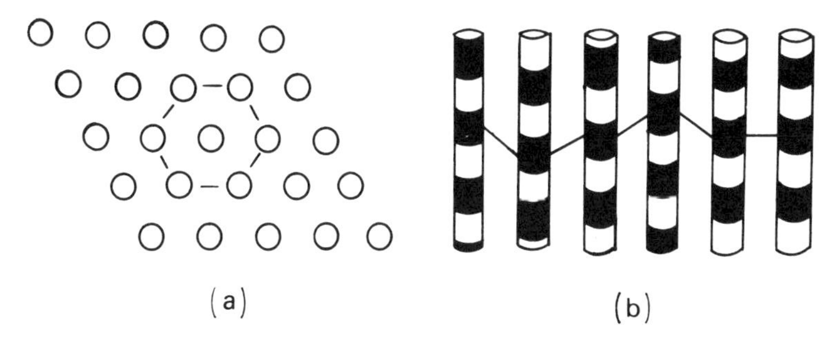

Fig. 7.7 – A regular lattice (a) with an irregular axial packing (b).

7.1.5 Crystal lattice transitions

Like other substances, polymers may exist in more than one crystal form. Thus polytetraflurorethylene exists in the form already mentioned below 19°C; between 19°C and 30°C, there is another helix with 15 -CF_2- groups per turn; and above 30°C, the vibrations increase so much that no specific helix can be defined. These changes are all a consequence of differences in free energy at different temperatures.

More interesting is the fact that in some transitions there is a very marked change in dimensions. Crystals with helical chains can open out to crystals with extended chains. A particular example is keratin, the protein in wool, which normally has a crystal structure with a chain conformation approximating to the α-helix of Fig. 7.5, but which can also form extended chain β-crystals. The transition is illustrated in Fig. 7.8. In the α-form the hydrogen bonds are between turns within a molecule; in the β-form, they are between neighbouring molecules. The change in length accompanying the transition is about 100%. Some synthetic polymers show a similar effect.

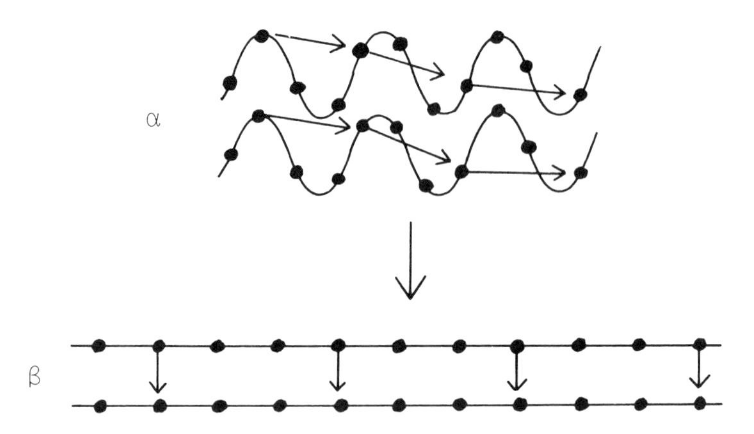

Fig. 7.8 – Schematic representation of the $\alpha \rightarrow \beta$ crystal lattice transition found in keratin and other polypeptides. The arrows ↦ represent hydrogen bonding.

7.2 DEFORMATION OF POLYMER CRYSTAL LATTICES

7.2.1 Elastic deformation

The equilibrium form of a crystal lattice (ignoring thermal fluctuations) occurs when the atoms, under the influence of various interatomic forces, have settled down to the positions which give the minimum internal energy of the lattice as a whole. If the unit cell contains m atoms, then the number of co-ordinates needed to characterise the unit cell is $(6m - 6)$ since each atom is characterised by three co-ordinates of position and three of direction, but six of the total can be dropped as being due to translation or rotation of the unit cell as a whole.

Consequently, the packing in the unit cell is given by the solution of the equations:

$$\frac{\partial U}{\partial x_1} = ; \frac{\partial U}{\partial x_2} = 0 ; \ldots \frac{\partial U}{\partial x_{6(m-1)}} = 0 \tag{7.1}$$

where U is internal energy, and $x_1, x_2 \ldots x_{6(m-1)}$ are the co-ordinates.

If the unit cell is deformed in some way, the atoms will adapt to new positions of minimum energy, given by a solution of (7.1) reduced in number by the restrictions imposed by the deformation. By definition, the minimum energy in a deformed state must be greater than the minimum energy in the undeformed state. The force F resisting the deformation will be given by:

$$Fdl = \mathrm{d}U \tag{7.2}$$

where $\mathrm{d}U$ is the change in internal energy resulting from the change in deformation $\mathrm{d}l$.

In principle, it is thus easy to see how both the undeformed state and the elastic moduli of a crystal can be theoretically calculated. All that one needs to know is the energy of each atom in the field generated by the other atoms in the neighbourhood. Unfortunately, while this is quite comprehensible as a problem, the exact solution is impossibly difficult. A few detailed treatments have been reported. For example calculations have been made of the form which chain folds take in polyethylene.

In the calculation of elastic moduli of crystals, a number of simplifications can be made. Firstly, the starting-point, the crystal lattice itself, is usually known, at least approximately, from experimental X-ray diffraction studies. Secondly, it is usual to consider only the interactions of near neighbours. Thirdly, some simple form of internal deformation of a unit cell resulting from a given external deformation is usually assumed. The energy changes resulting from the deformation can then be calculated. Fourthly, the strains considered can be made vanishingly small.

7.2.2 Modes of deformation

The simple forms of internal deformation of polymer crystal are as follows:

(1) *Within the polymer chain*

The intramolecular forms of distortion of a polymer chain, such as Fig. 7.9(a), are:

(a) Stretching of covalent bonds, leading to an increase in the spacing between atoms as in Fig. 7.9(b). This is strongly resisted, and can often be assumed to be negligible in extent. It would be the only mode of extension possible in a truly linear chain.

(b) Change of the angle between adjacent covalent bonds, as in Fig. 7.9(c). This is easier, and will be the most important mode of deformation along the length of a planar zig-zag chain.

(c) Rotation of one bond about an adjacent bond, with respect to the direction of the next adjacent bond. This is easier still, and will be an important mode of extension of non-planar helical chains, Fig. 7.9(d).

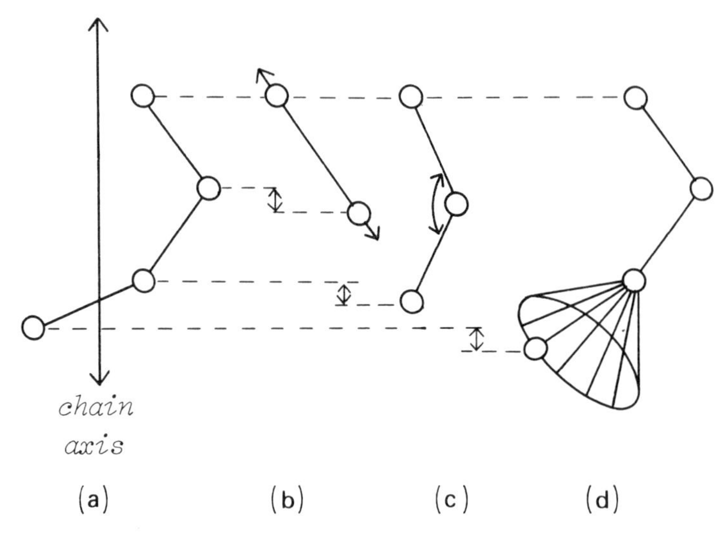

Fig. 7.9 – Forms of deformation within a chain in a polymer crystal: (a) undeformed chain. (b) bond extension; (c) change in bond angle; (d) bond rotation.

(2) *Between chains*

Deformation perpendicular to the chain axis will be dominated by the intermolecular separation, which is usually less strongly resisted then deformation within a molecule. The forces involved can be somewhat arbitrarily classified as:

(a) weak van der Waal's forces, due to the interaction of non-polar groups;
(b) rather stronger forces, occurring when strongly polar groups are present.
(c) special bonds, most notably the hydrogen bond, which resist deformation quite strongly.

Fig. 7.10(a) illustrates deformation by change in the distance between chains. The separation of chains can, of course, occur in a variety of directions, allowing different forms of deformation, as indicated for example in Fig. 7.10(b). The magnitude of resistance to intermolecular deformation will depend on the geometry and scale of the energy contours.

(3) *'Inter-molecular' forces within chains*

Where the chain is in a helical or other form so that some atoms are physically adjacent to atoms which are not chemically near neighbours, it is necessary to consider the resistance to deformation coming from forces of intermolecular character between turns. This is particularly important when there are hydrogen bonds between the coils of a helix, as as in Fig. 7.5.

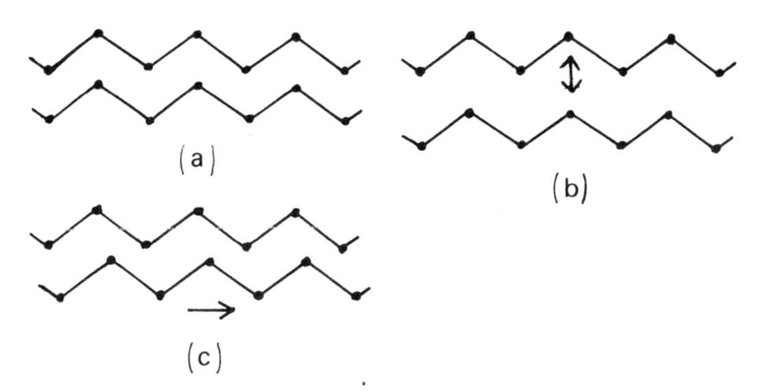

Fig. 7.10 – Deformation due to change in relative position of neighbouring chains: (a) undeformed; (b) change in distance between chains; (c) relative shear of chains.

7.2.3 An illustrative calculation

As an example of how these fundamental ideas can be applied, we take the calculation of the axial elastic modulus of the polyethylene crystal, using a method based on a derivation by Treloar. In this crystal as shown in Fig. 7.3 the chains are planar zig-zags, all parallel to one another. We therefore need consider the extension of only one chain, and this must occur through stretching of C–C bonds and change of bond angle at $\mathrm{C}\overset{\mathrm{C}}{\frown}\mathrm{C}$. The force constants for these modes of deformation are available from the results of measurements of the corresponding vibration frequencies of molecules.

The calculation may be tackled in various ways by the analysis of either force or energy relations in the system. For the force method, we take, as a model of the molecule, a set of rods, which are able to stretch but not bend, linked by joints, which can change angle but not stretch. This is illustrated in Fig. 7.11(a). The consequences of the application of a force F are then worked out, as a problem in statics. This method is perhaps the simplest method of dealing with such a simple linear chain; but it becomes much more difficult when applied to more complicated polymer crystals and more complicated modes of deformation.

It is therefore preferable to use an energy method in a form which is best for more complicated problems. It is also unnecessary to introduce an actual mechanical model: we merely specify the state of the chain in terms of the spacing x between the atoms and the angle 2θ between the bond directions, as shown in Fig. 7.11(b). The length along the direction of the chain axis for one repeat unit is taken to be $2l$. Consequently, we have:

$$l = x \sin\theta \tag{7.3}$$

$$\mathrm{d}l = \sin\theta\,\mathrm{d}x + x\cos\theta\,\mathrm{d}\theta \tag{7.4}$$

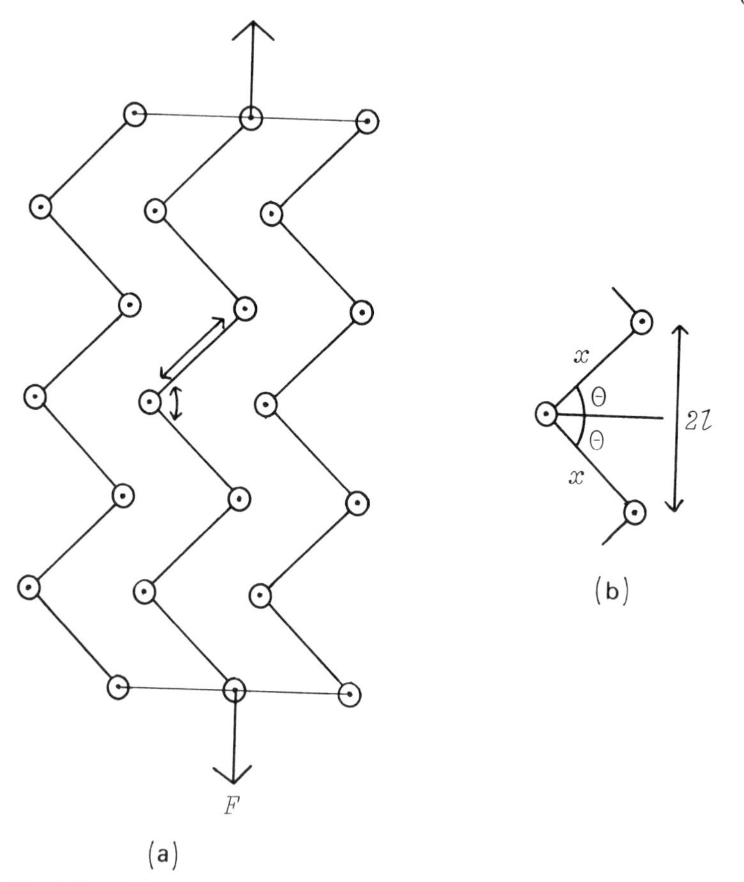

Fig. 7.11 – (a) Model for analysis of axial deformation of polyethylene crystal lattice by force method. (b) Basic diagram for energy method.

If we define strains as $\epsilon = \mathrm{d}l/l$, $\epsilon_x = \mathrm{d}x/x$, $\epsilon_\theta = \mathrm{d}\theta/\theta$, we get:

$$\begin{aligned}\epsilon = \mathrm{d}l/l &= \mathrm{d}x/x + \cot\theta\,\mathrm{d}\theta \\ &= \epsilon_x + \theta\cot\theta\epsilon_\theta\end{aligned} \tag{7.5}$$

We postulate that the strain energy U (internal energy of deformation) can be written as the sum of two terms in the usual form:

$$U = U_x + U_\theta = \tfrac{1}{2}k_x\epsilon_x^2 + \tfrac{1}{2}k_\theta\epsilon_\theta^2 \tag{7.6}$$

where k_x and k_θ are force constants for the deformations.

It is now necessary to determine how the total strain ϵ is divided between ϵ_x and ϵ_θ. Substituting from (7.5) in (7.6) we get:

$$\begin{aligned} U &= \tfrac{1}{2}[k_x(\epsilon - \epsilon_\theta \theta \cot \theta)^2 + k_\theta \epsilon_\theta^2] \\ &= \tfrac{1}{2}[k_x \epsilon^2 - 2k_\theta \epsilon \epsilon_\theta \theta \cot \theta + (k_x \theta^2 \cot^2 \theta + k_\theta)\epsilon_\theta^2] \end{aligned} \tag{7.7}$$

Since the equilibrium condition must be one of minimum energy (subject to the total deformation ϵ), we must have:†

$$\left(\frac{\partial U}{\partial \epsilon_\theta}\right)_\epsilon = \tfrac{1}{2}[-2k_x \epsilon \theta \cot \theta + 2(k_x \theta^2 \cot^2 \theta + k_\theta)\epsilon_\theta] = 0 \tag{7.8}$$

$$\epsilon_\theta = \frac{k_x \theta \cot \theta}{k_x \theta^2 \cot^2 \theta + k_\theta} \epsilon \tag{7.9}$$

Substitution in (7.7) then gives:

$$U = \frac{1}{2} \frac{k_x k_\theta}{k_x \theta^2 \cot^2 \theta + k_\theta} \epsilon^2 \tag{7.10}$$

We can write an alternative expression for U in terms of the crystal modulus E_c:

$$U = \tfrac{1}{2} A l E_c \epsilon^2 \tag{7.11}$$

where A is the effective cross-sectional area occupied by the chain. Hence:

$$E_c = \frac{1}{Al} \frac{k_x k_\theta}{k_x \theta^2 \cot^2 \theta + k_\theta} \tag{7.12}$$

It remains to relate k_x and k_θ to the force constants reported form spectrographic data (vibration frequencies). The force constant k_1 for extensional vibration is usually quoted in terms of actual separation of bonded atoms by a force f_x, as illustrated in Fig. 7.12:

$$f_x = k_1 \delta x \tag{7.13}$$

giving

$$U_x = \tfrac{1}{2} k_1 (\delta x)^2 = \tfrac{1}{2} k_1 x^2 \epsilon_x^2 \tag{7.14}$$

$$k_x = k_1 x^2 \tag{7.15}$$

The angular vibration data is usually reported in terms of the constant k_p which relates the force f_p perpendicular to a bond to the movement of the atom perpendicular to the bond direction, as illustrated in Fig. 7.13. It is assumed that the next bond remains fixed in position, and so we have $\delta p = x\delta(2\theta) = 2x\delta\theta$.

† In this particular instance of a simple linear chain it is also easily possible to deal with the relations between energy and deformation associated with changes in x and θ separately. However the method used above is more easily capable of generalisation to cover the situation when there are several parameters to adjust to reach the minimum energy state.

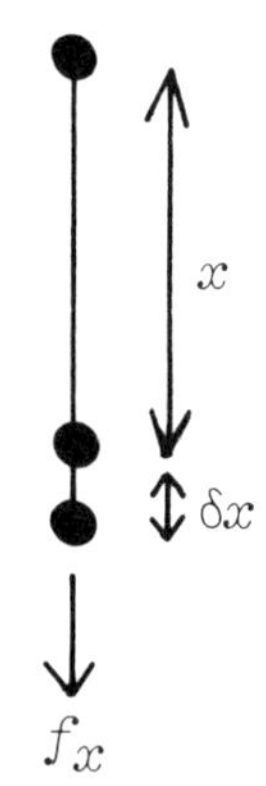

Fig. 7.12 – Deformation due to bond stretching.

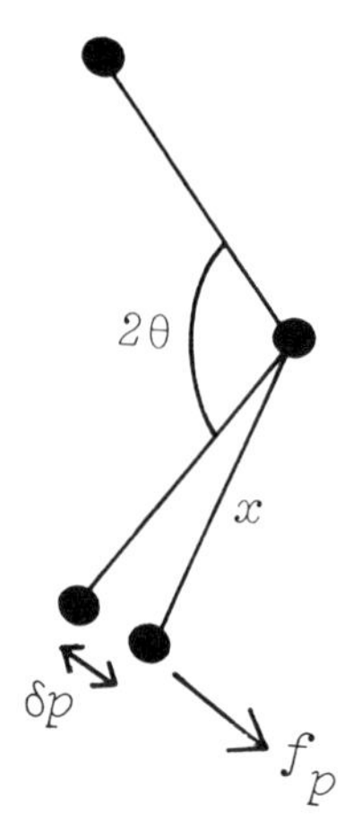

Fig. 7.13 – Deformation due to change in bond angle.

This gives:

$$f_p = k_p \delta p \tag{7.16}$$

$$U_\theta = \tfrac{1}{2}k_p(\delta p)^2 = 2k_p x^2 (\delta\theta)^2$$
$$= 2k_p x^2 \theta^2 \epsilon_\theta^2 \tag{7.17}$$

$$k_\theta = 4k_p x^2 \theta^2 \tag{7.18}$$

Substitution in (7.12) gives, from (7.3), (7.15) and (7.18):

$$E_c = \frac{l}{A} \; \frac{4k_1 k_p}{(k_1 \cos^2\theta + 4k_p \sin^2\theta)} \tag{7.19}$$

Treloar calculated a value of $E_c = 1.82 \times 10^5\,\text{N/mm}^2$ on the basis of the the following values taken from the literature for polyethylene:†

$$l = \tfrac{1}{2}(c\text{-axis of unit cell}) = 1.267\,\text{Å}$$

$$2\theta = 112^\circ\,; \quad \theta = 56^\circ$$

$$A = 18.24 \times 10^{-16}\,\text{cm}^2$$

$$k_1 = 4.36\,\text{N/cm}$$

$$k_p = 0.35\,\text{N/cm}$$

It is also interesting to calculate the relative contributions of bond stretching and change of angle to the deformation. This is given by rearrangement of (7.5) and (7.9):

relative contribution of angular change to total strain

$$= \theta \cot \theta \epsilon_\theta / \epsilon$$

$$= \frac{k_1 \cot^2 \theta}{k_1 \cot^2 \theta + 4k_p} = 0.74 \tag{7.20}$$

Thus three-quarters of the total extension can be attributed to change of bond angle, and the remaining quarter to bond stretching.

7.2.4 Other calculations and experiments

Further development of this subject involves the following features:

(a) the calculation of moduli in different directions
(b) the inclusion, where appropriate, of bond rotation
(c) the inclusion of intermolecular modes of deformation, which will dominate transverse deformation
(d) the adoption of more rigorous methods of analysing the energy changes in deformation, including more remote interactions
(e) the choice of use of *either* theoretical values of inter-atomic potentials *or* experimental data from vibration studies
(f) the study of polymers with more complicated unit cells.

However the principles are essentially as outlined in the previous section.

As one example of the results of a more elaborate calculation, Table 7.1 gives estimates of the full set of elastic constants of the polyethylene crystal: calculation by others show some difference in values, but the general pattern is similar. It can be noted that the resistance to extension in the direction of the chain axis (33) is an order of magnitude greater than any other resistance.

† On an atomic scale, it might be more realistic to give the k_1 and k_p values as 4.36 and $0.35 \times 10^{-8}\,\text{N/Å}$.

Experimental estimates of the crystal modulus can be made by applying a known stress to a semicrystalline fibre or other specimen and then measuring the crystal strain by means of the change in spacing shown in the X-ray diffraction diagram. The result will be correct if crystalline and non-crystalline regions are strictly in series, but will be in error for other morphologies. Table 7.2 gives some values. For polyethylene the values are of the same order as the theoretical estimates in Table 7.1. It is worth noting (a) that one of the transverse moduli of polyvinyl alcohol is higher, due to hydrogen bonding between chains, (b) that the axial modulus of polypropylene is much lower, since the helical chain conformation is easily deformed by bond rotation in a way which is not possible in an extended chain.

Table 7.1 – Theoretical estimates of elastic constants of the polyethylene crystal, in kN/mm^2 (after Wobster and Blasenbrey (1970)).

Direction of stress	Direction of strain					
	1	2	3	4	5	6
1	14	7	2.5	0	0	0
2	7	12	4	0	0	0
3	2.5	4	325	0	0	0
4	0	0	0	3	0	0
5	0	0	0	0	2	0
6	0	0	0	0	0	6

The direction of 3 corresponds to the length of the chain (c in Fig. 7.3).

Table 7.2 – Experimental estimates of crystal moduli in kN/mm^2 (after Sakurada, Ito and Nakamae (1966)).

Polymer	Axial modulus (33)	Transverse moduli (11 and 22)	
Polyethylene	240	3	4
Polyvinyl alcohol	255	9	5
Polypropylene	42	3	3

7.2.5 The mechanical promotion of lattice transitions

The crystal lattice transitions described in section 7.1.5 can be caused to occur by changes in other conditions, as well as by changes in temperature. Thus as with many substances, there will be some polymers in which one crystal form is stable at low pressure and another at high pressure due to the additional $(p\delta v)$ contribution to the free energy shifting the balance in favour of a lower volume lattice.

More interesting, and peculiar to particular polymer systems, are the transition from a helical to an extended chain crystal lattice shown in Fig. 7.8. Since this is accompanied by a large axial extension, it will be induced by an axial tension. A possible form of variation of energy† per unit volume U with axial strain ϵ is shown in Fig. 7.14(a). The changes in the energy diagram when stress is applied are shown in Fig. 7.14(b) and (c). An alternative representation is given by the stress-strain diagram, Fig. 7.14(d). At low tension, the α-form is

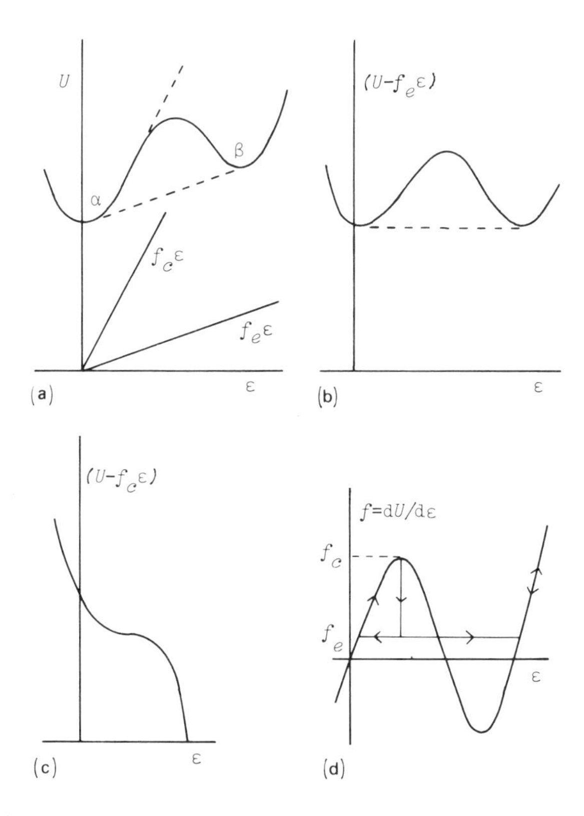

Fig. 7.14 – (a) Free energy in a polymer crystal lattice transition, together with energies associated with applied stresses f_e and f_c. (b) Energy diagram with equilibrium stress f_e applied. (c) Energy diagram with critical stress f_c which just removes the first trough. (d) Alternative representation given by plotting stress-strain relation obtained by differentiation.

† There is a tacit assumption that no entropy changes are associated with the transition. Strictly the free energy, and not the internal energy, should be used.

stable and suffers elastic deformation; but for stresses greater than f_e the β-form will be preferred since it leads to a lower total free energy. At f_e, the two forms will be in equilibrium, and may be present in any proportion. It is apparent from the diagram that the equilibrium stress f_e is the stress given by the slope of the common tangent to the two troughs of the free energy curve: it thus, almost inevitably, corresponds to a very low strain in the helical form. It is not, in fact strictly necessary to have a second minimum in the free energy curve: all that is necessary is a point of inflection so that a common tangent can be drawn.

Although it is the β-form which is stable at any stress greater than f_e, the problem of nucleation remains. If there is some fault in the crystal, or a free end where chains emerge, then the β-form may appear immediately the stress reaches f_e, and increase in proportion as extension increases, at constant stress, as in any other phase change. But, with a large perfect crystal, metastable extension of the α-form will continue up to the critical stress f_c which is at the point of inflection in the free energy curve. The region of decreasing stress, which follows, is unstable and the slightest fluctuation will lead to a portion opening out to the β-form while the rest contracts, and the stress falls to the equilbrium stress f_e. The equilibrium change between the forms will then continue as before, with the length of the β-zone increasing as extension proceeds.

If the crystal is small, or if the transition can start in a small part of it, then the absolute height of the energy barrier is reduced and a thermal vibration may help to carry the system past the critical point. The critical stress will thus be time and temperature dependent, and will also be affected by stress concentration. A stress-induced crystal lattice transition is a special case of the topic treated in A4.4.

The whole process has analogies with crystal fracture. In a simple crystal the free energy and force relations are as shown in Fig. A.3. Here the critical stress is the theoretical fracture stress, since when it is passed infinite extension will occur at one point. The critical stress in the α-β transition is also a fracture stress in the sense that is the point at which the α-crystal lattice ruptures by breaking the hydrogen bonds between the coils of the helix; but the polymer chains are not ruptured and can recrystallise in a new form. Provided the new form has a higher rupture strength than the low tension form, the process will go as described. If its rupture strength is less, then it can only appear when the critical stress is passed in some other way in order to establish equilibrium between the two forms.

There will also be a critical stress f_c less than the equilibrium stress and usually negative, for the reverse transition $\beta \rightarrow \alpha$. Once the α-form reappears, recovery would continue at the equilbrium stress f_e until all the β-form had disappeared.

While the above outline is generally correct, there are other factors to be taken into account in a full treatment. Firstly, there must be continuity of polymer chains between the α- and β-forms, and this can only be achieved by the presence of an intermediate, energetically unfavourable γ-region. The energy contribution of the γ-region must be included: it is analagous to surface free energy, which makes nuclei unstable. In the same way small α-zones will be unstable at stresses somewhat below f_e, and small β-zones at stresses above f_e. Secondly, there will be lateral contraction accompanying the transition, which will therefore be influenced by transverse stress also. Thirdly, the changes from α- to β- unlike many lattice transitions involves comparatively large movements of polymer chains, since the extensions are typically about 100%. There will be viscous resistance to the dragging of chain units over one another, and to the rupture of individual hydrogen bonds, and so the equilibrium stress may be significantly time dependent.

7.3 SINGLE CRYSTALS

7.3.1 Diversity of form

Crystallisation – of any substance – is notable for the diversity of forms which it produces. Ice crystals provide a myriad of patterns, dependent on the local conditions of freezing. This is true, also, of polymers, and many different forms of polymer crystal have been reported in the literature during the last fifteen years. And even this probably minimises the real diversity, because microscopists can only examine a few individual crystals among the clouds that form – and, understandably, there is a strong tendency to select for examination and photography those crystals which show the most clearly defined forms. Unfortunately, too, in the presentation of the subject, there has been a tendency to regard certain simple forms as universal in occurrence, and to relate all polymer crystallisation to these forms. We must remember that complexity is likely to be the rule, and simplicity of form the exception.

Nevertheless, in a single chapter, only the simplest basic forms can be described; but these must be regarded as merely indicating structures which commonly occur, often modified and elaborated, and as bringing out certain basic features just as a child's drawing of a flower gives the basic features but does not do justice to the complexity of a rose.

The simplest mode of crystallisation is slowly from a quiescent dilute solution on to a limited number of nuclei, and, except where otherwise indicated, this is the way in which the crystals described in sections 7.3.2 to 7.3.9 have been obtained. Greater complexity arises with faster crystallisation or a proliferation of stable nuclei due to supercooling, with more concentrated solutions, with solutions in motion, and, especially with crystallisation of bulk material from the melt. Other structures occur in the special circumstances of crystallisation at the time of polymerisation or when liquid crystals form as a precursor.

7.3.2 Simple monolayer crystals grown from solution

On cooling a dilute solution of a polymer, a finely dispersed precipitate of crystals appears and can be collected and examined, roughly in an optical microscope and in more detail in an electron microscope. Sometimes the crystals are simple platelets (lamellae) as shown in Fig. 7.15. In polyethylene, these are diamond-shaped and resemble the crystals of paraffins, which are, of course, analogous in chemical structure.

Fig. 7.15 – Single crystals of polyethylene: mainly monolayer but some spiral overgrowths (Holland and Lindenmeyer, *J. Pol. Sci.*, 57, 589, 1962).

With different polymers, or in different conditions of crystallisation different characteristic shapes may appear: for example, polyoxymethylene shows a hexagonal form. However these differences in shape which reflect the form of the crystal lattice and the preference for growth on certain faces are of no great general significance and will not be discussed further.

The monolayer platelets are always about 100Å in thickness, but their lateral dimensions vary considerably. Growth must therefore occur at the edges. Examination by electron diffraction gives very sharp patterns, which indicates a high degree of perfection in the crystal. Furthermore, the diffraction pattern gives conclusive evidence that the axis of the chain molecules lies perpendicular to the plane of the crystals. In paraffins, with molecules of limited and constant length, such crystals would be easily formed by stacking molecules side-by-side as suggested in Fig. 7.16(a). But, in the polymer, the molecules are very much longer (10,000Å or more), and could only be accommodated in the crystal by some form of chain folding as in Fig. 7.16(b). The simplest model of the mode of formation of a single platelet is shown in 7.16(c), in which the chain molecule is adding on by folding along one edge, giving a step which moves round the crystal as growth proceeds. When the end of one molecule is reached, the step (with attractive forces on two sides) would be a preferred place for another molecule to start crystallising.

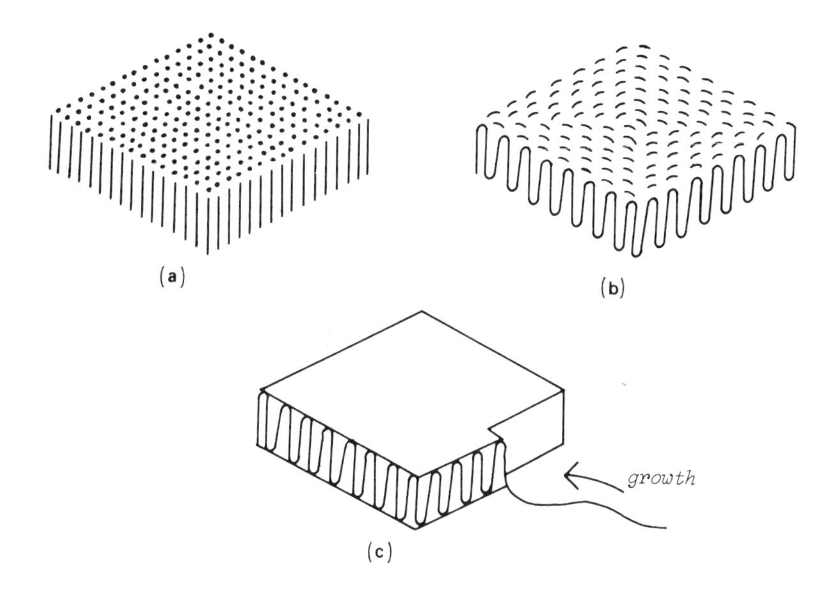

Fig. 7.16 – (a) Short chain paraffin crystal. (b) Folded chain crystal. (c) Growth of crystal.

7.3.3 Sectors in polymer crystals

Regular folding of chains on a growing face has an important consequence; the crystal splits up into sectors. This feature is unique to polymer crystals. In any ordinary simple crystal, for example the paraffin crystal of Fig. 7.16(a), structure is the same everywhere; but in the polymer crystal, the folds are in different directions in different sectors. This is apparent in Fig. 7.16(b) and is shown in more detail in Fig. 7.17 in relation to the unit cell. The folds are along each face and so split the crystal up into four sectors with differently oriented folds.

In the simple diamond-shaped polyethylene crystals, all the fold directions are equivalent, since they are on the diagonals of the rectangular unit cell. However growth can occur on other faces as well. Fig. 7.18 shows a crystal where growth on two of the edges (as well as the diagonals) of the unit cell gives a truncated diamond form. If this happens, the crystal splits up into six

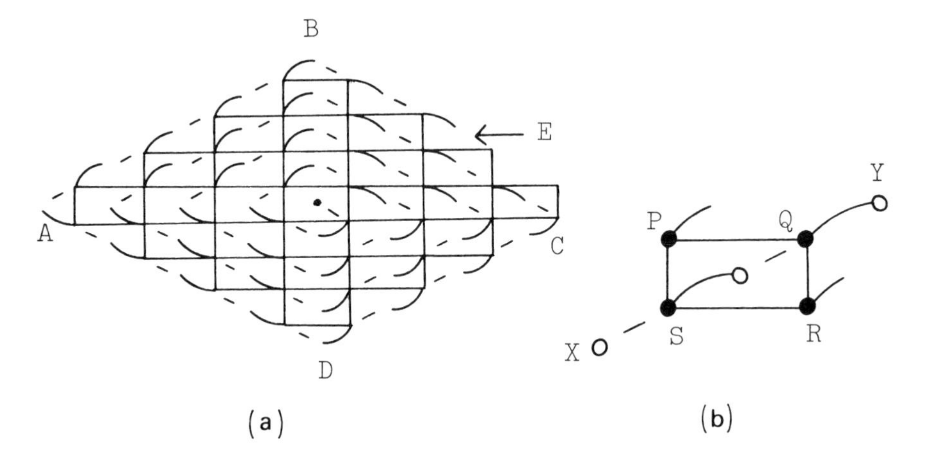

Fig. 7.17 – Relation of folding, unit cell and growth face: (a) general view with faces A B C D and growing chain at E; (b) detail showing a chain XY passing through a unit cell PQRS – note chains coming up at four corners and going down in centre.

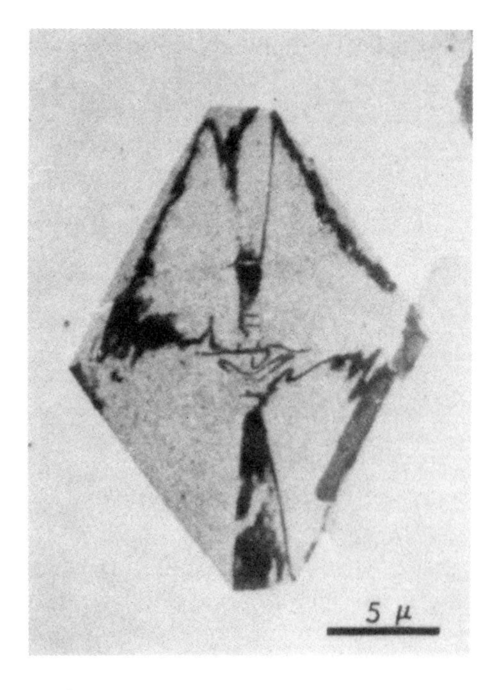

Fig. 7.18 – Truncated single crystal of polyethylene, showing differential extinction in optical microscopy between crossed polars (Bassett, Frank and Keller, *Phil. Mag.*, 8, 1739, 1963).

crystallographic sectors as shown in Fig. 7.19. In two of these the folding is not equivalent (in relation to the unit cell) to that in the other four, and is in fact a slightly less stable form, so that these sections can be preferentially melted as shown in Fig. 7.20.

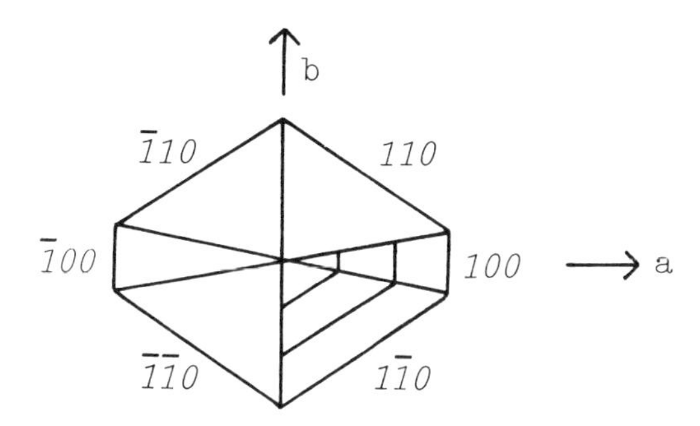

Fig. 7.19 – Growth faces and folding in the truncated crystal.

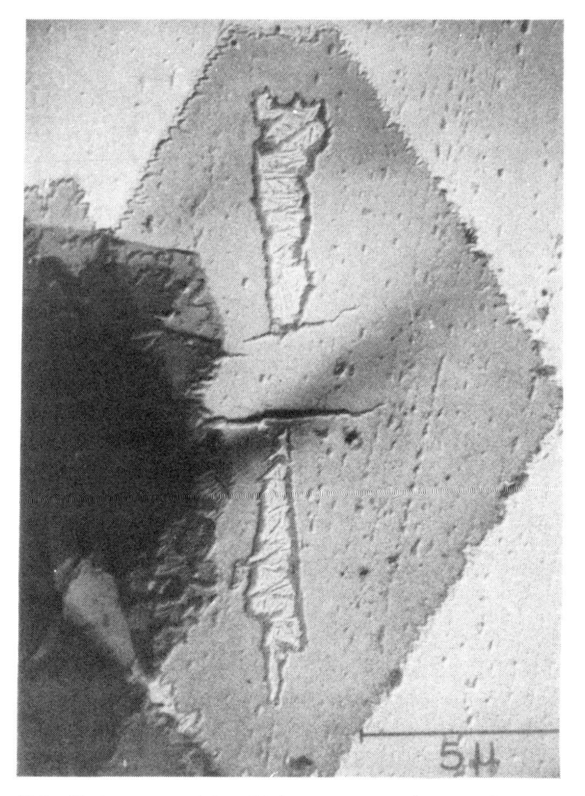

Fig. 7.20 – Polyethylene crystal in which two sectors have preferentially melted (Keller and Bassett, *Proc. R. Mic. Soc.*, 79, 243, 1960).

Where the unit cell has rectangular symmetry, four equivalent sectors occur, and it is only when additional sectors are formed that there is a lack of equivalence of folding. But with unit cells that are not rectangular, non-equivalent sectors would be inevitable.

In the above discussion, it has been assumed that the sectors differ only in the form of folding on their surfaces. However it has been found that the folds which represent a strained molecular conformation, also cause a slight distortion of the interior of the crystal, as indicated in exaggerated form in Fig. 7.21. This can be detected by slight differences in the electron diffraction patterns from different sectors, or by changes in the direction of Moiré fringes (see section 7.3.5) at the sector boundaries in overlapping crystals. The distortion of the unit cell means that the sectors would not meet in a single plane at the centre and the platelet would be transformed into a shallow pyramid. The effect is a very small one, and is not to be confused with the much more pronounced pyramidal form discussed below, but it has been observed in polyoxymethylene crystals in situations where the other mechanism is not occurring. A repetitive occurrence of this change of direction has been used to explain the existence of some curved crystal forms.

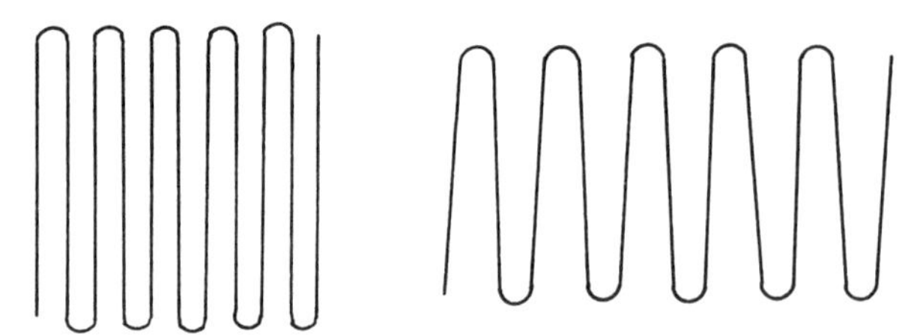

Fig. 7.21 – Schematic illustration of distortion of unit cells in interior of crystal due to folding on surface: (a) ideal lattice; (b) distorted lattice.

7.3.4 The influence of screw dislocations

The above discussion relates to the simplest form of regularly folded monolayer polymer crystal, and brings out some effects which are characteristic of polymers. But it is rare to find such simple crystals, and we must now turn to more complicated forms.

Studies of crystals in general have shown that growth occurs most easily at a dislocation, and particularly at a screw dislocation, such as is shown in Fig. 7.22. The dislocation provides a step which can continue growing indefinitely: it goes round and round, and so there is no necessity to nucleate another layer to give a new step when a layer becomes complete. Thus, if a dislocation forms for any reason, it will develop as a preferential line of growth.

There is a great deal of evidence that polymer crystals commonly grow by a screw dislocation mechanism, and Fig. 7.23 shows some of the spiral forms which result. Instead of the single layer of the simple platelet, the crystal is now composed of a number of overlapping lamellar layers. The separate layers have a thickness which is regular and equal to that of the corresponding platelets.

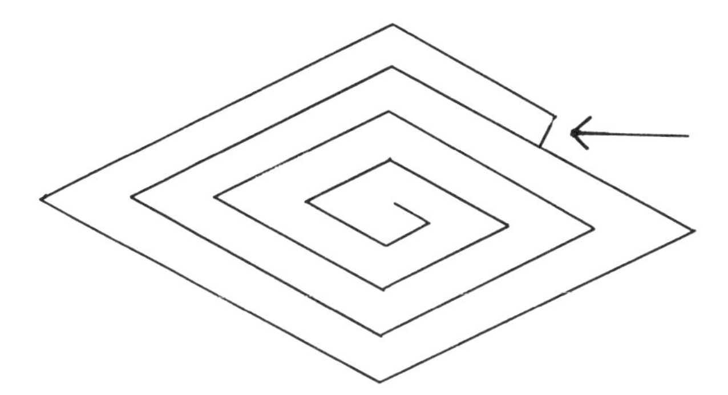

Fig. 7.22 – Schematic illustration of a screw dislocation in a crystal.

Fig. 7.23 – Spiral growths on a polyoxymethylene single crystal (Reneker and Geil, *J. Appl. Phys.*, **31**, 1916, 1960).

It is common for the separate layers to be slightly displaced relative to one another. This results in a rotation of the lateral orientation of the unit cell, and causes the stack of crystal layers to be twisted. A particularly beautiful example of this is shown in Fig. 7.24. A simple model of platelets crystallographically

Fig. 7.24 – Twisted multilayer crystal (Keller, *Polymer*, **3**, 393, 1962).

stacked on top of one another would not give any reason for the displacement of one layer relative to the next, and it is difficult to work out the precise cause and magnitude of slight displacements due to secondary effects. Nevertheless it is reasonable to expect that the distortion caused by the folds and by the screw dislocation would cause the layers to be slightly displaced in a regular manner which would give twisting. Indeed, it would need a very thorough and detailed argument to prove that the layers should remain in precise register in any particular instance, although the simple argument is valid as a first approximation and shows that there cannot be large displacements if the crystallographic orientation is maintained.

7.3.5 On Moiré-fringes and dislocation networks

The occurrence of multilayer crystals leads to two special techniques of examination: Moiré fringes due to the relations between the main lattices in the two crystals; and dislocation networks due to the packing together of two fold surfaces.

Moiré fringes appear in a simple form when two regular gratings with different spacing or different orientations are superimposed. For crystal lattices, they can be detected by dark field electron micrography. If there are defects in the lattice, this is shown by distortion or termination of the fringes. For example, Fig. 7.25(a) shows fringes formed between overlapped polyethylene crystals, terminating at various points due to lattice dislocations.

In many sorts of crystals, particularly metals, it has been shown that disorder appears as a network of defects separating regions which are not in exact crystallographic register with one another. Very similar networks have been observed when two polyethylene crystals are laid on top of one another as shown in Fig. 7.25(b). These differ from the Moiré fringes in that they are associated with the fold surfaces and not with the main crystal lattices in the interior. Essentially what happens is that if two crystal surfaces are brought together in such a way that they are not exactly matched, then, rather than remaining uniformly mismatched, the fold surfaces, which can be regarded as two-dimensional crystal lattices, prefer to distort so that they split up into regions which are crystallographically matched and are separated by a network of defects. Obviously a complete network must be formed since each region of match must be completely separated from its neighbours in a different register. Thermodynamically the reason for the formation of the dislocation network is that the mixture of regions of high order separated by a network of high disorder must be a state of lower free energy than a uniform spread of intermediate lack of order.

More detailed examination of dislocation networks, and associated theoretical calculations, leads to information on the nature of defects, the energies of dislocations, and forms of folding.

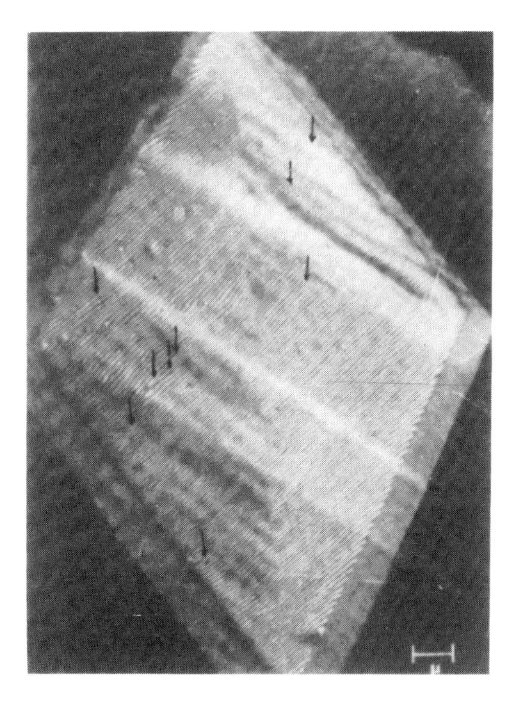

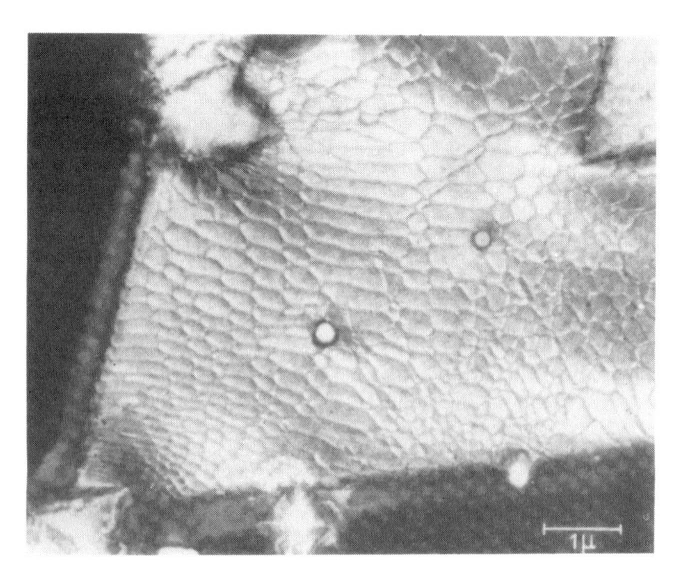

Fig. 7.25 – (a) Moiré patterns between two layers of polyethylene crystals. Arrows indicate termination of fringes due to dislocations. (b) Dislocation network between two polyethylene crystal layers (Holland, *J. Appl. Phys.*, **35**, 3235, 1964 and **36**, 3049, 1966).

7.3.6 Pyramidal form

Another observed feature, is that the crystals, whether platelets or spiral forms, are often markedly pyramidal in form. The separate sectors make up the faces of a hollow pyramid, as indicated in Fig. 7.26(a) and (b). It may be noted that

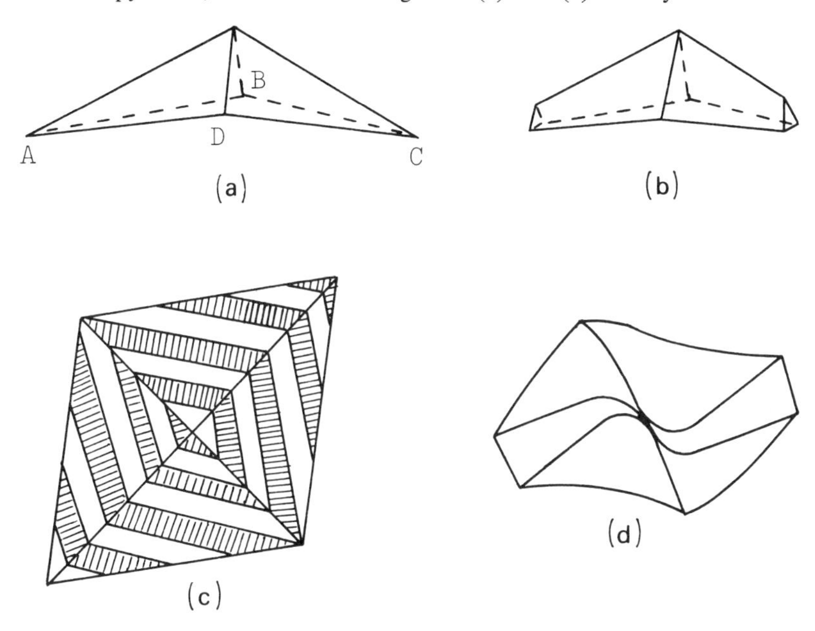

Fig. 7.26 – Pyramid forms in which polymer crystals commonly occur. (a) Simple hollow pyramid with four sectors. Note that the base A B C D is not planar: B D lies above A C. (b) Truncated pyramid with six sections. (c) Ridged pyramid. (d) Chain form.

the lower edges of the crystal are not planar as they would be in an ordinary pyramid. This form can be seen directly when rather large crystals are examined in an optical microscope while still floating in the solvent, Fig. 7.27(a). For detailed examination in the electron microscope, the crystals must be collected and dried and then the pyramids usually collapse and show a fold as indicated in Fig. 7.27. Other more complicated forms which are obviously derived from the basic pyramidal form are also observed. These include rigid structures as shown diagramatically in Fig. 7.26(c) and in reality in Fig. 7.28, and the 'chair' forms such as Fig. 7.26(d).

The pyramidal forms can be explained as a consequence of chain folding. The folds themselves are somewhat unfavourable chain conformations. Even so, they may in some circumstances pack preferentially side-by-side as in Fig. 7.29(a) and give a planar crystal (or a very shallow pyramid due to the distortion mentioned earlier). But it seems more likely that the folds will often be bulky, and will more easily pack in the staggered from shown in Fig. 7.29(b). If the fold length is to be maintained, this would cause the crystal to grow in the hollow pyramidal form already described. A reversal in the direction of staggering would lead to the more complex ridged and chair forms.

There is an interaction between sectorisation and staggering. Thus with growth on two types of face, giving six sectors, there is only one ratio of size of

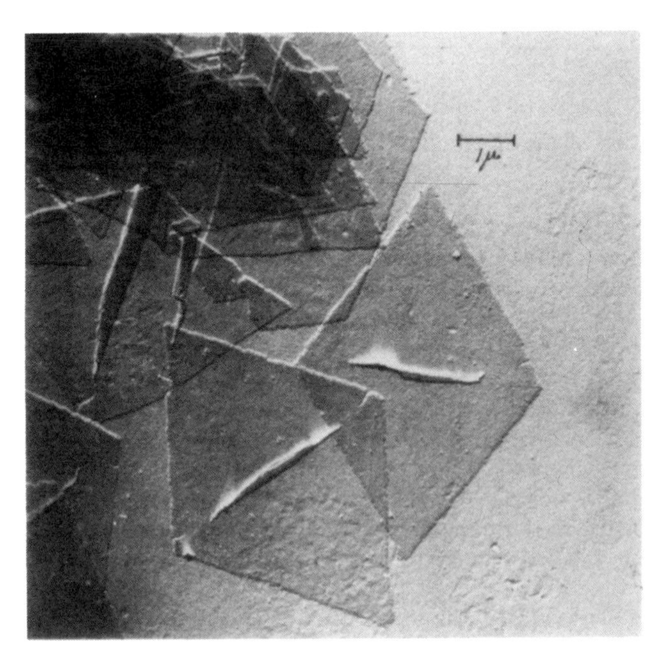

Fig. 7.27 – Pyramidal crystal of polyethylene collapsed with formation of a fold. (Holland and Lindenmeyer, *J. Pol. Sci.*, 57, 589, 1962).

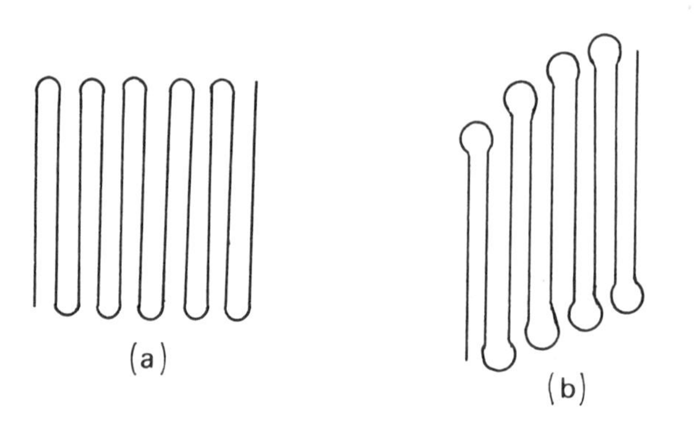

Fig. 7.29 – (a) Simple packing of folds. (b) Staggered packing of folds.

Fig. 7.28 – Rigid monolayer crystal of polyethylene (Bassett, Frank and Keller, *Phil. Mag.*, 8, 1739, 1963).

faces which maintains the preferred obliquities, because the boundaries between the sectors must be the lines of intersections of given planes. For polyethylene, this geometry is shown in Fig. 7.30(a) and a crystal which has grown in this form

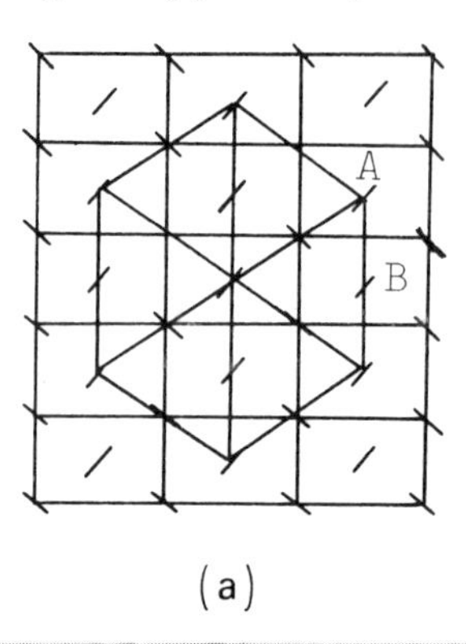

(a)

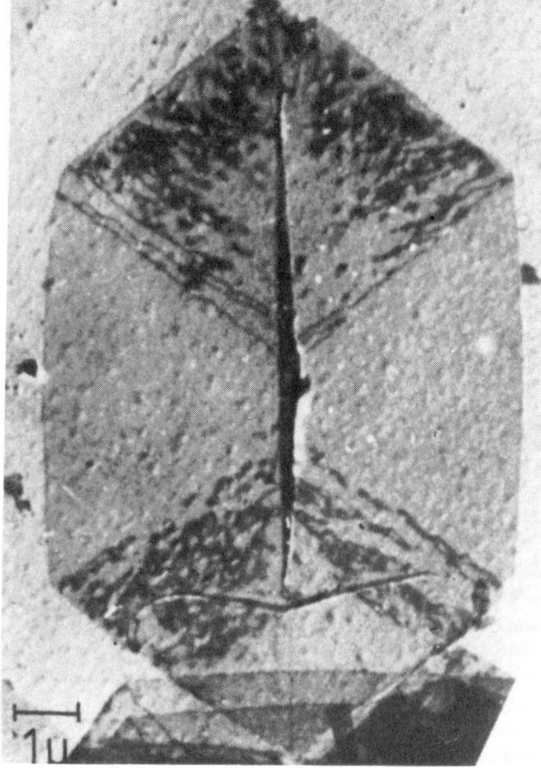

Fig. 7.30 – (a) Preferred geometry of polyethylene crystal in order to match obliquity and folding: face A is 312 at an angle of 29° and face B is 201 at −34° 20. (b) Collapsed hollow pyramidal crystal which has grown in this shape. (Kawai and Keller, *Phil. Mag.*, **11**, 1168, 1965).

is shown in Fig. 7.30(b). Where crystals have other forms due to preference for growth on particular faces, as in Figs. 7.19(a), 7.20 or 7.27(a), this must be accommodated either by alteration of obliquities or by distortions in matching of sectors. The occurrence of large 100 sectors, as in Fig. 7.30(b) is encouraged by high crystallisation temperatures and more concentrated solutions: it may be noted that there is then no growth on the usual diagonal faces, such as 110, as shown in Fig. 7.19(b).

7.3.7 Diverse crystal forms

Other complications in crystal form arise due to causes which are typical of crystallisation in general: fresh nucleation on a growing crystal; variations in solute concentration near a growing crystal either as a result of removal of solute by crystallisation or of other changes in the environment; temperature gradients; and so on. Some overgrowth is seen in Fig. 7.15, and extreme dendritic crystallisation is shown in Fig. 7.31(a). Sometimes, the overgrowth may develop with the symmetrically opposite, twinned crystal lattice: Fig. 7.31(b) shows a rosette growth due to this cause. Fig. 7.31(c) shows what has been called a quadritc in which polypropylene has formed a dendritic growth by twinning so as to give a cross-hatched structure with an angle of about 80° between the branches.

In another form of twinning, a double pyramidal crystal forms as illustrated

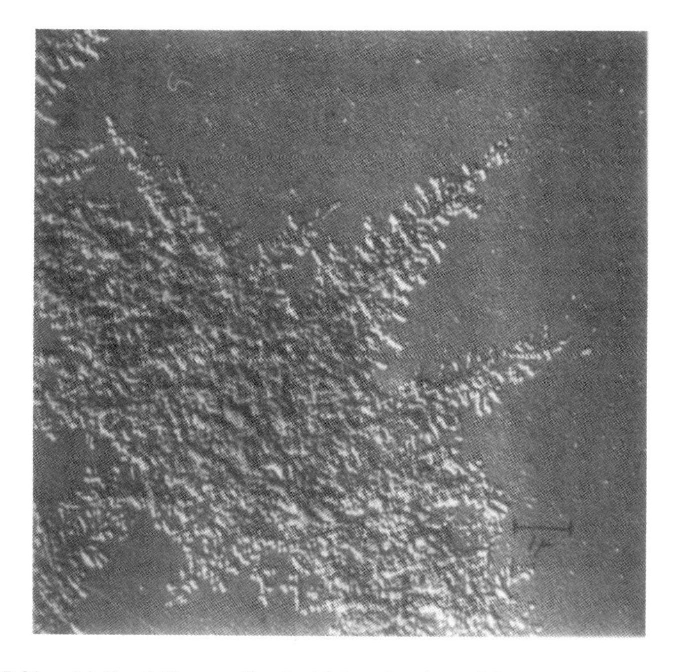

Fig. 7.31 – (a) Dendritic growth of a high molecular weight polyethylene crystal (Holland and Lindenmeyer, *J. Pol. Sci.*, 57, 589, 1962).

Fig. 7.31 – (b) Rosette of twinned polyethylene crystals (Lindenmeyer, *J. Pol. Sci.*, *C*, **1**, 5, 1963). (c) Quadrite of polypropylene (Khoury, *J. Res. Nat. Bur. Std.*, *A, Phys. and Chem.*, **70A**, 1, 1966).

in Fig. 7.32. This is the simplest example of a rather common form of growth, namely overgrowths of lamellar crystals splaying out as they grow. Fig. 7.33(a) shows a simple multilayer polyethylene crystal, while a more elaborate multilayer structure which has grown from a more concentrated solution in a sheaf-like form, is shown in Fig. 7.33(b). When collapsed on drying, the structures appear rather different.

In general, the forms which occur will be more complicated the faster the rate of growth of the crystal, the more concentrated the solution and the larger the crystal. The simplest forms will be small crystals, slowly grown from dilute solution. In crystallisation from concentrated solution, rather larger objects are obtained: these have been termed **hedrites** or **axialites**. When seen from some directions they appear to be solid polygonal crystals, but, when viewed from

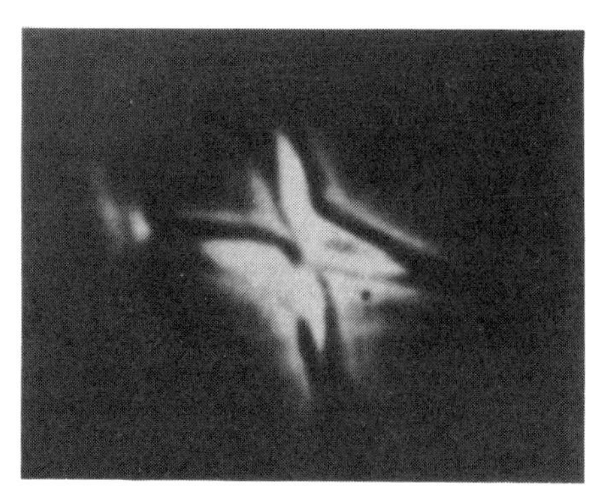

Fig. 7.32 – Double pyramid polyethylene crystal seen edge-on in liquid (Bassett, Keller and Mitsuhashi, *J. Pol. Sci., A,* **1**, 763, 1963).

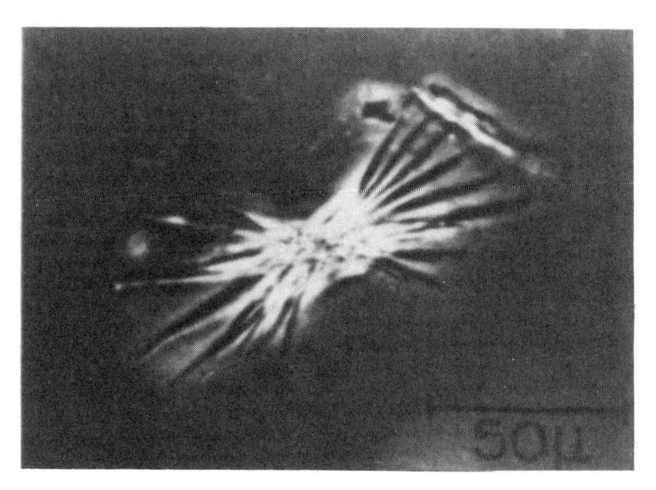

Fig. 7.33 – (a) Multilayer polyethylene crystal is suspension (Mitsuhashi and Keller, *Polymer*, **2**, 109, 1961).

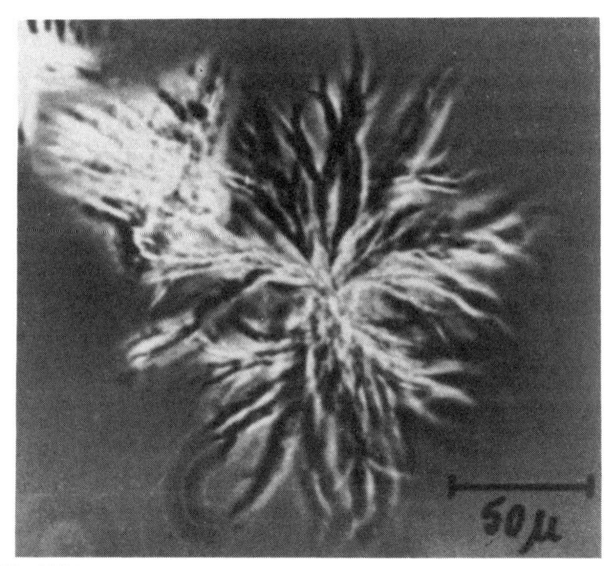

Fig. 7.33 – (b) Sheaf-like structure observed edge-on in suspension (Bassett, Keller and Mitsuchashi, *J. Pol. Sci.*, A, **1**, 763, 1963).

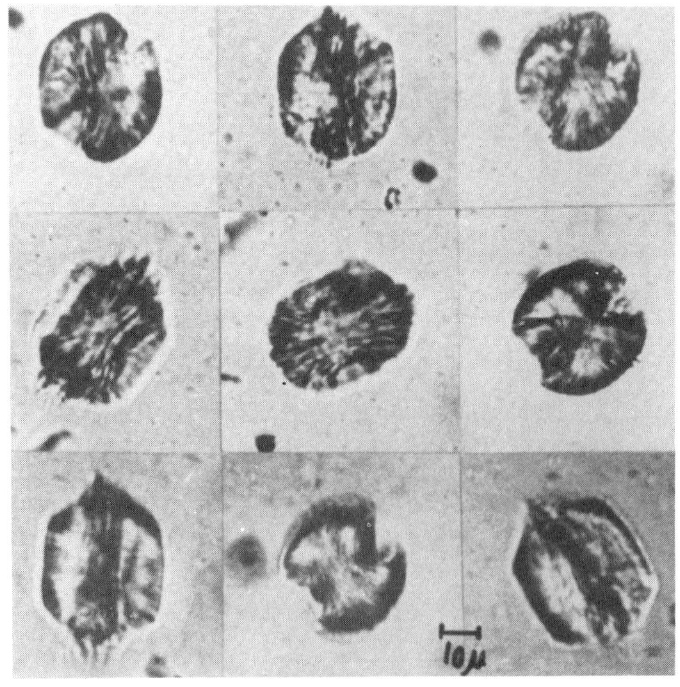

Fig. 7.34 – A crystal of polyethylene grown from concentrated solution. This 'axialite' is viewed from different angles (Bassett, Keller and Mitsuhashi, *J. Pol. Sci.*, A, **1**, 763, 1963).

other directions it is apparent that they are thick sheaf-like assemblies of lamellar crystals. Fig. 7.34 shows examples of the same axialite viewed from different directions.

In compact structures of this form, which are rather like two slightly open books placed back-to-back, it has been suggested that there are tie molecules (or crystalline portions) between the lamellae so that the extent of splaying is limited. Fig. 7.35 shows a portion of material crystallised from a rather concentrated (5%) solution: here there are obvious thread-like crystal strands between the layers, and, while it may be that these have been pulled out during drying, it is likely that they are fibrous tie crystals.

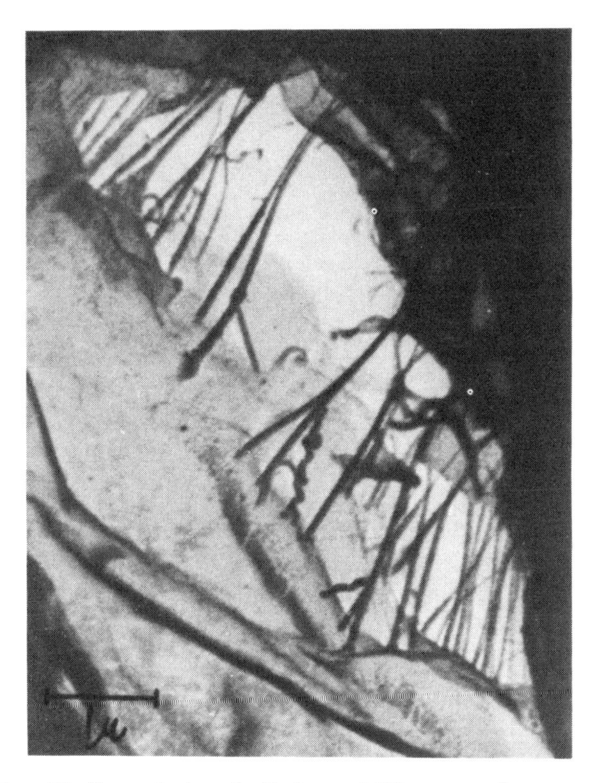

Fig. 7.35 – Fibrillar units in polyethylene axialities grown from concentrated solution (Bassett, Keller and Mitsuhashi, *J. Pol. Sci.*, A, **1**, 763, 1963).

7.3.8 On the formation of crystals of uniform size

Polyethylene crystals will dissolve in xylene at 97°C. However over the range 97°C to 110°C some residual effects of the crystal form remain, with subsequent recrystallisation becoming progressively slower as the dissolution temperature is increased. After dissolving within this range all the crystals subsequently grow to a uniform size, as in Fig. 7.36, indicating that they are forming on pre-existing nuclei. Furthermore because there are many residual nuclei, there is a very large crop of small crystals, which have not developed many complications. After

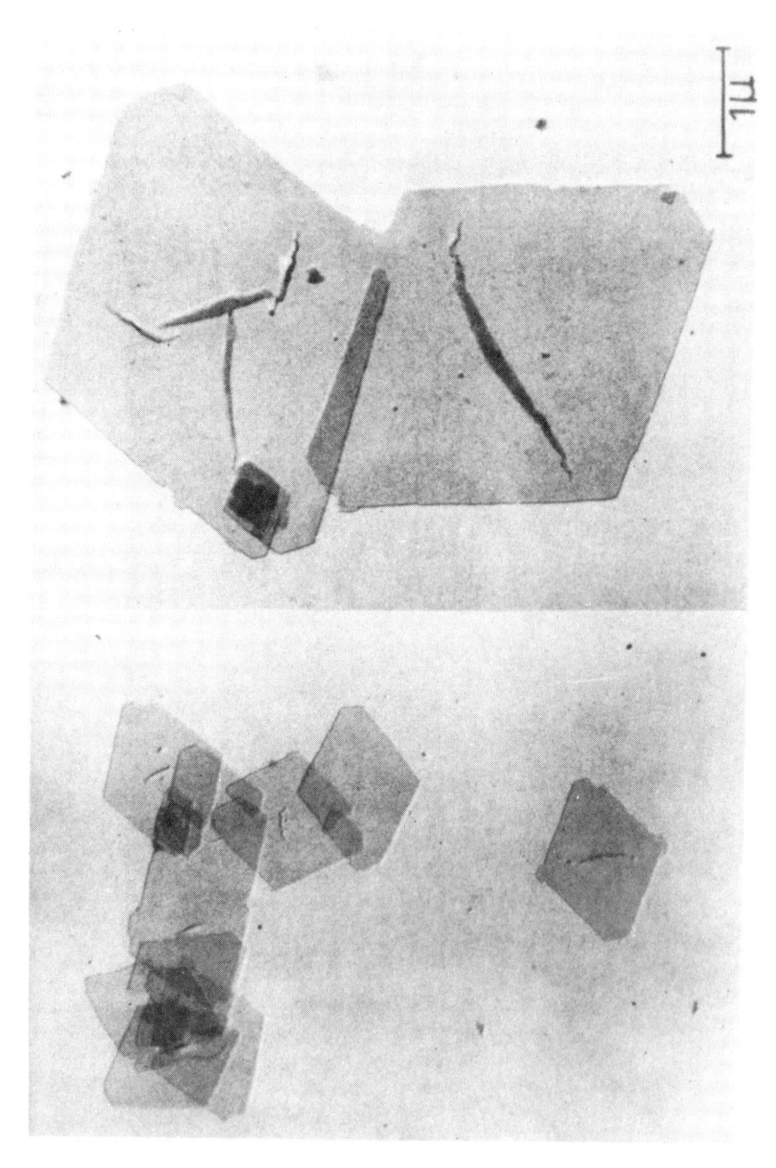

Fig. 7.36 – Monolayer crystals grown by recrystallisation after dissolving at (a) 99°C, (b) 101.5°C (Blundell, Keller and Kovacs, *J. Pol. Sci.*, B, **4**, 481, 1966).

solution at a higher temperature, the residual nuclei disappear and crystallisation has to take place on a few nuclei giving large complicated crystals with dendritic growth, sheaving and so on. As indicated by the illustration in Fig. 7.36 it is possible to control very close the size and form of crystal by controlling the conditions of formation.

It has been suggested, on the basis of light scattering studies of the solutions, that the residual nuclei are individual monomolecular crystals formed of particularly long chains which have been refolded to a higher fold length, and are stable when less perfect crystals have melted. They appear as disc-like thickening in the middle of the crystals subsequently formed. It seems probable that effects of this sort are responsible for other memory effects in crystallising polymers.

7.3.9 Fracture of single crystals

If single crystals are deposited on an extensible film to which they cohere, then it is possible to fracture the crystals by stretching the film. Fig. 7.37 illustrates what happens. If the extension is perpendicular to the line of folding, the crystal fractures by a clean break as shown in Fig. 7.37(a) but if it is along the line of the fold, the continuity of chains is maintained and threads appear across the gap, Fig. 7.37(b). Presumably the latter are fibrillar crystals formed as neighbouring groups of folded molecules pull into extended form and crystallise together.

Fig. 7.37(a) – Rupture of polyethylene crystals in two perpendicular directions: parallel. (Lindenmeyer, *J. Pol. Sci.*, C, 1, 5, 1963).

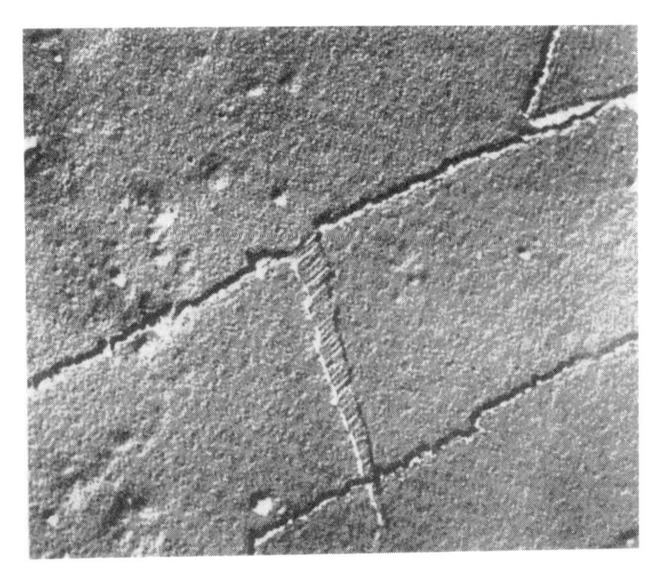

Fig. 7.37(b) – Rupture of polyethylene crystals in two perpendicular directions: perpendicular to the (110°) plane bounding the sector (Lindenmeyer, *J. Pol. Sci.*, *C*, **1**, 5, 1963).

7.3.10 Fibrous crystals and shish-kebabs

There is another effect in crystallisation from solution which is a result of the polymeric form. If the solution is under stress (that is to say, if it is in a state of non-uniform flow) then as we have discussed in Chapter 3, the molecules will be pulled into a somewhat extended chain conformation. With small stresses, this might be expected to cause some modification of crystal habit. But in extreme cases where the force of extension was strong enough to prevent any folding, the only way in which crystallisation could occur would be by the formation of long fibrillar crystals with fully extended chains. With less severe stresses,

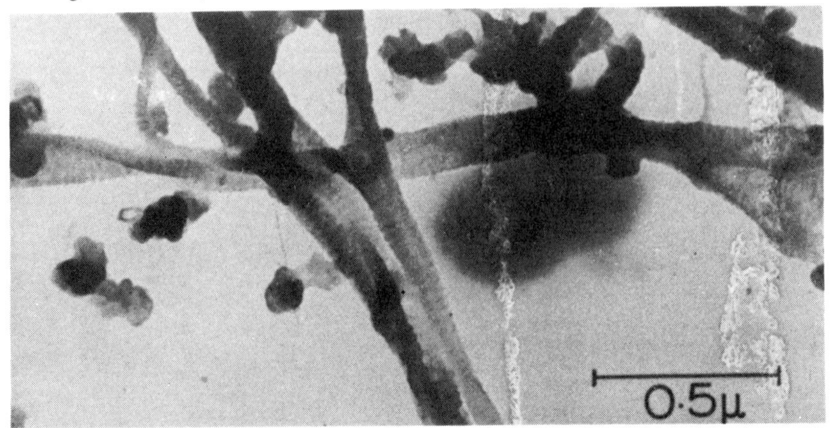

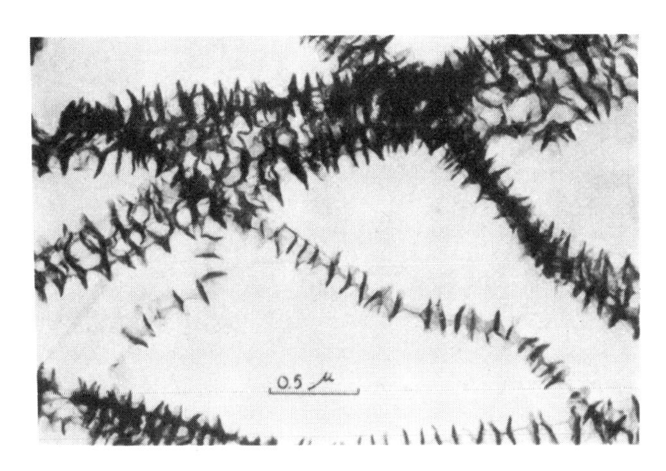

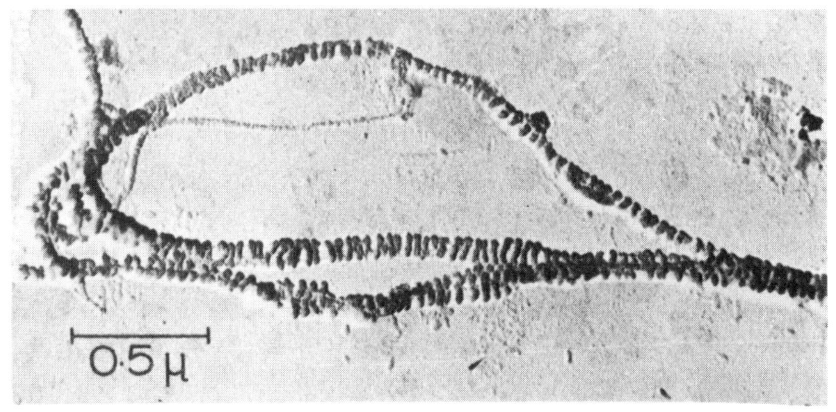

Fig. 7.38 – (a) Polyethylene crystallised from an agitated xylene solution (Keller and Machin, *J. Macromol. Sci. B*, 1, 41, 1967). (b) Shish-kebabs of polyethylene (Pennings, *Proc. Int. Conf. on Crystal Growth*, Boston, June 1966, Pergamon). (c) A crystal similar to Fig. 7.38(a) after attack by fuming nitric acid (Keller and Machin, *J. Macromol, Sci. B*, 1, 41, 1967).

some folding might occur (due to the driving force of crystallisation) but it might be biassed by the stress, and still give fibrillar crystals. Fibrillar forms are, in fact obtained on crystallisation from agitated (stirred) solutions, where there are strong stress gradients. An example is shown in Fig. 7.38(a); it is faintly streaked indicating the probability of some folding. A form which is more often observed is the 'shish-kebab' structure shown in Fig. 7.38(b). It is believed that the platelets surrounding the main stem form by crystallisation during the cooling of the solution after the fibrillar stems have formed: the fibrils act as nuclei for the chain-folded growth. Simple fibrils, as in Fig. 7.38(a), can be obtained by replacing the solution by solvent at a suitable temperature, or by high-temperature washing of shish-kebabs.

round in the solution. There are, in fact, indications of some microstructure within the main stem. Treatment with nitric acid, which preferentially attacks folds, breaks the stem up into the 'micro-shish-kebab' shown in Fig. 7.38(c). A central thread remains and resists a long period of attack. This suggests that there is an ultimate fibrous strand at the centre with some folded material as a continuous sheath around it. The main stem, both core and sheath is more stable than ordinary folded chain crystals. On top of this stem come the lamellar overgrowths of ordinary stability.

It is interesting to note that crystallisation occurs more readily in stirred solutions; thus polyethylene can be crystallised as fibrils from a solution in xylene up to 108°C, although in a stationary solution the limit is 90–92°C, or 95°C with self-seeding. Faster stirring encourages crystallisation. Furthermore it is the longest chains which crystallise first as fibrils.

All these observations are consistent with the view that dragging chains, particularly long chains, through the solution by the stirrer leads to an extended form which both crystallises more readily, because the free energy barrier is reduced, and also establishes a lower free-energy state. Some detailed theories incorporating hydrodynamic work have been attempted.

An oriented assembly of fibrous crystals can be collected like a sheet of paper round the stirrer. Such a sheet shows a high modulus (200 N/mm^2), but an extensibility of only 10%.

In recent years, there has been much research on crystallisation in more controlled forms of flow than crude stirring or agitation. For example Couette apparatus may be used. This work confirms the general pattern of fibrillar crystallisation under conditions of shear flow. The subject will be taken up again in Volume 2 as an example of particular structural forms. The sequence which leads either to (a) fibrillar crystals or extended chains or (b) shish-kebabs with overgrowths is illustrated in Fig. 7.39.

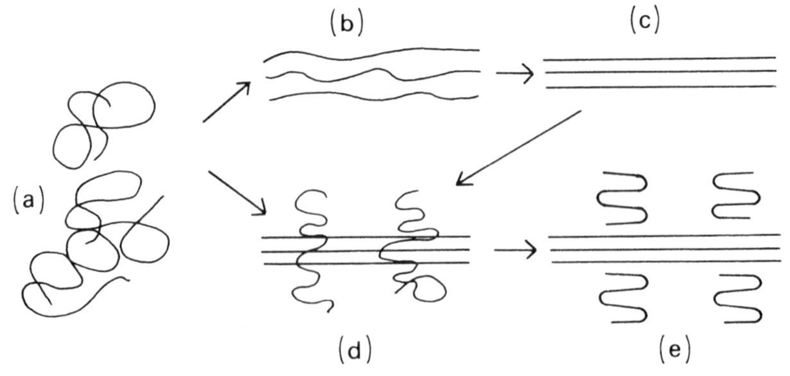

Fig. 7.39 – (a) Molecules in solution in random conformation; (b) become oriented by flow; and (c) crystallise as extended chain fibrillar crystals, possibly with some folding. (d) They then collect additional molecules; and (e) these form crystalline overgrowths.

7.3.11 Behaviour of different polymers

Although the great majority of the observations presented above have been on polethylene, other crystallisable polymers behave similarly. Single crystals have been grown from solution for a great variety of polymer types including many polyolefins, polyamides, cellulose and cellulose derivatives and so on. Even polyacrylonitrile with a pseudo-crystal lattice will form rather irregular ellipsoidal crystals.

There is only one general point to mention. Where the crystal lattice shows a staggering between groups, as for example in nylon, Fig. 7.4, then the fold surface must be staggered to fit it. The formation of lamallae with oblique faces is thus a consequence of the form of the crystal lattice, and not merely of the most favourable arrangement of the fold itself, as in Fig. 7.29.

7.4 CHAIN FOLDING

7.4.1 The nature of folding

The phenomenon of chain folding is so important in the behaviour of crystalline polymers that it is necessary to look carefully at this particular aspect of crystallisation. Firstly, one can say that in many polymers there is no great difficulty about chain folds as such. The examination of models of polymer molecules shows that most will fold back on themselves relatively easily to give a fold which would be compact enough to fit into a crystal surface, though this might not be true of polymers with more complicated chemical structures, such as those with very bulky side-groups, rings in the main chain, or ladder or girder structures. Even when the molecule naturally takes up a helical form, suitable paths from one helical portion to the neighbouring anti-parallel portion can usually be found.

Of the many possible forms of folding, certain forms will lead to a minimum free energy in the crystal, and will thus be favoured. Because of their regularity, these can be termed crystallographic folds. Some calculations have been made in order to find out what these forms are. This involves a minimisation of the combined energy of chain distortion and inter-chain forces, based on appropriate potential functions. In polyethylene the most favoured fold contains about 5 main chain atoms. Because of the different directions of folding needed to fit with the main crystal lattice, all folds will not be identical, but they will themselves form a regular pattern, which can be regarded as a form of crystal.

Secondly as discussed in section 2.4.5 it is clear that a long isolated molecule will fold up in order to minimise its free energy by crystallisation. The balance between the increase of energy in folding and the decrease of energy due to interaction of chains (or, to use the complementary quantity, the reduced free surface) will determine the equilbrium fold length.

The situation in polymer crystals composed of many molecules is more complicated. Observation of single crystals in the electron microscope indicates that the thickness of the crystals which is a measure of fold length, is a well-defined quantity, which is constant for any given polymer crystallised from a given solvent at a given temperature. Furthermore when a crystalline precipitate settles with all the platelets parallel, the resulting coherent mat gives a narrow-angle X-ray diffraction pattern with four orders of discrete reflections visible. The Bragg spacing was identified with the crystal thickness, and the sharpness of the pattern indicates a high degree of regularity of thickness. This is an accurate method of measuring thickness. Interference optical microscopy can also be used.

The variation of fold length with temperature and solvent is shown in Fig. 7.40. The increase with temperature has been widely observed, but, as illustrated in Fig. 7.41, the relation is really to the degree of supercooling, independently of the solvent. If the temperature is changed during crystallisation, a step appears as the fold-length changes.

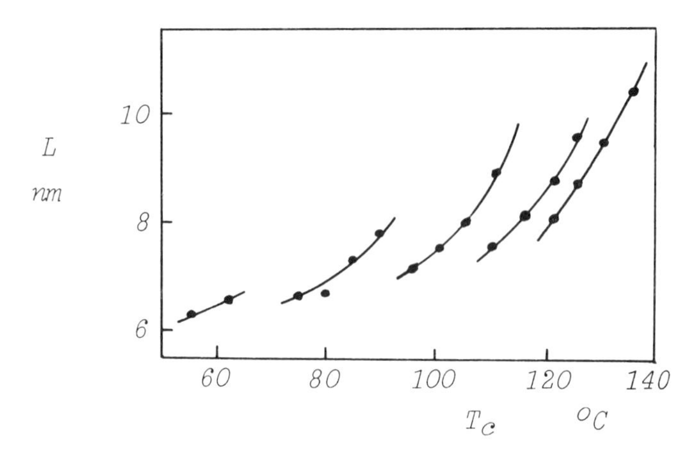

Fig. 7.40 – The dependent of lamellar thickness L of polyoxymethylene crystals on crystallisation temperture, for crystallisation from different solvents. After Nakajima and Hamada (1972).

Even more remarkable is the fact that when a crystal is annealed at a temperature higher than the crystallisation temperature, it refolds to the thickness characteristic of the new temperature. In doing so, portions of the crystal disappear while others get thicker to give the broken up appearance of the crystal in Fig. 7.42 or the picture frame of Fig. 7.43. The dependence of the fold length on annealing temperature, as indicated by the annealing of crystalline mats, is shown in Fig. 7.44.

As with all annealing processes, this is a coarsening of texture: the crystals get thicker.

Fig. 7.44 shows that the annealing process continues indefinitely at a steadily decreasing rate, with the long-period spacing continuously increasing. It has been

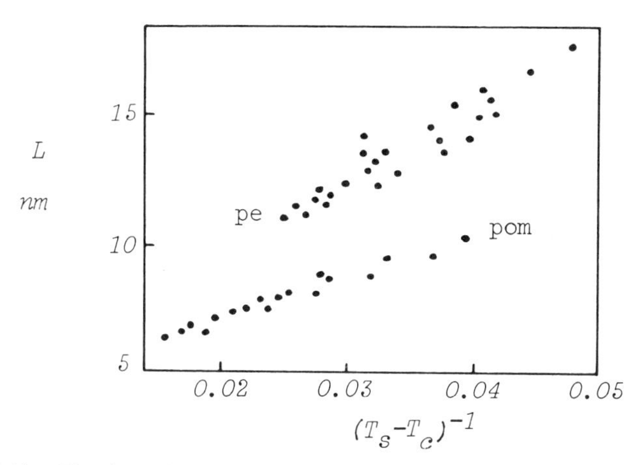

Fig. 7.41 – The data of Fig. 7.40 – and other data for polyethylene – replotted against reciprocal of degree of cooling, that is, the difference between temperature of dissolution T_s and temperature of crystallisation. After Makajima and Hamada (1972).

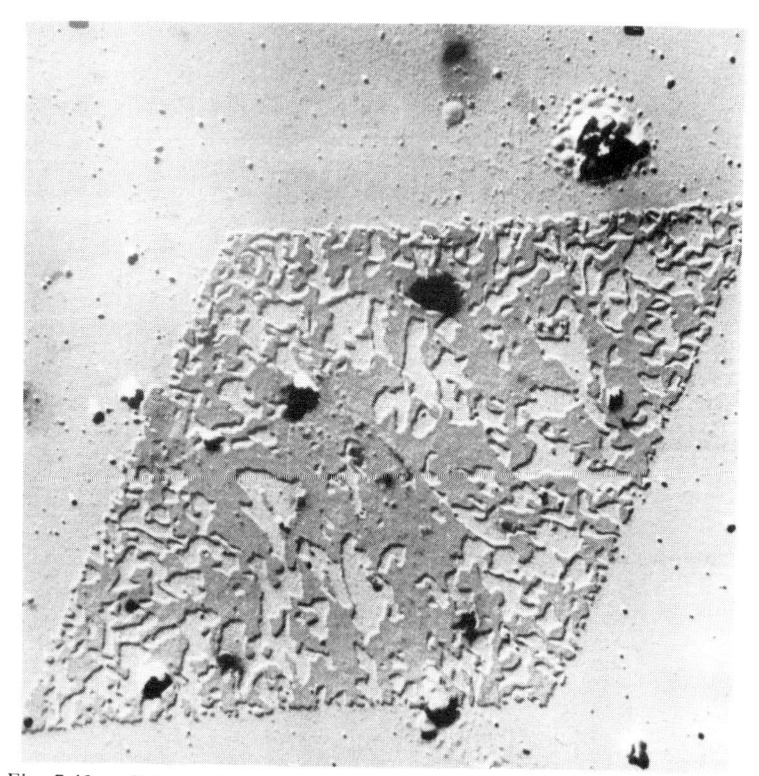

Fig. 7.42 – Polyethylene single crystal formed at 85°C and then annealed at 125.7°C for 35 minutes (Statton and Geil, *J. Appl. Pol. Sci.*, **3**, 357, 1961).

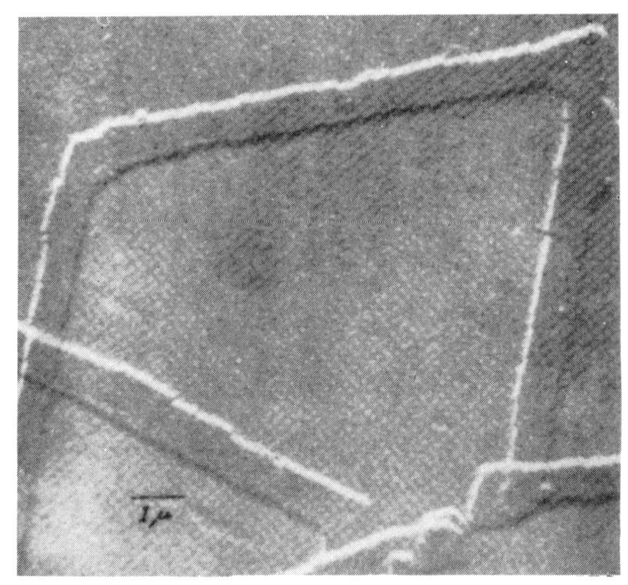

Fig. 7.43 – Picture-frame crystal resulting from annealing in suspension (Holland, *J. Appl. Phys.*, 35, 59, 1966).

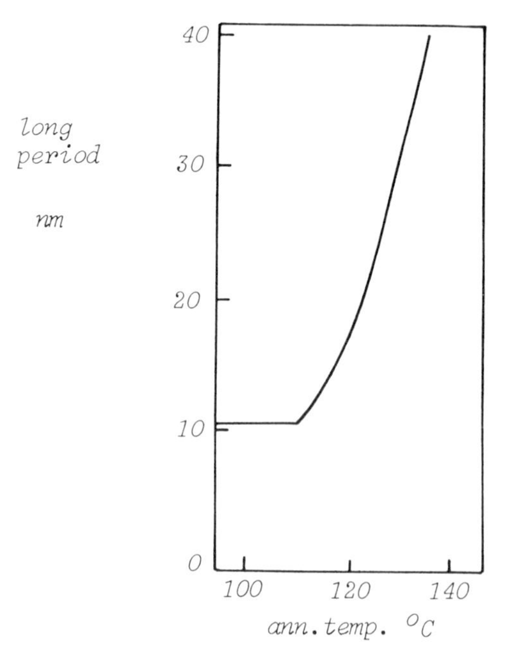

Fig. 7.44 – Change of fold length as shown by long period of polyethylene crystal mats, after annealing for 15 hours at different temperatures. After Statton and Geil (1960).

found that experimental results commonly fit the empirical equation:

$$l = l_0 + B(T) \log (t/t_0 + 1) \tag{7.21}$$

where l is long period at time t, l_0 at time t_0, and $B(T)$ is a proportionality factor which is dependent on temperature.

However more detailed investigation shows that the initial increase in crystal thickness occurs more rapidly than predicted by (7.21), while at very long times the change becomes slower. There is also a dependence on chain length, with short chains annealing more rapidly and long chains showing very little change beyond the first sudden increase. The annealing rate is also affected by the crystallisation conditions, and by the environment during annealing.

If crystals are annealed while in suspension, a combination of solution and recrystallisation occurs. Molecules from the surface of crystals dissolve, and then recrystallise around the edges of crystals to give a thicker rim or frame as shown in Fig. 7.43. If the temperature is high enough, the inner crystal may dissolve completely.

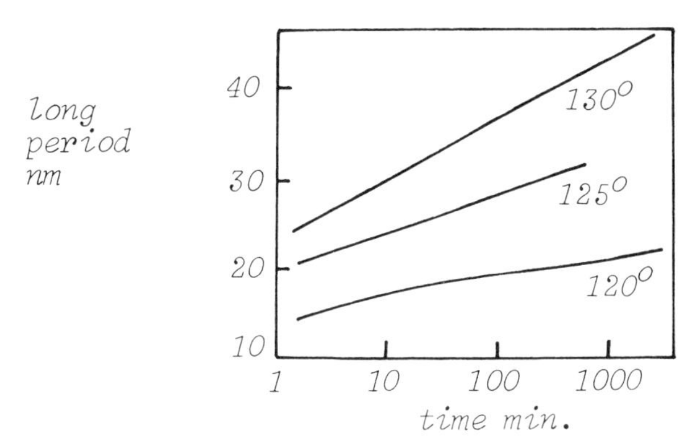

Fig. 7.45 – Change of long period of polyethylene crystal mats with time at different temperatures, °C. After Fischer and Schmidt (1962).

The theoretical situation on fold length is still confused and none of several complicated theories is really satisfactory. In these circumstances, the mathematical details of the theories will not be given. The discussion will be limited to qualitative fundamentals.

The subject will be approached on the assumption that essentially the same theory should account for both the thickness of crystals as grown and the changes on annealing. There are four classes of theory which may be considered (a) equilibrium theories (b) nucleation theories (c) kinetic theories (d) some combination of the other theories.

7.4.2 Equilibrium theories

Equilibrium theories assume that the system is able to reach its position of minimum free energy. In fact this is inherently improbable, since a structure as large as a single crystal is certain to have reached a metastable state. It is also incompatible with the irreversible change of thickness on annealing: a true equilibrium condition could only lead to a reversible change.

But since systems are likely to move in the direction of equilibrium it is worth seeing what these are, and a detailed theory could be based on the right choice of metastable equilibrium. A few conditions of minimum energy can be described. A single chain would fold up as shown in Fig. 7.46(a), at a length

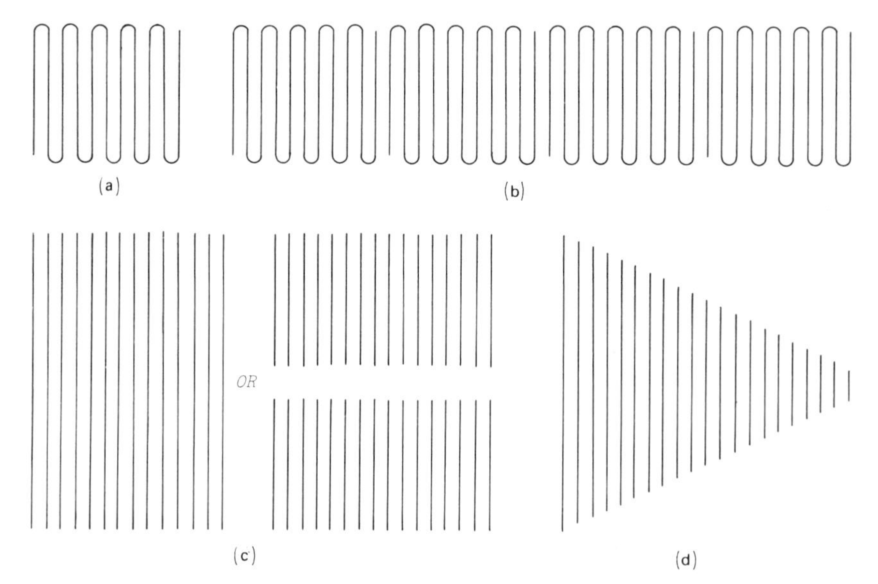

Fig. 7.46 – (a) Equilibrium folding of isolated chain, either in space, or on a substrate, with reduction of surface energy to balance fold energy. (b) With many chains, folding will lead to undue increase of surface energy. (c) Extended chain crystal, minimising fold energy and surface energy. (d) Segregation of variable chains lengths.

governed by the balance between folding energy and surface energy. For reasons given in section 2.4.5 this length would be temperature dependent and might also depend on interactions with solvent. Equation (2.6.2) shows that the equilibrium fold length would increase as the length of the molecule increases. A limited number of chains will also achieve lower energy by folding. Furthermore, Lindenmeyer has shown that isolated chains crystallising on a substrate will fold up in order to reach their equilibrium state, and indeed this is obviously favourable. Thus there is ample reasons why folding should start.

But, with more than a certain number of molecules as indicated in Fig. 7.46(b) folding would increase and not decrease, the surface area, and so would be of no benefit. The equilibrium crystalline form for a large number of separate chains of equal lengths would therefore be an extended chain crystal as shown in Fig. 7.46(c). If there were a range of chain lengths, they would segregate into separate parts of the crystal as indicated in Fig. 7.46(d). Thus, on a simple complete equilibrium theory, as a crystal grew larger its thickness would increase until it grew as an extended chain crystal. On the basis of some more sophisticated theories it is suggested that a folded chain crystal would be preferred because thermal vibrations of the lattice would give an increased energy in the extended chain crystal. However the general opinion is that this type of theory does not explain the experimental results.

7.4.3 A metastable equilibrium theory

There is a particular metastable equilibrium theory which should be mentioned, because while adding to our understanding it also illustrates the difficulties which arise when attempting a detailed analysis. Lindenmeyer has computed the free energy of a polymer crystal as a function of crystal thickness. His expression for internal energy can be simply written as:

$$U = n_f u_f - n_i u_i \tag{7.22}$$

where n_f is the number of folds, u_f is the energy per fold, n_i is the number of pairs of segments which interact as neighbours, and u_i is the cohesive interaction energy between segments. In order to obtain the values of n_f and n_i, we have to consider how the chains will fold. One difficulty is to know what to do with the tails of the chain which is left over after the insertion of a number of regular folds. Lindenmeyer considered two possibilities: (a) the chain ends are excluded, as in Fig. 7.46(a); (b) chain ends can be included within the crystal, but give rise to strings of vacancies as in Fig. 7.47(b). The latter alternative will give an additional fold, and a change in the number of paired segments.

In counting up n_i, it is necessary to exclude from the total those segments which are in folds or which are unpaired at the ends. The total number of odd segments in Fig. 7.47(a) (or the total vacancies in Fig. 7.47(b), may be distributed

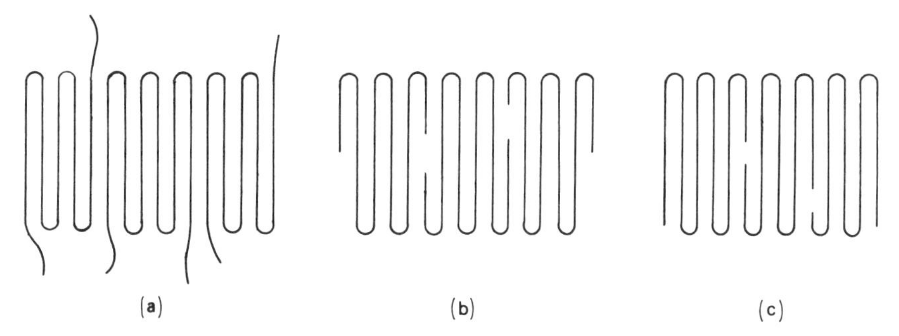

Fig. 7.47 – (a) Folded chain with excluded ends. (b) Folded chain crystal with included ends. (c) Folded chain crystal with matched ends.

in any proportion between the tails at either end of the chain. This gives an additional entropy contribution to the free energy, which is included in Lindenmerer's computation.

Lindenmeyer excludes from consideration the possibility that chain ends might pair in the crystal, as in Fig. 7.47(c), because this leads to complicated statistics. His calculation can also be criticised on the grounds that the model is somewhat simplified, and that the energy and entropy expressions are not precisely correct; but the main trends of the predictions can be expected to be valid.

Fig. 7.48 shows the results of the computation for a crystal composed of chains containing 1000 segments, with 10 segments per fold. As predicted above, the lowest energy state is for an extended chain crystal, although it can be noted that if this state has to be approached through a gradual increase of fold length, the barriers get progressively higher. Of course, it is possible that the lowest energy state could be reached by some other route. More significant, perhaps, is the fact that the benefit involved in increasing crystal thickness becomes slight as the number of folds gets less, and thus little is gained by moving too far down the line of successive minima, which correspond to regular folding with no odd segments as tails or vacancies.

An interesting situation arises when we have a distribution of chain lengths and we assume that these have to crystallise together in a crystal of given thickness, and that it is not possible for them to fractionate and reach the form shown

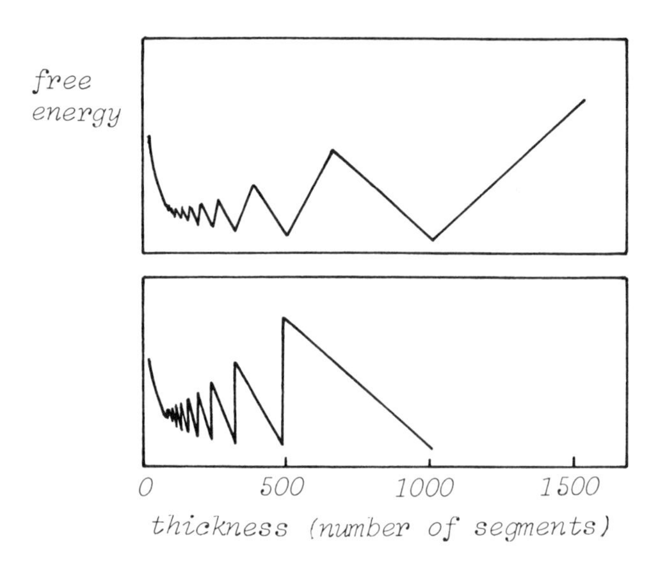

Fig. 7.48 – Computed variation of Helmholtz free energy with fold length for a molecule of length 1000. (a) Chain ends in crystal. (b) Chain ends excluded. After Lindenmeyer (1967).

in Fig. 7.47(c). It is at this point that Lindenmeyer's approach becomes a *metastable* equilibrium theory – the possibility of going to the true equilbrium of Fig. 7.46(d) is somewhat arbitrarily, but reasonably, excluded. Because of the irregular length of chains, there must always be some odd segments whatever the crystal thickness. At any given thickness, the free energy can be found by averaging over the values for the separate individual lengths. This is illustrated in Fig. 7.49 for a discrete distribution of lengths. Because of the unfavourable

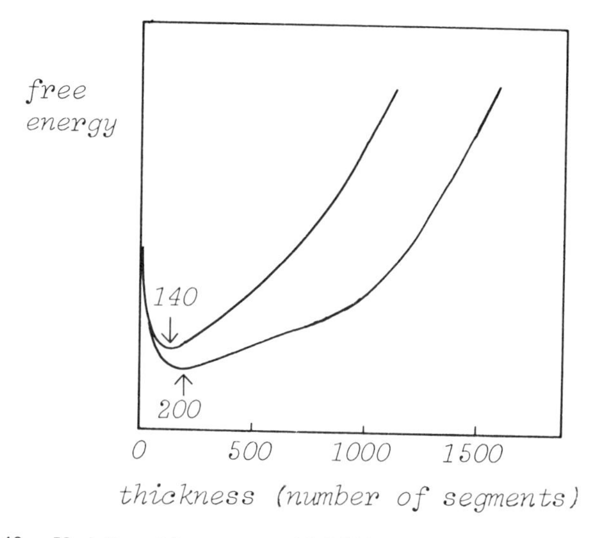

Fig. 7.49 – Variation of free energy with fold length for a discrete distribution of chain lengths, ranging from 200 to 4000 segments. (a) Chain ends in crystal. (b) Chain ends excluded. After Lindenmeyer (1967).

values resulting from the long tails which occur when the chain length does not equal, or form a multiple of the crystal thickness, the lowest free energy corresponds to a folded chain and not an extended chain crystal, since the tails will be generally shorter in multi-folded crystals. Fig. 7.50 shows the results of a computation of the free energy for a distribution which can be regarded as typical of a particular commercial polyethylene sample (Marlex 50). When crystallised from the melt as discussed later, this material forms both extended chain and folded crystals. It is reasonable to assume that the short chains are able to fractionate out and reach the lower state of the extended chain crystal, but that the longer ones get caught up in metastable folded chain crystals. Since the barrier to escape from the folded chain state will increase with chain length, we see that the lower the temperature the lower will be the length at which escape is possible. In the computation, therefore, short chains should be excluded from the distribution used to calculate the effect of crystal thickness. On the basis of the observed length of extended chain crystals, Lindenmeyer puts the cut off at

550 segments for crystallisation at 130°C and 325 segments at 120°C. With this added restriction, the computation indicates lowest free energies at crystal thicknesses very close to the observed values for the folded chain crystals of 280 segments at 130°C and 150 segments at 120°C. The influence of temperature comes in partly through the entropy term and partly through the change in exclusion of short chains.

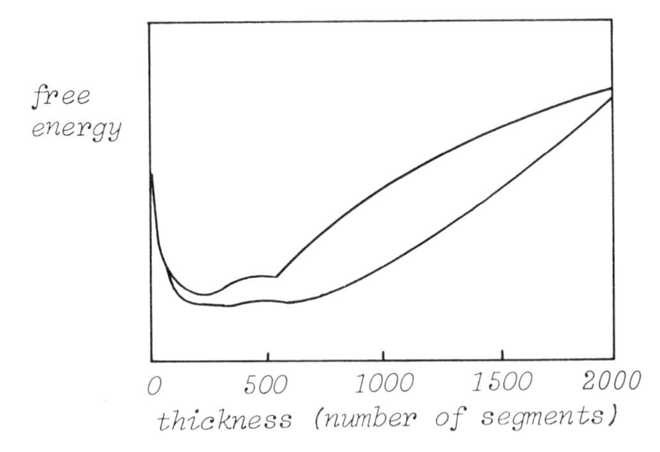

Fig. 7.50 – Variation of free energy for a distribution which simulates a commercial polyethylene. (a) Chain ends in crystal, (b) chain ends excluded. After Lindenmeyer (1967).

7.4.4 Nucleation theories

Crystals must grow from nuclei, which are small but are nevertheless large enough to be static and grow. An obvious possibility, suggested early on, is thus that the size of the nucleus determines the crystal thickness. It would be possible then for the thickness to be determined by calculating the equilibrium folding of the limited number of chains (or the limited length of a single chain) needed to establish a stable nucleus. The difficulty with this simple approach based on primary nucleation, is two-fold: (a) the crystal thickness is obviously not decided by the thickness of the original nucleus, since it can change during crystallisation or on annealing, and also subsequent crystallisation on extended chain crystals can lead to folding; (b) a more exact kinetic argument is needed to show why a particular nucleus (in a polymer crystal) is not only large enough to lead to continued growth, but also stable enough to prevent any readjustment of the fold length as growth proceeds.

However it is worth discussing one of the more advanced nucleation theories because (a) it is the basis of the more sophisticated kinetic theories mentioned in the next section, and (b) the continued secondary nucleation of additional polymer chains could be a controlling mechanism. It may be noted that there are two forms of secondary nucleation in polymers: the first, as in all crystals, is crystallisation on a new face, after the previous face has been filled up (this necessity disappears when there is a screw dislocation); the second is the start of crystallisation of a polymer chain.

In general, a kinetic approach rests on the view that, whatever may be the ultimate position of lowest free energy, what actually happens is that the system changes in such a way that the energy decreases as rapidly as possible (just as a ball rolls down a line of steepest slope) and that this may well lead to a metastable state cut off from the true equilibrium state by an impassable barrier.

In order to obtain a model which is amenable to simple analysis, we consider the behaviour of a chain which is crystallising on an existing substrate as indicated in Fig. 7.51. Suppose it folds back. Will the folded length be stable against detachment, and thus give a nucleus for continued folded-chain crystallisation?

The change in free energy ΔF on formation of the fold would be:

$$\Delta F = F_{\mathrm{f}} - Al\Delta F_{\mathrm{c}} \tag{7.23}$$

where F_{f} is the free energy associated with the fold, l is the length of the fold, ΔF_{c} is the decrease in the free energy for crystallisation of unit volume, and A is the effective area of cross section of a chain.

Approximately we can put:†

$$\Delta F_{\mathrm{c}} = h(T_{\mathrm{m}} - T)/T_{\mathrm{m}} \tag{7.24}$$

where h is the heat of fusion, T_{m} is the melting point, and T is the temperature in the system.

This leads to:

$$\Delta F = F_{\mathrm{f}} - Alh(T_{\mathrm{m}} - T)/T_{\mathrm{m}} \tag{7.28}$$

If $\Delta F > 0$ (positive energy of formation) there would be a decrease of energy on detachment of the fold length, and so the addition would not be stable. But if $\Delta F \leqslant 0$, it would be stable against detachment and so would be a possible nucleus for further growth. Rearrangement of the above equation leads to an expression for l_{min}, the shortest stable fold, corresponding to $\Delta F = 0$:

$$l_{\mathrm{min}} = F_{\mathrm{f}} T_{\mathrm{m}} / Ah(T_{\mathrm{m}} - T) \tag{7.29}$$

† The general argument is as follows. At the melting point, melt and crystal are in equilibrium, and so there is no change in free energy on crystallisation.

$$\Delta F_{\mathrm{c,m}} = 0 = \Delta U - T_{\mathrm{m}} \Delta S \tag{7.25}$$

If at some other temperature, we assume that ΔU and ΔS remain unchanged, we have:

$$\Delta F_{\mathrm{c}} = \Delta U - T\Delta S \tag{7.26}$$

Subtracting (7.25):

$$\Delta F_{\mathrm{c}} = (T_{\mathrm{m}} - T)\Delta S = (T_{\mathrm{m}} - T)h/T_{\mathrm{m}} \tag{7.27}$$

since $h = T_{\mathrm{m}} \Delta S$.

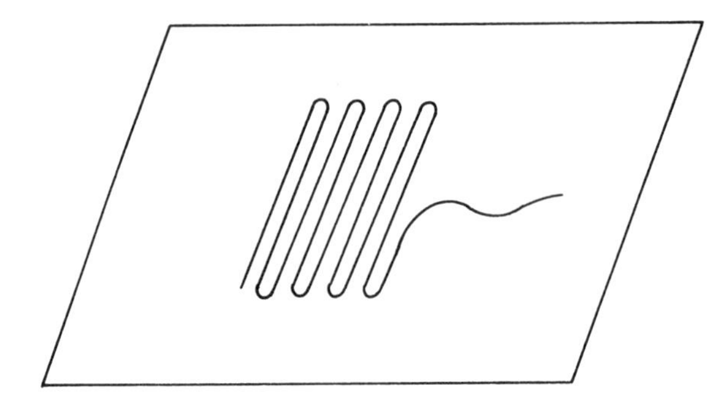

Fig. 7.51 – Model of chain crystallising on a substrate.

One point of immediate interest is the way in which the fold length l_{min} is dependent on temperature through the degree of supercooling $(T_m - T)$ as required by experiment; and, if we consider crystallisation from solution, dependent on the heat of solution, which replaces h, thus giving the right dependence on type of solvent. The difference from the equilibrium theory comes in because we are considering stability against detachment into the melt or the solution and not against recrystallisation in another form.

All that the calculation of l_{min} indicates is that folds longer than this will be stable for further growth. This is not to say that refolding to a longer length will not occur, without complete detachment into the melt, as growth proceeds.

The next stage in the argument is to consider what fold length would grow fastest. There are three things which can happen to a newly growing fold of the type shown in Fig. 7.51: it may detach itself; it may lead to continued growth at the same fold length, or it may form a new fold at another length. The probabilities of these effects are illustrated schematically in Fig. 7.52. The growth rate as an independent effect is independent of fold length, but the probability of detachment decreases and the probability of refolding increases as the fold length increases. We thus find a maximum probability at a certain fold length l^* greater than l_{min}; and this can be regarded as the type of nucleus which is likely to grow most rapidly and thus give the mean (or strictly the modal) crystal thickness.

A more detailed calculation which is too involved to present here, leads to an expression of the form:

$$l^* = [2F_f T_m / Ah(T_m - T)] + \delta l \tag{7.30}$$

Since the refolding is just a reflection of the situation discussed previously, it is reasonable that there should be a symmetry in the situation which leads to the first term equalling $2l_{min}$.

The second term δl derives from a more detailed consideration of the contributions of bulk and surface energy effects. It is usually small, but increases rapidly when there is high supercooling: the effect has not been observed experimentally.

The original theory by Lauritzen and Hoffman gave the expression:

$$l^* = [2F_f T_m / Ah(T_m - T)] + kT/A\delta_s \tag{7.31}$$

where δ_s is the surface free energy.

This expression was taken as indicating the thickness of the crystal nucleus which would grow most rapidly and set the pattern for further crystallisation.

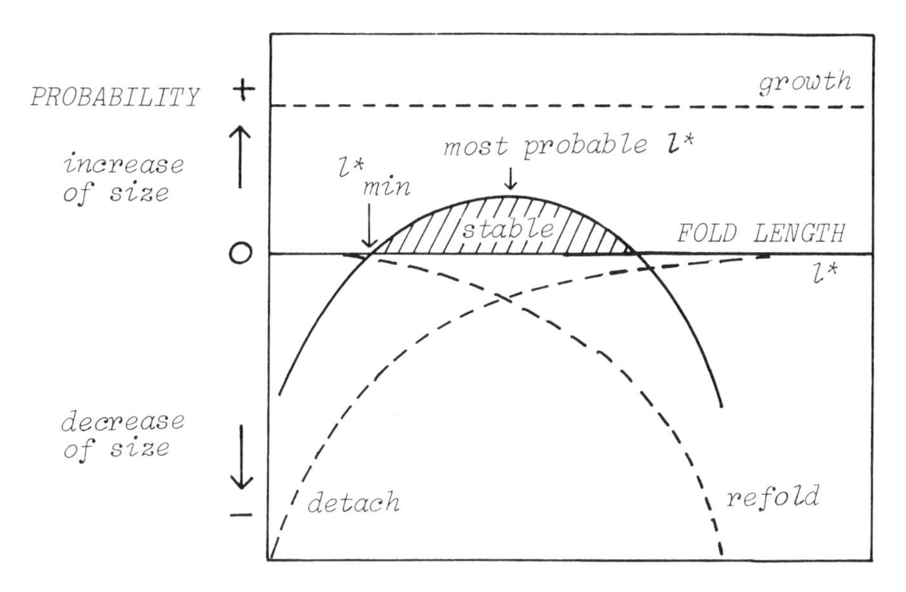

Fig. 7.52 – Relative probability of various events. Note that increased probability of detachment and refolding are shown negatively.

Viewed as a theory of secondary nucleation of each newly attached chain this treatment has its attractions; and, fitting values, which are not unreasonable, for the parameters, it gives good agreement with experimental results. In concept, its main defect is that it has to be based either (as in the simple argument given here on crystallisation on a substrate, or (in the more detailed treatments) on a crystal which is already growing at the required size! In the latter method – and this makes the approach inadequate as an explanation of changes of thickness – the increase of fold length would obviously be unfavourable since it leads to additional free surface, as shown in Fig. 7.53.

Nevertheless the basic ideas are relavant to the more complicated situation which exists in reality.

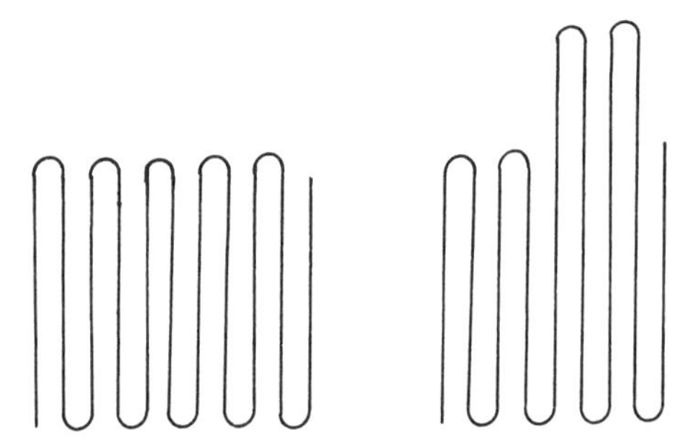

Fig. 7.53 – Unfavourable consequence of refolding to a longer length.

7.4.5 Kinetic theories

The kinetic theories have developed from the Lauritzen–Hoffman theory discussed in the last section. Mathematically they become even more complicated. Essentially, instead of considering just the initial nucleation (either primary or secondary), they consider the stability of each successive fold length. They do this allowing fluctuation in the fold length. This enables the crystal to change to a new fold length at any stage during its growth. It is equivalent to saying that fold lengths may take any value in the stable area of Fig. 7.52 although this should be modified to allow for the fact that projections beyond the length of the previous chain will be less favourable and will give rise to a sharper cut-off.

One difficulty with theories of this sort is that they are based on the idea of a fluctuating fold length, and will thus give rise to a rough surface on the crystal. This is contrary to experimental observation. A possible way out of this difficulty would be to invoke a two-stage process, with the initial formation of a rough surface, followed by a slower rearrangement to give a lower-energy regular folding.

On the substitution of given values of the parameters, the various detailed theories of the above types give rather widely different predictions. However the uncertainty of the values of the parameters is so great that they can all claim agreement with experiment on the basis of fitted values which are not unreasonable.

7.4.6 An overall view

The rather unsatisfactory nature of the above discussion, which indicates that a lot of complicated, but still inadequate, theories can all explain the experimental results, is due to the complexity of the real situation. In reality, we should consider all the possible movements which would occur in a volume of the growing crystal and its surroundings. This encompasses a vast number of polymer chain segments, and is clearly an impossible procedure. So far it has not proved possible to find a model which adequately represents reality and yet is simple enough to be formulated and analysed. The various approaches all have some validity, and the reality combines all of them and others. In this section, an attempt will be made to give a qualitative account of what may be happening in order to bring out some other features of the situation. In addition to what is included here, the other features, already mentioned, must also be taken into account.

An attempt at an integrated view of the situation would start from the assumption that the crystals will be driven towards their equilibrium form. But three difficulties arise: (i) the equilibrium form changes as the system grows: (ii) at some points the move towards lower energy states will be prevented by energy barriers which are too high to be surpassed; (iii) other changes, towards metastable equilibrium states, may lead to a more rapid immediate decrease of free energy and so be preferred.

Consider what might happen in a single very long molecule. Crystallisation would start when a small portion folded back and forth on itself to an extent sufficient to give a nucleus greater than the critical size. By definition, the lowering of internal energy would then only just have overcome the decrease of entropy. The crystal would thus be in a highly mobile state, since it would be on the verge of dissolving into disorder. In a low molecular weight substance, this would result in a rapid dynamic equilibrium between molecules entering and leaving the crystal: in the polymer chain it would mean that, in addition, the crystal was moving about on the chain and varying its pattern of folding (thus achieving an entropy greater than a crystal nucleus which was static on the chain).

With the nucleus over its critical size, crystallisation would proceed. But the equilbrium fold length would then increase; and since the crystal would still be mobile, this fold length would increase towards the new equilibrium valve. However as the chain length grew longer the crystal would become more stable, since more units would have to be moved over an intermediate unfavourable energy barrier in order to adjust the fold length. Eventually the crystal would reach a thickness where the adjustment of fold length could not keep pace with the rate of growth and a stable thickness would have been achieved.

Much the same argument would be valid for crystallisation on a substrate. Folding would start as an equilibrium form for a short length, but fold length would increase as the crystal grew until stability was achieved. These arguments are, in fact, similar to those used in the kinetic theories.

Annealing would be explained. On raising the temperature, the crystal would become more mobile, and the thickness would increase until a new condition of stability had been achieved. However, since the process would be controlled by jumps over an energy barrier it would be both time and temperature dependent, and thickening would continue indefinitely at a reducing rate since the ultimate equilibrium form would remain far away.

It must be noted that a rather large co-operative change is needed to lead to stability of a new fold length. Several lengths must change together in order to give a large enough region of the new length. An isolated projection would not be stable: a larger change would be necessary. It seems likely that this would occur only by simultaneous jumps of whole fold lengths.

7.4.7 The idea of a disordered fold surface

Some workers have suggested that the folds are not regular, but are disordered loops at the surface, as indicated, for example, in the switchboard model of Fig. 7.54 which was proposed by Flory. The weight of evidence, which has been described, is against this view as a general rule. However it is certainly possible that some irregularity may occur, and that the regularity of folding may not be perfect everywhere.

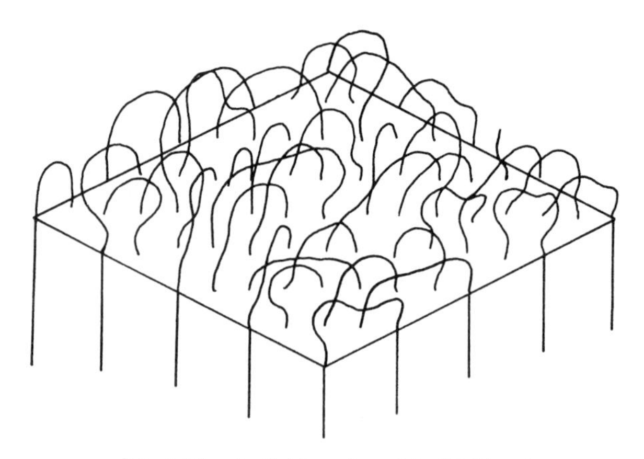

Fig. 7.54 – Switchboard model of fold surface.

In the same way it is possible that another chain may form a new secondary nucleus ahead of the main growth point, leading to a situation where two chain ends were left projecting. The two ends would then be likely to crystallise leading to a dendritic overgrowth. The probability of this happening will depend on the relative rates of growth and secondary nucleation, but will clearly become greater as the crystal grows larger.

7.5 BULK CRYSTALLISATION

7.5.1 The different situation

There are a number of important differences between crystallisation in the melt and crystallisation in dilute solution. In the former, the environment both of the open faces of the crystal and of any growing polymer chain consists of segments of other polymer molecules and not of solvent. In a very dilute solution, there will be time for a growing molecule to be incorporated fully into the crystal before another molceule arrives and nucleates; and crystals will be able to grow separately at a distance which remains remote from other crystals until all the polymer has crystallised. In the melt there will be mutual interference. Crystallisation from concentrated solution, which will only be mentioned incidentally, would be expected to show an intermediate pattern of behaviour.

The fact that each chain is tangled up with other chains in the melt means that it becomes more difficult to drag the whole length of the chain into a single crystal; and with long molecules, it is likely that different parts of the same chain will have joined on to different growing crystals: consequently a more complicated arrangement with less regular folding and with more links between crystals can be expected. The presence of other chains close to the crystal is likely to lead to more branching and confused growth as additional secondary nucleation occurs. And the growth of crystals together will lead to a particular larger-scale morphology.

Much of the complication which arises will be typical of all crystallisation in confused conditions, but it is worth pointing out two reasons why the polymeric nature of the material leads to special features.

The first is that while the unit cell of a polymer crystal and the unit cell of an ordinary crystal are in many ways comparable, there is the difference that any molecule of an ordinary crystallising substance can fit into any unit cell, but only a particular unit in the polymer system (the next unit along the chain) can fit correctly into the unit cell of a polymer crystal. This means that whereas additional growth points can be assimilated into an ordinary crystal without discontinuity of the crystal lattice this is not so in polymer crystals.

The second special feature is that, while ordinary crystallising molecules only interact with near neighbours, a polymer molecule has a long-range interaction along the chain. This gives rise to an additional free energy term in the crystallisation situation: apart from the decrease in free energy due to the near neighbour environment, there is also an increase in free energy due to the fact that the rest of the chain in the melt will be pulled into a more extended conformation. The extended conformation will not persist in the melt, because the slippage of the molecule past its neighbours will lead to a more disordered conformation with lower free energy. Nevertheless the increase in free energy is likely to last long enough to influence the pattern of crystallisation. The effect is really a manifestation of visco-elasticity in the crystallising material, and since crystallisation from the 'melt' often occurs when the material has cooled down so that it is in a quasi-solid, rubbery-flow state, not far above the glass transition temperature, it is likely that the development of rubbery-elastic free energy in chains due to the extension consequent upon crystallisation will be appreciable.

Other, more commonplace features that lead to complication in crystallisation will be thermal gradients, stress gradients, nucleation and growth rate phenomena, memory in the melt, presence of impurities or short polymer chains, influence of chain ends, and so on. In this complicated situation, it is surprising that it has been possible to recognise a form which is typical of most bulk crystallisation of polymers.

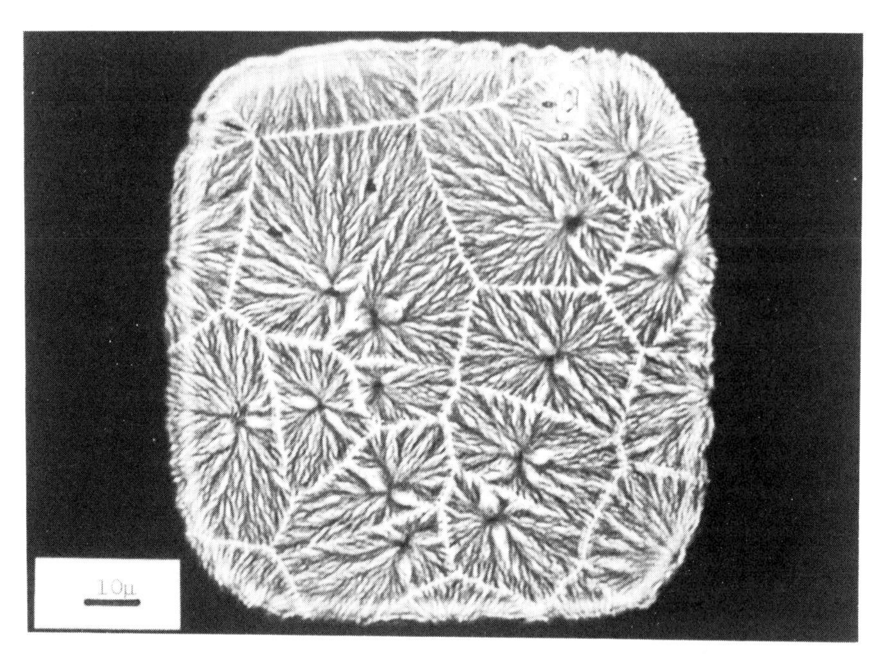

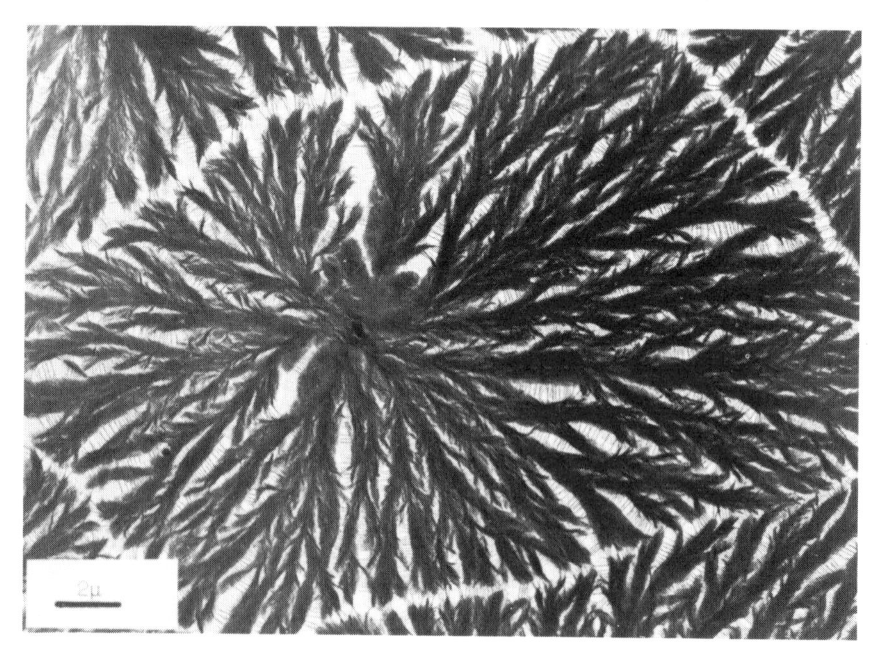

Fig. 7.55 – Spherulitic crystallisation of a polyethylene film. (a) Phase contrast optical microscopy. (b) Electron microscopy (Keith, Padden and Vadimsky, *J. Pol. Sci.*, A-2, **4**, 267, 1966).

7.5.2 Spherulitic crystallisation

The common from of crystallisation of polymers cooling in bulk or in thin films from the melt is shown in Fig. 7.55. The characteristic pattern is of crystals radiating out from a central nucleus. Crystalline growths of this type, which have been observed over the years in many crystallising substances, are known as *spherulites*.

The common features of growth can be mentioned briefly. If growth is arrested, perhaps by quenching before crystallisation is complete, the overall spherical form of growth can be seen as in Fig. 7.56(a) to (c). But if growth is allowed to proceed to completion, then boundaries form between the spherulites as seen in Fig. 7.56(d). If all the spherulites start growing at the same time and

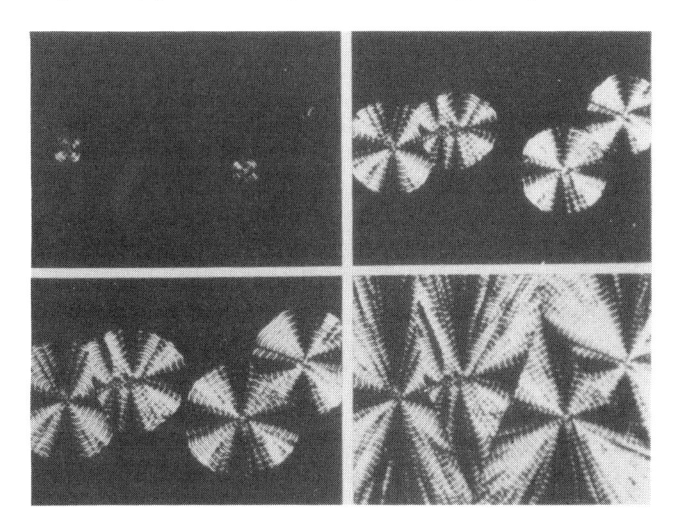

Fig. 7.56 – Spherulites at different stages of growth (Lindenmeyer, *J. Pol. Sci., C*, 3, 163, 1963).

continue to grow at the same rate, then, as shown in Fig. 7.57, the boundary will be the perpendicular bisector, or in three dimensions the bisecting plane, of the line joining the original nuclei: consequently, the material will be split up into zones which are irregular polygons, or polyhedrons in three dimensions. On the other hand, if the growth starts at different times, or is at different rates, then the boundaries will be hyperbolic.

Spherulitic growth may appear whenever crystallisation develops with repeated branching, as illustrated schematically in Fig. 7.58. Indeed an oak tree is a good example of spherulitic form, except that the bias against downward branching means that only half the spherulite develops. It is also necessary that the branching in crystal growth should be non-crystallographic, since regular crystallographic branching leads to preferred growth in particular directions and so to

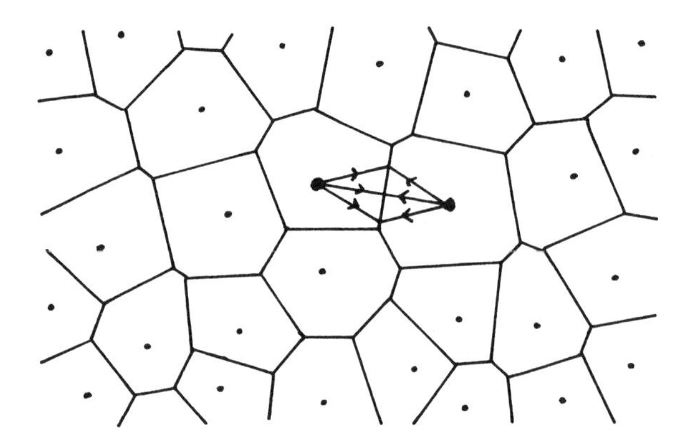

Fig. 7.57 – Boundaries between spherulites growing at same rate from same starting times.

regular forms, such as are found in ice-crystals or the dendritic polymer crystals shown in Fig. 7.31, and not to the uniform growth in all directions which is characteristic of spherulites. The fine structure of the spherulite will depend on the nature of the branching and on the form of the crystal between branches. In practice, most spherulites appear to be a radiating array of fibrous crystals branching at small angles, although, for convenience, Fig. 7.58 was drawn with rather larger angles. It is also usually found that one particular crystal axis lies parallel, or nearly parallel, to the radius of the spherulite thus showing that the crystal orientation is unrelated to the orientation in the original nucleus, but is

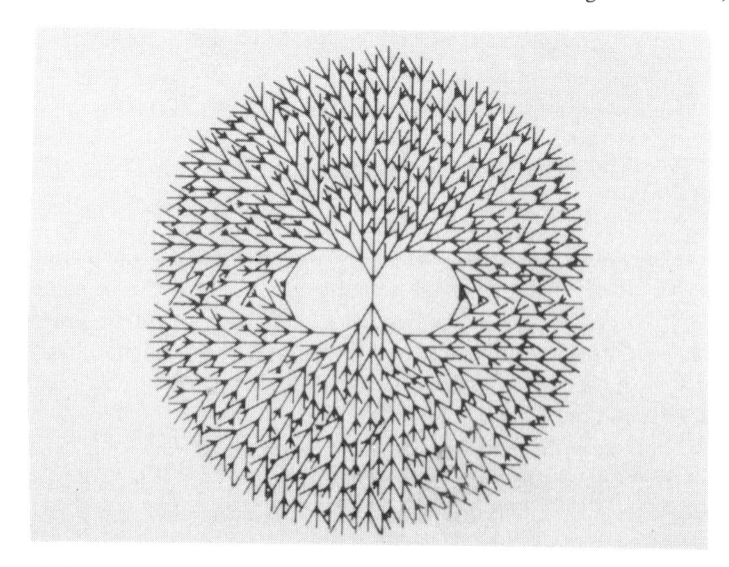

Fig. 7.58 – Schematic illustration of spherulitic growth by branching (Hearle, *J. Appl. Pol. Sci.*, 7, 1175, 1963).

determined by the form of overall growth of the polycrystalline aggregate. The typical pattern of spherulitic growth is thus of repeated splitting of crystals into separate parts at some arbitrary, but usually small angle.

The centre of the spherulite is the characteristic sheaf-like form which can be seen in Fig. 7.58 and in Fig. 7.55(b) and Fig. 7.59. This form is obviously reminiscent of some of the more complicated forms found in crystallisation from solution, illustrated in Fig. 7.32 and 7.33, when single crystals developed into splayed multilayer crystals or sheaf-like bundles. It is therefore possible to propose a probable sequence leading to spherulitic crystallisation in bulk. Initially, a folded chain, lamellar, single crystal would start to grow on a nucleus. As growth proceeded, secondary nucleation would lead to branching; and, in some manner, this would go beyond the two dimensional 'open-book' splaying of the solution-grown crystals into three-dimensional branching, so that growth was ultimately proceeding equally in all directions. As the surface of the sphere expanded, further branching would occur to fill in the additional space.

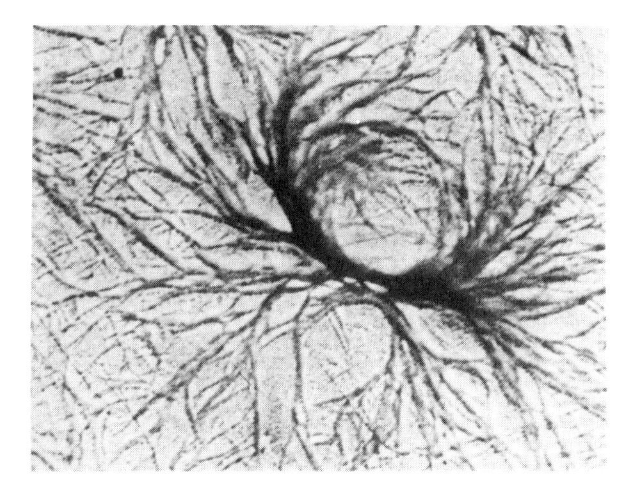

Fig. 7.59 – Sheaf-like form in early spherulitic crystallisation of nylon (Cooper, Keller and Waring, *J. Pol. Sci.*, **11**, 215, 1963).

7.5.3 The theory of spherulitic branching

In the common sequence of crystallisation, the original nucleus determines the crystallographic orientation, so that the whole growth from each nucleus is a single crystal, even though it is often complex in its external form. In contrast to this, spherulites are polycrystalline aggregates formed from a single nucleus. They are commonly formed in rather viscous systems in which an impurity has to be rejected: in addition to polymers, this includes certain minerals and some organic compounds crystallised from melts containing modifiers. Keith and

Padden have formulated a theory which is generally applicable to all these systems. They postulate, that, after secondary nucleation, the presence of impurity, which cannot diffuse away because of the viscosity of the system, may lead to a divergence of the two growths. This is illustrated schematically in Fig. 7.60, and is clearly a mechanism by which repeated small angle branching can occur. The operation of this branching all over a surface will keep the growth moving outwards radially, and, due to the preference for growth on a certain crystallographic face, keep each individual crystal aligned close to the radial direction.

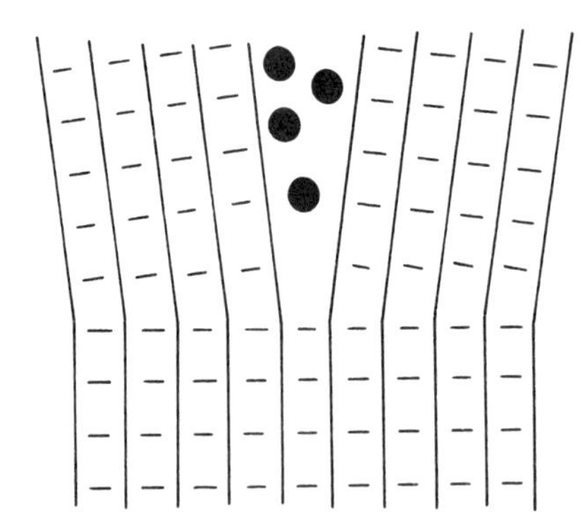

Fig. 7.60 – Entrapment of impurities leading to branching in crystal growth.

It is also possible to predict the circumstances in which branching will occur. The theory will be presented here in rather simple terms in two different ways. One approach considers a needle-like crystal growing into an impure melt as indicated in Fig. 7.61(a). As crystallisable material is removed ahead of the growing crystal, a zone which is rich in impurity will be formed. This zone will be smaller in size, the greater the crystal growth rate G cm/sec, since the faster the growth the more rapidly the point moves into an unaffected region. And the zone will increase in size as the diffusion coefficient D cm^2/sec increases, allowing the excess impurity molecules to diffuse over a wider area. Since the ratio $\delta = D/G$ has the dimensions of a length, it is reasonable to assume that the size of the zone is proportional to D/G, in the absence of any other controlling mechanism. A more rigorous analysis of the diffusion problem does, in fact, show that the excess concentration $(C_x - C_\infty)$ decreases as $\exp(-x/\delta)$ with distance x from the growing point. Thus δ represents the distance over which the excess impurity also serves to slow down the rate of growth, because, as a result of the depression of the melting point, the degree of supercooling is reduced.

If now instead of a growing point, we consider growth moving across a face of width W as indicated in Fig. 7.61(b) and (c), then if $W \ll \delta$ the environment all over the growing face will be effectively uniform and there will be no reason for it to break up into separate zones. But if $W \gg \delta$ then it is likely that a new growth in a region of low impurity content (and thus greater supercooling) will establish itself successfully and lead to branching. The excess impurity trapped between the two growing points will lead to the divergence of the branches at a small angle, as already discussed.

Obviously, a rigorous analysis of the whole situation would be very complicated, demanding a thorough examination of the geometry of the growing region, of the changes in concentration and temperature, of the stability of growth, and of the free energy differences which determine the dynamics of the whole process. The treatment by Keith and Padden is essentially a somewhat more detailed version of the analysis given above. It confirms that $\delta = D/G$ is the quantity which determines the maximum stable size of a growing unit, and that above this size branching will occur.

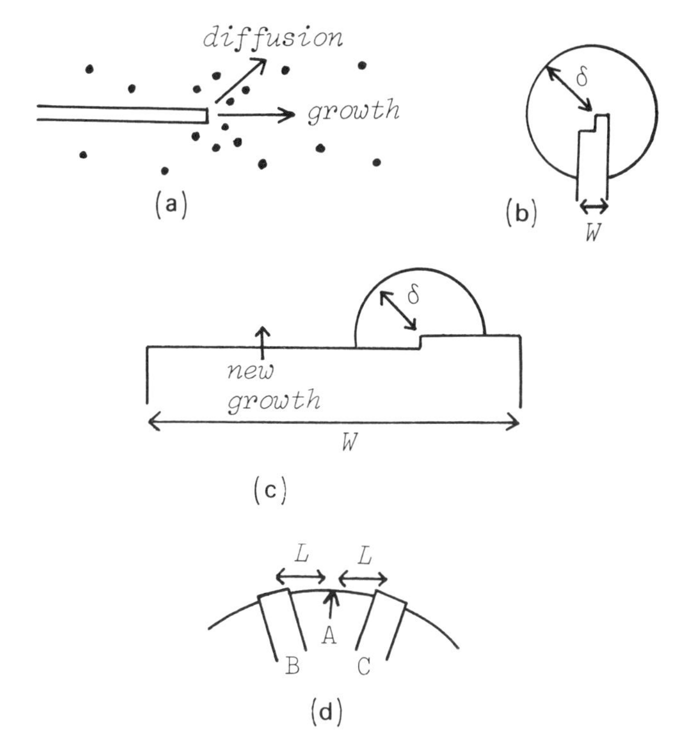

Fig. 7.61 – (a) Needle like crystal growing into an impure melt. (b) Growth at narrow face. (c) Growth at a wide face. (d) Growing surface of spherulite.

The alternative approach to the problem, which is complementary rather than conflicting, concentrates attention on the crystallising material rather than the impurity, but since what happens is the inter-diffusion of two substances it does not really matter which diffusion coefficient one adopts. Consider the growing surface of a spherulite, as in Fig. 7.61(d), with fibrillar crystals originating from its surface. Suppose a new nucleus appears at A, at a distance L from existing fibrils at B and C. We ignore the awkward geometrical question as to how this can arise as a result of secondary nucleation from B or C, and merely concentrate on the stability problem. If the nucleus is to grow it must collect most of the crystallisable material in the zone ahead of A; otherwise it will break up faster than it grows and will be unstable. The rate of advance into this zone will be G cm/sec, but the rate of diffusion of crystallisable molecules away towards the established, larger preferred crystals at B and C will be proportional to D/L: the length L enters the expression because the greater the value of L the smaller will be the diffusion gradient. The growth of the nucleus is thus likely to be stable only if G is large enough in relation to D/L, or conversely if L is large enough in relation to $\delta = D/G$. Thus once again we see that a quantity related to D/G, and likely to be of the same order of magnitude, appears as the dimension which characteristises the distance between the radiating crystals at which further branching is likely to be stable.

The rather awkward geometrical aspects of the problem, avoided in the above arguments, are perhaps best dealt with by postulating that, due to fluctuations, incipient branches are perpetually being established by secondary nucleation, but that they can only become the basis of a new continuing growth when they are able to move into a region which is farther than $\delta = D/G$ from any existing growth.

It must also be noted that there are two characteristic dimensions in spherulite growth: the width of the crystals and the separation between them, and it is not really clear which one is most likely to equal δ. If there is very little impurity, the spacing and thickness will be identical, but where there is a lot of impurity the spacing will be greater. All that one can say is that one would expect both spacing and width to be proportional to D/G, and of the same order of magnitude. Keith and Padden use the terms **compact** and **open** to describe the spacing between fibrils and **coarse** and **fine** to describe the thickness of the fibrils.

There is a good deal of experimental evidence to support the above theoretical account of what happens. While values of G are easily available, there is a lack of data on values of D. However values of δ between 0.2 and 0.4 μm predicted for polyethylene are in reasonable agreement with estimates that the width of crystalline ribbons obtained in electron microscope studies of replicas lie between 0.5 and 1 μm. The thickness of the ribbons will depend on the fold period.

The openness of spherulites will be influenced by the amount of impurity and if, for example, isotactic polypropylene is crystallised in the presence of atactic polypropylene, then the greater the proportion of impurity the greater the openness of the spherulites. Even with 100% crystallisable polypropylene, spherulites still develop, but this has been interpreted as due to the fact that short polymer chains are rejected and act as impurity. There is experimental evidence in support of the view that low molecular weight polymer is rejected.

However, it should be pointed out that there is no evidence that a homogeneous polymer fraction would not crystallise in spherulitic form, and it may be that extraneous impurity is unnecessary. In effect, once a given polymer chain has started to crystallise, it is different from all its neighbours and so the remainder inevitably act as impurities.

7.5.4 Form of the radiating units in polyethylene spherulites

The last two sections were a digression applicable to all spherulitic crystallisation. In this section, we return to a consideration of particular aspects of polymer spherulites.

The symmetrical arrangement of the crystallographic axis of the polymer crystal about the centre of the spherulite is shown by the characteristic Maltese cross, due to birefringence effects, observed when spherulites are viewed in polarised light, as in Fig. 7.56.

Almost invariably, the polymer chain axis is perpendicular to the radius of the spherulite, and thus also perpendicular to the 'fibrils'. This was unexpected, at first, since fibrils are usually regarded as units with polymer molecules aligned along their length. However, it is what would be expected, if the spherulites grow by chain folding on the surface of the 'fibril' tips, as illustrated in Fig. 7.62. Closer examination of polyethylene spherulites reveals that the 'fibrils' are in fact ribbon-like units, and are thus related to lamellar single crystals.

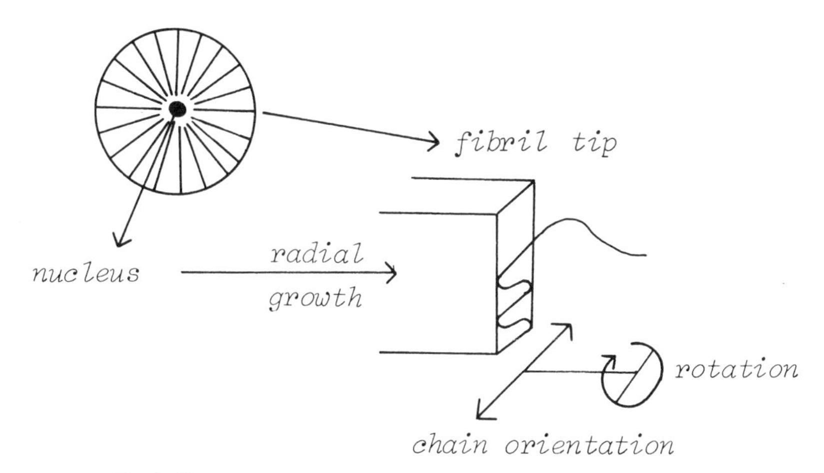

Fig. 7.62 – Crystallisation at tip of a growing fibre in spherulite.

Very often, spherulites show a banded structure in polarised light, as in Fig. 7.56, and this can be interpreted as due to a rotation of the ribbon, which thus rotates the direction of the polymer chain axis of the crystal in the plane perpendicular to the radius as indicated in Fig. 7.62. In other words the ribbon twists as it grows, and this can be regarded as related to the twisting which was

Fig. 7.63 – (a) Lamellar crystals in a fractured polyethylene sample with spherulitic crystallisation. (b) Interlamellar link between lamellae in a polyethylene sphereulite (Keith, Padden and Vadimsky, *J. Pol. Sci.*, A-2, **4**, 267, 1966).

observed in single crystals, though the geometry is not quite the same and needs closer study. Direct evidence of twisting can be seen in the surfaces of polyethylene spherulites, since the lamellar ribbons are seen in bands alternatively edge-on and flat. The rotation is more or less in phase throughout the spherulite.

It should be noted that the structure in any given small region is a lamellar structure, as shown in Fig. 7.63(a). The extent to which the fold surfaces of the lamellae are irregular, the extent to which chains emerge from the lamellae and interconnect them, and the extent to which there is disordered chain packing between lamellae is still speculative. Furthermore as we have seen the lamellae are really parts of long ribbon-like crystals growing out from the nucleus of the spherulite. The orientation of the lamellae thus vary with radial position in the spherulite and, where there is twisting, with the distance from the nucleus.

Between the lamellae in spherulites, there may be fibrous tie crystals, as shown in Fig. 7.63(b).

7.5.5 Variations of spherulitic crystallisation

Although the occurrence of spherulitic crystallisation is widespread in crystallisation of polymers from the melt, there are many important variations of the basic theme, dependent on the nature of the polymer and the conditions of crystallisation.

Whereas in many polymers besides polyethylene the spherulites consist of densely packed ribbons, there are others where the nature of the fibrils is not well understood but may be more complicated. Sometimes the branching is more profuse. If the fibrils are less ribbon-like, they need not all lie locally parallel and twist in unison: as a result, the only pattern in polarised light is the simple cross coming from the orientation of the chain axis perpendicular to the radius. Some thread-like fibrils may be rolled up or helically coiled lamellae. There have also been some reports of spherulites in which the orientation of the polymer chains is along and not perpendicular to the spherulite radius.

An obvious source of variation in spherulitic crystallisation is in the form of nucleation. The fewer the nuclei, the larger will be the spherulites. If the temperature changes during crystallisation, then different spherulitic forms of growth may be superimposed. Thus if crystallisation starts a high temperature on a few nuclei, and then continues at a low temperature on many nuclei we may find a few large spherulites embedded in a mass of small spherulites.

In the limit as the number of nuclei becomes very large, only the central portions of spherulites will have an opportunity to develop, before they run into one another, and the resulting structure will be a mass of small crystals separated by disordered regions: this is the same as the fringed micelle structures discussed in the next chapter.

Where the nuclei are arranged in some special way, other forms may appear. Thus if for some reason, there is a row of nuclei, then the spherulites will soon join up side-by-side, but can grow on farther at right angles to the row to give disc-like 'spherulites'. On the other hand, if nuclei are in a sheet, for example on the surface of material, then spherulitic growth will occur in cylinders at right angles to the sheet. This may give rise to a different structure on the skin of a material, and can be particularly important in fibres.

Spherulites will always be terminated at the edges of the material. In films, growth will be limited so that 'spherulites' will grow in only two dimensions.

If there is a temperature or concentration gradient, then spherulites may grow faster in one direction than another and become asymmetric. This can reach an extreme in a material which is cooling in such a way that an advancing edge of a spherulite can move through the material at a given contour of temperature and grow indefinitely without fresh nucleation. The growth then consists of sectors of giant spherulites.

Any particular crystallising system must be carefully studied to determine the precise form which it takes up.

7.5.6 On nucleation and growth rates

The kinetics of crystallisation have been rather extensively studied. Since the crystal is normally denser than the non-crystalline material, the increase of crystallisation is usually measured by changes in density, though other methods can also be used. A typical growth curve is shown in Fig. 7.64. The early part of crystallisation when the units are growing independently has frequently been interpreted in terms of a theory developed by Avrami, which indicated that the fraction crystallised should increase as t^n where n is an exponent which depends on the nature of nucleation and the number of directions in which crystallisation is proceeding. Values of n are given in Table 7.2. However later studies have cast a good deal of doubt on the applicability of the theory and it will not be discussed further here.

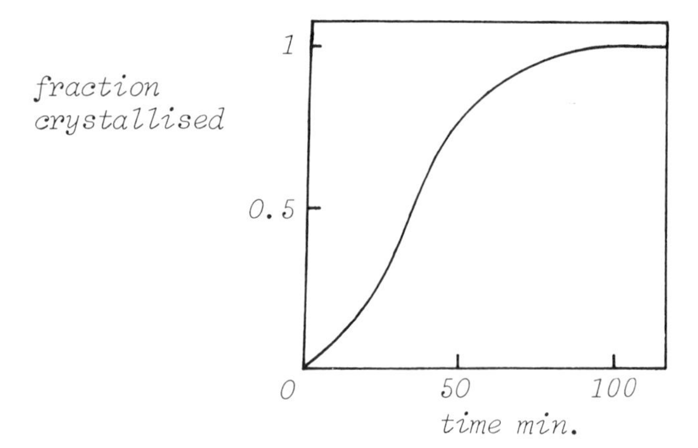

Fig. 7.64 – Typical crystallisation growth curve – for polyethylene terephthalate at 110°C. After Keller, Lester and Morgan (1954).

Table 7.2 – Values of Avrami index, n.

Growth	Nuclei	
	Formed $\propto t$	Available at $t = 0$
Spherulitic	4	3
Disc	3	2
Rod	2	1

The general view is that nucleation is most often heterogeneous so that all nuclei will be present at the start of crystallisation. This does not however mean that the number of nuclei is independent of temperature. Just below the melting point only very large nuclei will be effective, but at lower temperatures smaller ones will also be effective: the number of nuclei thus increases with decrease of temperature, as shown in Fig. 7.65(a). The heterogeneous nuclei are usually foreign particles of some sort or another.

An interesting example of heterogeneous crystallisation occurs when a polymer which has already crystallised is melted by heating slightly above its melting point, and the cooled again. Crystallisation occurs on many nuclei which are clearly remnants of crystalline order left in the melt. A particularly striking example, which confirms this, occurs with films in which only limited spherulitic growth has occurred. If these are carefully melted and cooled, the rapid recrystallisation occurs only within the regions of the previously existing spherulites.

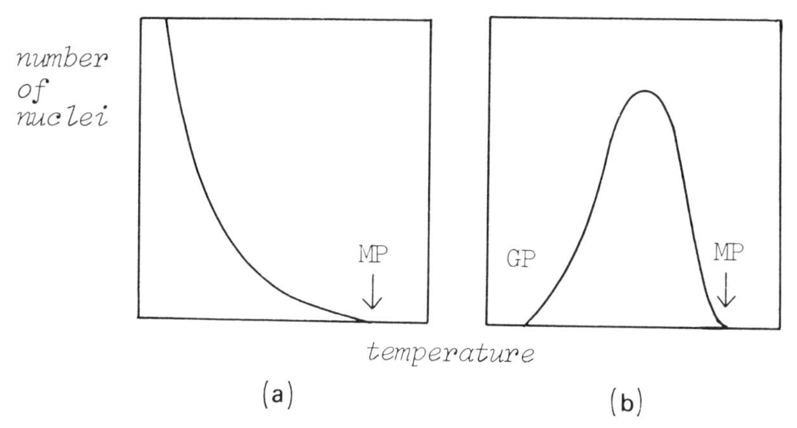

Fig. 7.65 – Nucleation density. (a) Heterogenerous (pre-existing) nuclei. (b) homogeneous (sporadically forming) nuclei. MP is melting-point; GP is glass transition.

Although heterogeneous nucleation is common, there are circumstances in which primary nucleation will occur sporadically as a result of the chance formation of a large enough ordered region. The variation of sporadic primary nucleation rate with temperature is shown in Fig. 7.65(b). Just below the melting point the tendency for nuclei to break up is nearly as great as the tendency to grow, so that the nucleation rate is small (above the melting point the rate of break up, of course, exceeds the rate of growth). However as the temperature falls the balance shifts in favour of growth and so the nucleation

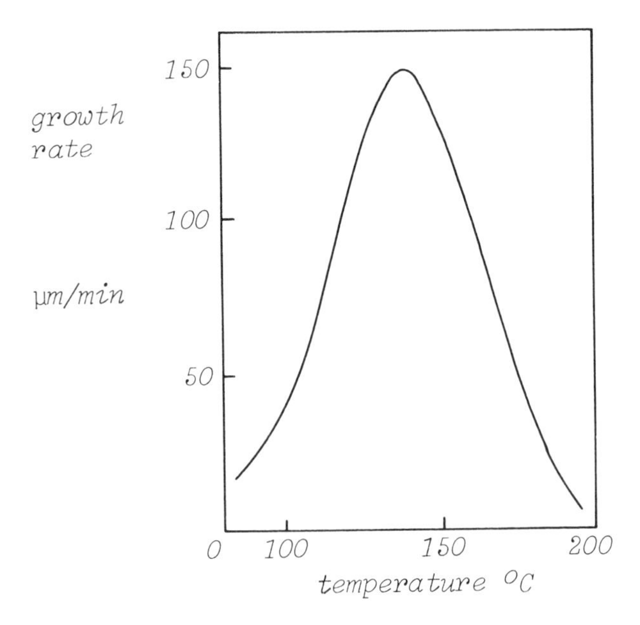

Fig. 7.66 – Growth rate of nylon 6 spherulites at different temperatures. After Magill (1962).

rate rises. Eventually, as the thermal fluctuations become less strong, the whole process slows down and the nucleation rate falls again, reaching zero at the glass transition temperature when all mobility is frozen out of the system.

Secondary nucleation is inevitably sporadic, and determined by random fluctuations. It will therefore follow the form of Fig. 7.65(b). Since our discussion of the nature of polymer crystallisation has shown that it is controlled in rate by the nucleation of new folds or fresh growing chains, it follows that growth rates must vary with temperature in the same way as homogeneous nucleation. Fig. 7.66 shows a typical plot of growth rate with temperature.

It is also possible to suggest the form of Fig. 7.66 purely in terms of general growth mechanisms. At high temperatures, just below the melting point, there is only a small driving force for crystallisation, but this increases with supercooling. At low temperatures, movement is sluggish and all changes are slowed down.

Numerical values of nucleation and growth rates depend considerably on the nature of the polymers. With some polymers such as polyethylene and nylon crystallisation is rapid, and it is almost impossible to prevent it occurring. With others, such as polyethylene terephthalate, it is possible to quench rapidly and obtain an amorphous solid: crystallisation occurs only when the material is held at a suitable temperature for some time.

7.5.7 Secondary crystallisation

The preceding discussion has been concerned with the development of the main spherulitic morphology in crystallisation. However, the initial crystallisation may well leave segments of polymer chains trapped between the crystallites, but capable of readjustment to give further crystallisation. A wave of secondary crystallisation may thus follow the initial crystallisation.

7.5.8 Extended chain crystallisation

Chain folds are an unfavourable form with increased volume. If a high pressure is imposed they will be even less favoured and extended chain crystals are formed. This is, in fact, the situation, and Fig. 7.67 shows extended chain crystals formed by crystallisation of polyethylene under high pressure.

Fig. 7.67 – Fracture of polyethylene crystallised under high pressure, showing extended chain lamellae, about 1μm thick (Lindenmeyer, *S. P. E. Trans.*, 4, 1, 1964).

7.5.9 Stress induced crystallisation

Some polymers which will not crystallise in an unoriented state will crystallise when the material is stretched. Natural rubber is a good example. Although it is normally non-crystalline, it will crystallise readily at around 0°C if it is stretched. The extended form is then stabilised, but, on warming, the crystals melt and the rubber springs back to its original length. Polyethylene terephthalate is another example: on rapid quenching an amorphous material is found, but on drawing crystallisation occurs.

What must happen in these materials is that the free energy barrier between disoriented amorphous material and the crystalline state is too high to be easily surmounted. But when molecules are pulled locally into alignment, then the chains fall into the crystalline lattice packing. This is a particular form of nucleation. If the overall alignment is high before crystallisation occurs, then thread-like crystals with rather extended chains are likely to be formed. However these may act as nuclei for further crystallisation in shish-kebab form.

Where less orientation has to occur before crystallisation, then separate nuclei rather than fibrous units may develop.

Even with polymers which will crystallise anyway, the presence of regions of deformation due to stress is likely to induce preferential crystallisation and to modify the forms which result.

7.5.10 Crystallisation on polymer formation

Another circumstance which leads to different forms is crystallisation as the polymer molecule itself forms. The controlled supply and the absence of interference from other polymer molecules leads to special types of structure. There are various ways in which this condition can arise. It happens frequently during the development of natural polymers in living cells. It can happen during synthetic polymerisation on a surface or in the solid state from a monomer crystal. It can also occur when a chemical derivative of a polymer is converted back to a form which will crystallise: this happens in the regeneration of some cellulose fibres.

7.6 MELTING OF POLYMER CRYSTALS

In ordinary chemical compounds consisting of small molecules, and especially for the diversity of organic substances where it is used as a means of determining identity and purity, the melting-point is a sharply defined temperature. This is not so for polymer crystals. Due to their small size, as well as to internal strains and defects, the melting-points, even of materials crystallised carefully in the laboratory, cover a range of temperatures. The subject will be taken up again in relation to bulk samples in the next chapter, but it is appropriate to include one set of results for well-characterised samples, including a mat of single crystals.

Fig. 7.68 shows how the melting-point of polyethylene increases as the fold length increases. Although the highest observed values are less than 140°C, and these are for the long fold period of 50 nm, the results suggest an extrapolation to 145°C for large crystals.

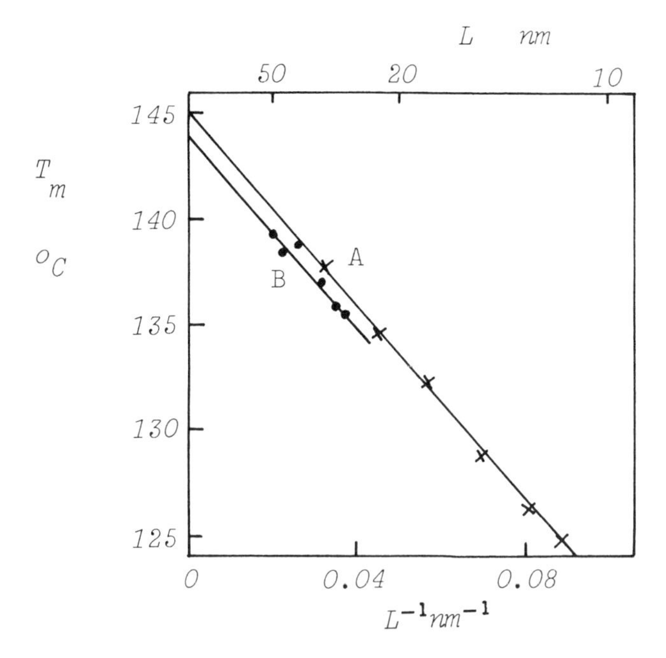

Fig. 7.68 – Variation of crystal melting-point with lamellar thickness, L for polyethylene: (A) dried single crystals; (B) bulk polymer. After Hoffman *et al.* (1969).

CHAPTER 8

Morphology and Thermomechanical Responses of Partially Crystalline Plastics and Fibres

Order! Order! – *Mr. Speaker, in the House of Commons.*

8.1 CHARACTERISATION OF REAL MATERIALS

8.1.1 The contrast between perfection and reality

There is a sense in which the study of the forms of polymer crystallisation in the laboratory, once it has demonstrated the important fundamental features of behaviour, may become merely an exercise in the art of crystal growing.

It is only necessary for the natural historian to look around in winter, with no more than a magnifying glass or a simple microscope, to observe the variety of forms of crystallisation of water: ice and snow can occur in a great diversity of morphologies, some small-scale and some larger, depending on the precise conditions of crystallisation. A similar diversity is inevitable in polymer crystallisation, but the subject is made more difficult, more mysterious, and more attractive to scientists for two reasons. Firstly, the scale of the morphology is much smaller, around 10 nm, so that it cannot be simply observed, but requires study by sophisticated scientific instruments and sophisticated interpretation. Secondly, the length of polymer molecules is greater than the scale of the fine structure, and separate parts of the same molecule get trapped in different crystalline regions, while other intervening parts remain in substantial regions of disorder which cannot be removed. Small molecules of simple shape can move independently, so that every molecule can slot into crystalline register with the neighbours at least on one side: disorder therefore exists only in the lack of register of atoms on either side of a boundary surface, or at a line or point defect. But polymer segments cannot move without disturbing neighbouring parts of the chain; and, therefore, once a system has reached a certain level of crystallisation, individual segments can only move into crystalline register by pulling other segments out of register into disorder. Substantial amounts of disorder are always present; it is common for 'crystalline' polymers to be no more than 50% crystalline; and so the characterisation of structure involves more than a description of crystalline shapes and sizes.

Because so much diversity of form is inevitable, and so dependent on the detailed circumstances of crystallisation that every case is a special case, there is a danger that laboratory studies might move from simple experiments which demonstrate some general principle into a game of complex artefacts, fascinating but unimportant.

Nevertheless there are difficult and complex systems which must be studied. Many natural and industrial materials are **partially crystalline polymers**, and, in order to understand their function or optimise their production and performance, we need to understand their structure and behaviour. These materials will have formed in a variety of ways: through biosynthesis and deposition in living cells; though solid-state polymerisation; through crystallisation from the melt after extrusion or moulding; through precipitation from ordinary solution after casting or extrusion; through an intervening stage of liquid crystals. And their structure may have been modified subsequently by thermal and mechanical treatments. In principle, structure could be determined either by prediction from their history or by deduction from observations on the final form. In practice, both approaches interact together, within the realms of science and technology, in attempts to characterise the important features of important materials. But these materials are not simple, and it is folly to expect too much by way of detailed structural description: skill and art lie in finding out what is possible and useful, and in finding ways of expressing it without obscuring knowledge by semantic controversy.

8.1.2 Degree of crystallinity

In some materials **degree of crystallinity** can be defined unambiguously. Consider, for example, a simple organic substance in which relatively large perfect crystals are forming from the melt. At any instant, the crystals could be removed and weighed, or they could be observed and their volume estimated, or the total evolution of latent heat could be measured, or the volume change could be observed, or the surface area of the crystals could be found by adsorption, or the X-ray diffraction pattern could be resolved into sharp reflections and diffuse scattering, or the infra-red absorption of liquid and solid could be separated, or any of a variety of differing responses of crystal and liquid could be monitored. All these methods would give, with minor differences, the same value for the degree of crystallinity. If, at some instant, the material was suddenly quenched so that the residual liquid turned to glass, there would be little more difficulty in estimating the degree of crystallinity of the solid.

But because the scale of the structure is close to molecular dimensions, the situation in polymers is less clear. Differing techniques which respond to different attributes of ordered and disordered material may yield widely different estimates of 'degree of crystallinity'. However, any given technique will enable a comparison to be made between different materials, and usually there is a correlation among the orders of ranking given by different methods. Values range from arbitrarily low values for a block copolymer containing only a small fraction of crystallisable material, through values of about 33% for regenerated cellulose precipitated from solution, 50% for nylon crystallised from the melt, 66% for natural cellulose fibres, to 80 or 90% for a carefully crystallised linear polyethylene or a fully aromatic polyamide (like the fibre Kevlar) formed from a solution with liquid crystals. The degree of crystallinity is thus seen to depend partly on how easy it is for particular polymer molecules to fit into crystalline register and partly on the nature of the crystallisation process.

It is also important to note that a certain degree of crystallinity seems to be natural for any particular system. The crystallisation of nylon from the melt is a good example. It is possible, by very rapid quenching of perfectly dry nylon, to get a glassy solid with zero crystallinity, but, once crystallisation occurs, it is difficult to stop at a degree of crystallinity less than about 45%, as measured by the density, and difficult to increase the value above about 55% by changing the process of crystallisation or by subsequent annealing.

8.1.3 The influence of molecular form

It will be useful to examine again the features of the constitution of polymer molecules which influence the ease and stability of crystallisation.

Regularity

Save in the exceptional circumstances where another unit is so similar that it can substitute in a crystal lattice, regularlity is essential for crystallisation. Any irregularity must either be accommodated as localised defects within a crystal or be segregated into disordered regions. The regularity has to be in atomic constitution of repeat units, and in the geometry of stereochemical form.

Shape and interaction

The crystallisation process is driven by the force of attraction between neighbouring chains or, in other words, by the reduction in internal energy which occurs on crystalline packing. This will be most effective when the shape of the molecules is such that they can fit closely together, with all the atoms in favourable positions, minimising internal energy, and when the nature of the chemical groups is such that they interact strongly with relatively high bond energies. Hydrogen bonding groups and strongly polar groups will be more effective than $-CH_2-$ or $-CF_2-$, but this may be countered by a simplicity of shape which enables inherently weak van der Waal's bonds to have a large cumulative effect.

Paradoxically, it must be noted that if the interactions between chains are too strong they may inhibit crystallisation by holding the disordered material together too firmly.

Repeat length

The repeat length does not obviously influence the internal energy of the crystal, but it does influence the ease of crystallisation, and through its effect on the entropy of the non-crystalline material, may affect the melting-point. If, in a simple chain such as polyethylene, a particular location for a chain segment in the crystal lattice is not favoured, then the chain can shift by one main chain atom, $-CH_2-$, or, if the zig-zag geometry is constrained, by two main chain atoms, CH_2CH_2, in order to find a more favoured position. There are thus many possible positions, and crystal size can vary in steps of 1.25 Å, which, for crystals of the order of 100 Å, can be regarded as continuously variable. But the situation is very different with a polymer like nylon 66. The repeat unit is $-CO.CH_2.CH_2.CH_2.CH_2.CO.NH.CH_2.CH_2.CH_2.CH_2.CH_2.CH_2.NH-$, with a length of 17 Å. Several consequences follow from this. If the chain does not fit in one place in the lattice, then the next available position involves a shift by 14 main chain atoms. The available crystal sizes are strongly quantised at 17, 34, 51, 68, 85, 102, 119 Å etc. An extra repeat unit will contain 38 atoms and so could not be accommodated as a localised defect within a crystal: it would inevitably give a substantial region of disorder. On the other hand the $-CH_2-$ repeat of polyethylene could be regarded as packing in crystal defects. As illustrated in Fig. 8.1, the molecule of polyethylene can reasonably be represented by a simple string of beads, or even by a continuous line, but this is inadequate for the other polymers illustrated in Fig. 8.1, which are used in high quality fibres and plastics. As we examine the mechanical and other properties of these materials, we shall see that it is performance, and not accident or perversity, which leads to the choice of the more complicated polymers, despite their higher cost.

Chain directionality

A similar, but less important, effect is that some polymer chains have direction, while others are symmetrical about the mid-point of a repeat unit. Thus, as shown in Fig. 8.1, nylon 6 differs from nylon 66 only in that a -CO- and a $-CH_2-$ group have changed places; but, in addition to halving the repeat length, this also means that the nylon 6 molecule has direction. As we go along the chain, from left to right in Fig. 8.1, we find a sequence of -NH.CO- continuing, but in nylon 66 -NH.CO- alternates with -CO.NH- to give symmetry. When there is direction, the chain molecules have to pack in a preferred way, either antiparallel ↑↓↑↓ or parallel ↑↑↑↑, and in the disordered regions the choice of

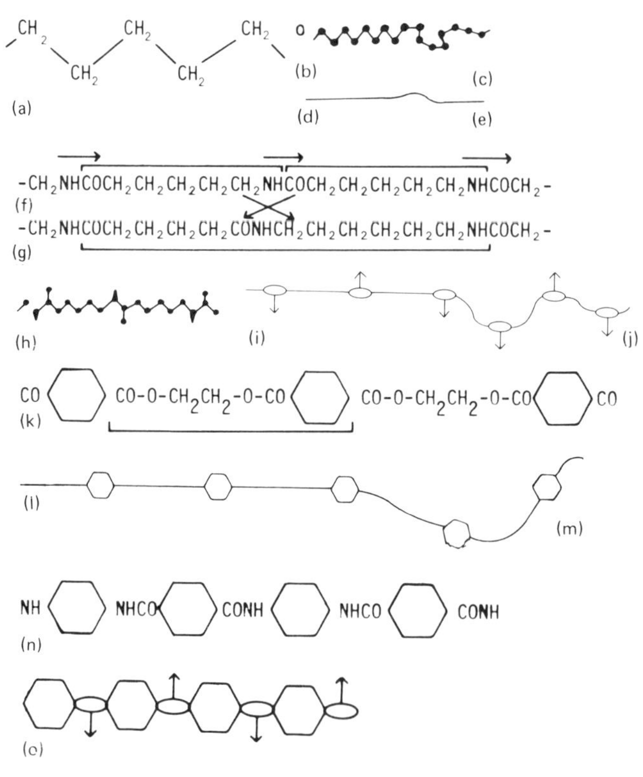

Fig. 8.1 – Polyethylene (a) is reasonably represented by a simple chain in regular crystalline form (b) or irregular (c), or even by a line (d, e). Nylon 6 (f) and nylon 66 (g) have longer repeats, and ought to be represented by more complicated models (h, i, j). Note shorter repeat and directionality of nylon 6, and slight rearrangement to give nylon 66. Polyethylene terephthalate (k) and the fully aromatic polyamide (n) also require complicated representation (l, m, o).

chain direction will contribute to the entropy. But in a symmetrical molecule it does not matter: ↕ is the same as ↕ (and the printer does not know which way up the author intended them to be!).

Conformation of single chains

There will be a preferred minimum energy conformation for an isolated segment of a single chain. It does not necessarily follow that this is the minimum energy for a collection of chains in the crystal lattice. However crystallisation will be helped if the chain segments which are adding on are in the required form or close to it. Thus compatibility between the preferred conformation of the single chain and the form in the crystal will favour crystallisation.

Chain stiffness

Paradoxically, crystallisation is easier when the degree of chain stiffness is either small or large. With very flexible chains, there is little resistance to the rearrangement of the chain from a disordered conformation to an ordered one. On the other hand, very stiff chains, provided there is a single dominant minimum energy form giving a regular structure of suitable shape, will occur in solution as stiff rods which then easily pack together in a crystal lattice. The difficulty comes with intermediate stiffness, which is inadequate to prevent the formation of an irregular chain but sufficient to oppose rearrangement.

8.1.4 Structural models

The classical approach to the problem of how to describe the structure of partially crystalline polymers is to draw a picture of an arrangement of chains and to add some verbal explanation. Provided they are treated with caution, such models are relevant and useful. But their inadequacies must be stressed. Essentially, they are imperfect representations of a model imperfectly imagined in the mind of the originator; and they are not firmly stated like mathematical theories, but are subject to variation in different representations by different authors. Hosemann has drawn an interesting example, reproduced as Fig. 8.2, in order to illustrate

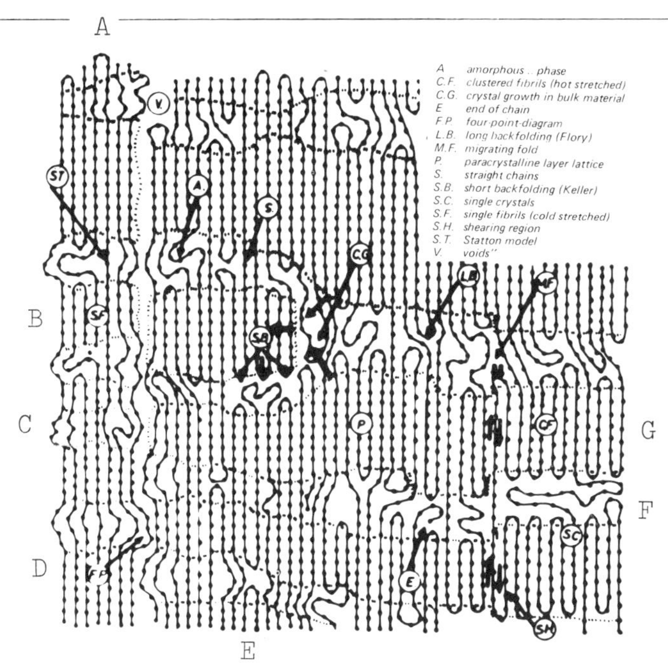

Fig. 8.2 – Schematic composition diagram prepared by Hosemann (1962) as a 'model of the different types of macro-lattices in stretched linear polyethylene' with additional identification of various forms of model. (A) pseudo-fibril; (B) modified fringed micelle, (C) amorphous with correlation; (D) fringed micelle; (E) crystal defects; (F) tie molecule; (G) fringed lamellar (Hosemann, *Polymer*, 3, 349, 1962).

the diversity of forms which have been suggested. There are also some specific limitations. First, the polymer molecules are usually represented by single lines with no attention to the volume, shape or stiffness of the chain, to the repeat length, or to other features indicated in Fig. 8.1. Secondly, the pictures are two-dimensional analogues of a three-dimensional structure. Apart from the general inadequacy, any strict two-dimensional system is limited in that it is not possible for chains to cross one another, and so it is always possible to find lines of split between chains. This limitation is clearly obvious in Fig. 8.2 but often it is partly avoided by drawing lines that do cross to give a quasi-three-dimensional form. Thirdly, the way in which the drawing is built up by successive lines to give a single static picture is not the same as allowing a collection of chains to move into a partially ordered structure. Fourthly there is inevitable artistic bias. Generally, polymer scientists do not have the artistic ability or patience of the old biological illustrators, so they do not attempt to make very detailed drawings and they have more freedom since they are drawing an imagined and not an observed form. The disordered regions are usually left open in order to help clarity, too few chains are included, and various forms may result from the carelessness or over-tidiness of the artist; only small areas are represented, and there may be more diversity over larger regions. Fifthly, we must remember that different materials – different polymers or different histories – will have different structures, and we are not looking for a single universal model. Finally, the behaviour at the molecular level is dynamic. Thermal vibrations will certainly be present, and the structure may be changing with time. All these limitations must be borne in mind when looking at the pictures, which should not be regarded as any more than crude guides to what the structure really is.

Apart from being a mode of communication, models may also be used as a basis for theroretical analysis or interpretation. When used in this way, the description of the features of the model should be clearly and explicitly stated; but the purpose is then to define the limitations of the assumptions of the theory implicit in the description of the model, and not try and describe reality. In this difficult subject area, the value of models depends not only on their validity as an attempt to represent reality, but in the extent to which they are capable of being used to increase understanding or prediction, either quantitatively or qualitatively. This leads, at the extreme, to the view that it does not matter whether a model is 'true', all that matters is whether it is useful.

8.1.5 Two-phase models

When it was realised, in 1930, that the lengths of polymer molecules were greater than the size of the crystalline regions, the situation was resolved by the **fringed micelle** theory, illustrated in Fig. 8.3. This is a two-phase model in the sense that one can recognise two types of structural arrangement: namely crystalline regions, which are small crystallites or micelles with the molecules regularly packed in a crystal lattice, distributed in a matrix which is the disordered (amorphous or non-crystalline) region. Individual chains pass through a number of crystalline regions, alternating with segments in the disordered region. At the edge of each crystalline micelle the chains form a fringe as they diverge into the non-crystalline material.

The fringed micelle structure explains many features of polymer behaviour. The crystals provide cohesion, stability and strength, and maintain the orientation in oriented structures. The non-crystalline material lowers the density, provides freedom for deformation, provides accessibility for water and additives, and

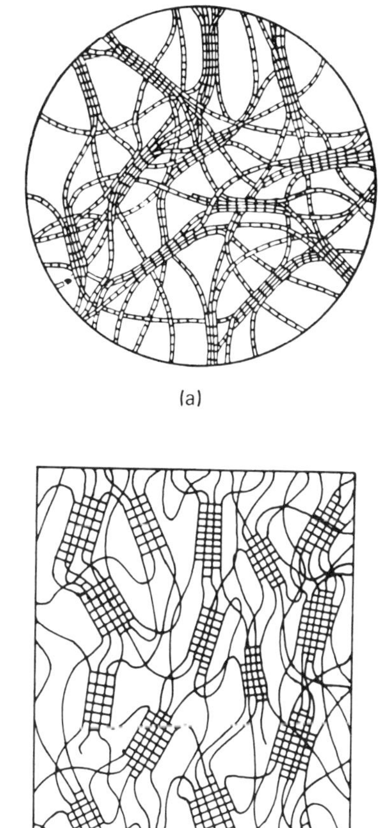

Fig. 8.3 – (a) View of fringed micelle structure as originally proposed by Herrmann and Gerngross (Herrmann and Gerngross, *Kautschuk*, 8, 181, 1932). (b) Typical view of fringed micelle structure in an oriented polymer (Hearle, *J. Pol. Sci., C*, **20**, 215, 1967).

allows a path for diffusion. Furthermore there is a plausible mode of formation. If one imagines crystallisation to start in many places, then the same molecules may get trapped in several growing crystals. Each crystal continues growing until the growth is blocked because the molecules are trapped in different places, and the disordered tangle left between them cannot be sorted out any further. The sequence is illustrated in Fig. 8.4.

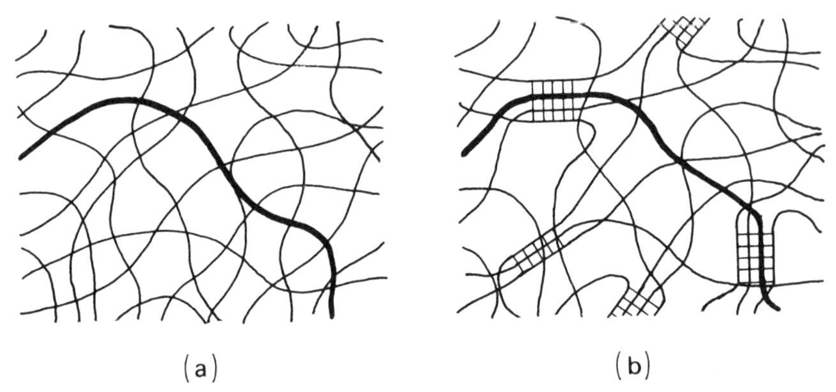

Fig. 8.4 – Crystallisation of amorphous polymer, (a) starting at various places leads to a fringed micelle structure (b). The chain XY is shown trapped in two micelles and is thus unable to crystallise any more.

Several developments of the fringed micelle model have been proposed. In the early days, there was a mind-block and chain folding was not considered. But with the clear demonstration of chain folding in single crystals, this barrier came down, and it was recognised that a **modified fringed micelle** structure should be adopted in which some chains folded back at the edge of a micelle while others fringed off and acted as tie-molecules to other micelles. This is illustrated in Fig. 8.5(a). We note four possibilities at the edge of the crystal: the molecule may fold back; it may act as a tie-molecule to the next micelle; it may link to a more remote micelle; or it may lead to a chain end. Fig. 8.5(b) shows that there can be many tie-molecules, even if folding predominates.

Finally it was recognised that a random arrangement of micelles, such as that in Fig. 8.6(a) can only accommodate a low degree of crystallinity. A more ordered arrangement such as that in Fig. 8.6(b) is needed. The micelles are arranged in stacks which may be regarded as **pseudo-fibrils**. A model of structure on these lines is shown in Fig. 8.7. Subject to all the limitations mentioned earlier, and regarding the model merely as a guide to a more varied structure, one has confidence in Fig. 8.7 as a working model for the structure of a nylon or polyester fibre, because several scientists, with different backgrounds, have come to very similar views. The structure can be regarded as a two-phase model, but some take it to be three-phase by emphasising a difference between non-crystalline regions within a pseudo-fibril and the non-crystalline material between

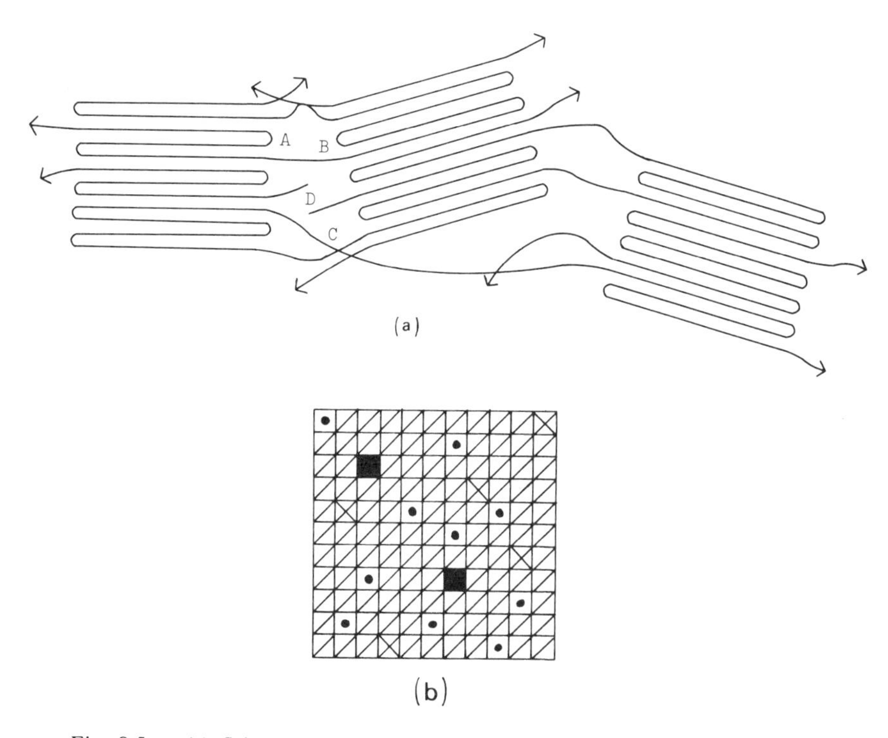

Fig. 8.5 – (a) Schematic view of modified fringed micelle structure with (A) chain folding (B) close tie molecules, (C) distant tie molecules, and (D) chain ends. (b) Note that an 'end-on' view, including the third dimension shows that there can be an appreciable number of near tie molecules (●) and some far ties (X) or terminations (■) even though folding (⁄) is predominant.

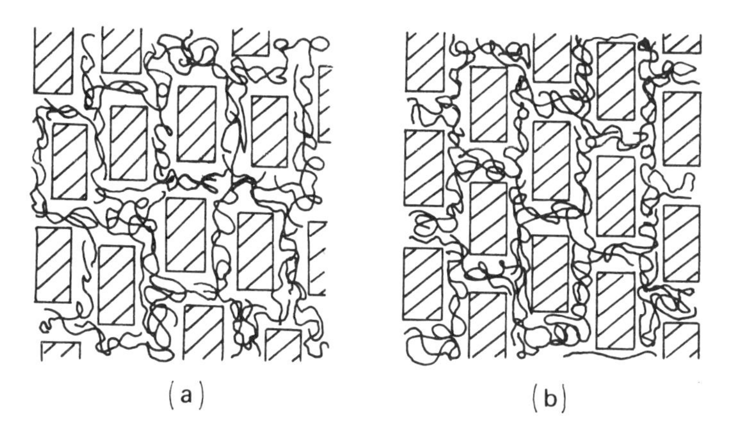

Fig. 8.6 – (a) Random arrangement of crystalline micelles in a amorphous matrix. (b) Ordered arrangement with higher crystallinity.

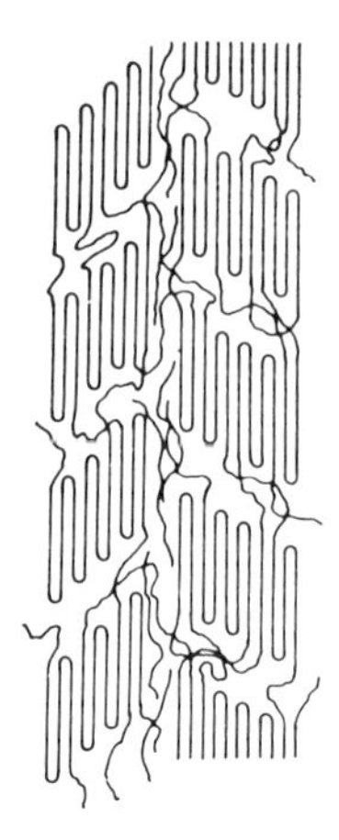

Fig. 8.7 – Typical current view of fringed micelle structure (Hearle and Greer, *J. Textile Inst.*, **61**, 243, 1970).

the pseudo-fibrils. In some materials there is evidence of a fibrillar fine structure, and the **fringed fibril structure**, illustrated in Fig. 8.8(a) was suggested. This is similar to the fringed micelle theory, except that the crystalline regions are long fibrillar strands, with individual chain molecules branching in and out to alternate between the crystalline fibrils and the disordered matrix. It is also possible to introduce chain folding into a fibrillar model, and modify Fig. 8.8(a).

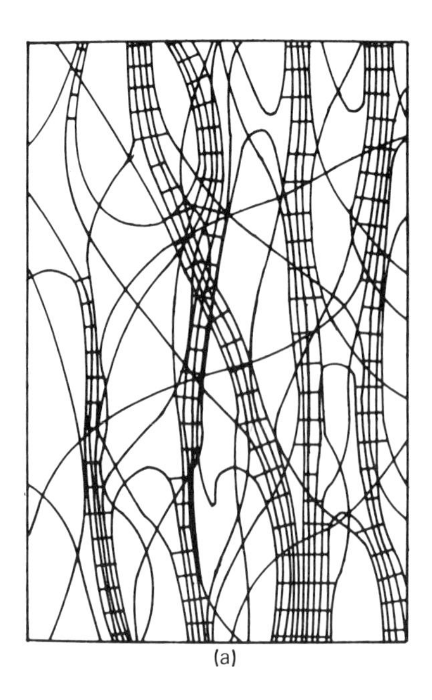

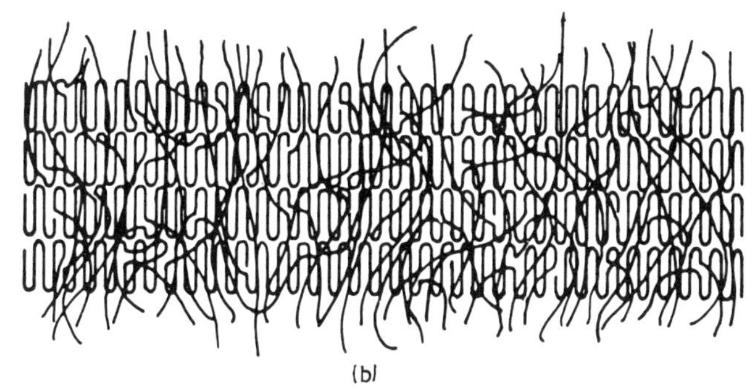

Fig. 8.8 – (a) Fringed fibril structure (Hearle, *J. Pol. Sci.*, 28, 432, 1958).
(b) Fringed lamellar structure.

Single crystals of polymers are usually lamellar with the chains crossing the platelets. This leads to the suggestion of a **fringed lamellar structure**, with thin, wide crystalline regions separated by disordered regions as illustrated in Fig. 8.8(b). Once again, there will be a mixture of chain folding, tie-molecules, and free ends.

We should note that all these two-phase models imply a continuous matrix of non-crystalline material, enclosing separate crystalline regions. This is a natural consequence of the view that crystals grow from separate nuclei within a disordered material. However, in terms of geometry, there is also the possibility that disordered regions could be included within a continuous crystalline block, or that there might be two intermingled continuous networks: the improbability of occurrence of continuous crystal matrices is that, although crystals easily divide into branches, they do not usually unite again.

Quantitative application of the fringed micelle models depends on the possibility (or the difficulty) of specifying the properties of the phases, which are not truly separate, and on the methods (and problems) of the mechanics of composites.

8.1.6 Continuous models

The fringed micelle and similar theories are based on a geometric distribution of separate known (or at least potentially knowable) forms of structure. There is a contrasting statistical view which regards the structure as uniform in the sense that, if a proton was to go and sit in the structure and look around, the environment would be everywhere the same, within the statistical variability: it would not be possible to identify the surroundings as crystalline or amorphous, as would happen in a two-phase structure.

Kargin viewed some polymer materials, such as cellulose, as essentialy non-crystalline, but with a moderately high degree of correlation between the positions of neighbouring chains, although not to the extent needed to recognise a crystal lattice. While this remains a possibility to bear in mind, it has not been developed further.

Hosemann, on the other hand, has proposed a **paracrystalline** structure, which can be explicitly defined, so that various consequences can be analysed mathematically. In a paracrystal it is assumed that any spacing x_i in the ideal crystal lattice takes instead a value $(x_i + \epsilon_i)$ where ϵ_i is a statistically variable deviation. Consequently, over the short lengths the structure appears to be a distorted crystal lattice, but over greater distances, as the random errors accumulate long-range order is lost, as illustrated schematically in Fig. 8.9. As the distance increases, the deviation in spacing from that of the perfect lattice will increase as $n^{1/2}\,\epsilon_i$, where n is the number of repeats.

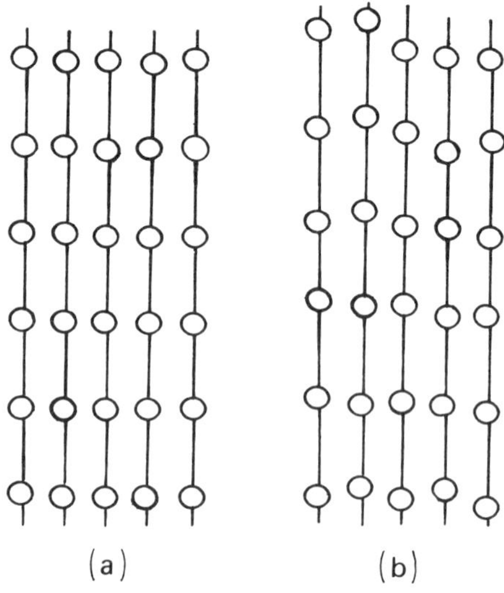

Fig. 8.9 – Neighbouring chains with constant axial spacing x_i. Neighbouring chains with variable axial spacing $x_i + \epsilon_i$. In the diagram ϵ_i is randomly $\pm\ 0.1\,x_i$.

It may be that over a given zone the deviations ϵ_i are all the same, or closely correlated: the paracrystalline structure then becomes identical with a strained crystal lattice. This might be due to external forces. But it could also be due to internal stresses or 'frozen-in' strains. Disorder could be interpreted as a variability in structure accommodated as a complicated pattern of internal strain, arising from difficulties in reaching perfect crystal packing or from variations in local orientation from place to place in an isotropic material.

8.1.7 Crystal defect and crystalline gel models

Disorder may be regarded as distributed within a crystal as localised defects. In its purest form, this model of a partially crystalline polymer would have the defects more or less uniformly distributed within a single crystalline material. The defects might be of various types. For example, polymer molecules have a finite length; and Fig. 8.10 shows a drawing which illustrates the disorder which can arise merely due to a random distribution of chain ends. Folds, kinks, jogs and other defects could be similarly distributed. Even a rather poorly ordered material might be regarded as a system with a high density of defects, though for the reasons given in section 8.1.3 this model would only apply to rather simple polymer chains. Parts of Fig. 8.2 can be interpreted as defect structures.

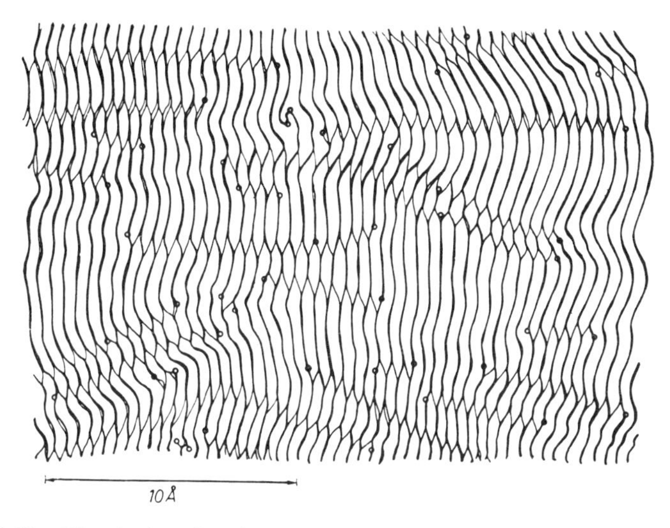

Fig. 8.10 – Disorder introduced into a crystal lattice solely by random location of chain ends as defects, after Predecki and Statton (1966) (Predecki and Statton, *J. Appl. Phys.*, **37**, 6053, 1966).

The defect model may be developed in various ways. For example, some correlation of particular geometries could lead to variations in orientation so that, on the larger scale the material is isotropic, even if a local crystal lattice direction can be identified everywhere. Alternatively, the defects may concentrate and act as boundaries between more ordered regions.

The quantitative application of crystal defect models derives from attempts to use the methods successfully used by metallurgists to characterise dislocations in metal crystals. The problems with polymer chains are more difficult.

The converse of the defect model is the crystalline gel model, in which segments in localised crystalline register are distributed more or less uniformly throughout generally disordered material. Perhaps the best way to visualise this model is to imagine its formation from the melt. In the liquid, just above the melting-point, many neighbouring segments of molecules will be in crystalline register, but the situation is dynamic, with the local order continually breaking-up and re-forming elsewhere to allow the liquid to flow. In polymer molecules, as distinct from small molecules, the situation can change when the density of segments in register increases above a certain level.

Consider a group of pairs of segments in register at some instant, as shown in Fig. 8.11. Suppose that the central pair A separates due to thermal vibration, but the other four remain in position. Because the separate segments are not free to move far, the same pair A will come together again and link up once more.

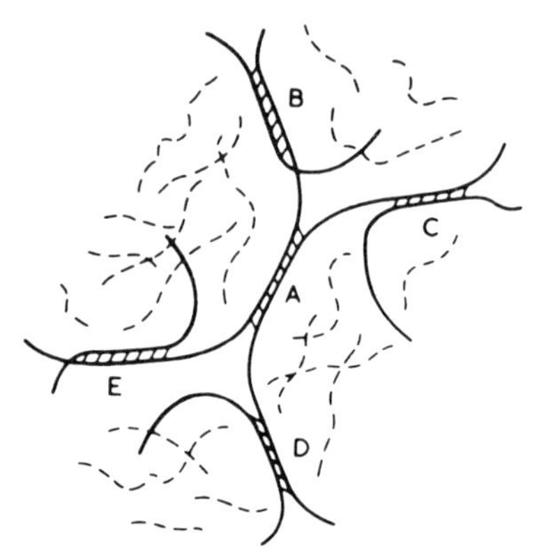

Fig. 8.11 – Neighbouring segments in register in a crystalline gel (Hearle, *J. Appl. Pol. Sci.: Appl. Pol. Symp.*, **31**, 137, 1977).

At some other instant the pair *B* might break, but the segments will be held in the same general position by their neighbours. Thus the structure has formed a gel. The local crystalline register holds the material together and prevents the relative movement of whole chain molecules which is necessary for liquid mobility. Close to the melting point the material will be a dynamic gel, since the segments will be continually breaking and re-forming, but only when this reaches a critical level will the freedom of the liquid appear. The discerning reader will detect the thermodynamics underlying this description of the dynamic crystalline gel.

Fig. 8.12 shows ways in which attempts have been made to represent the

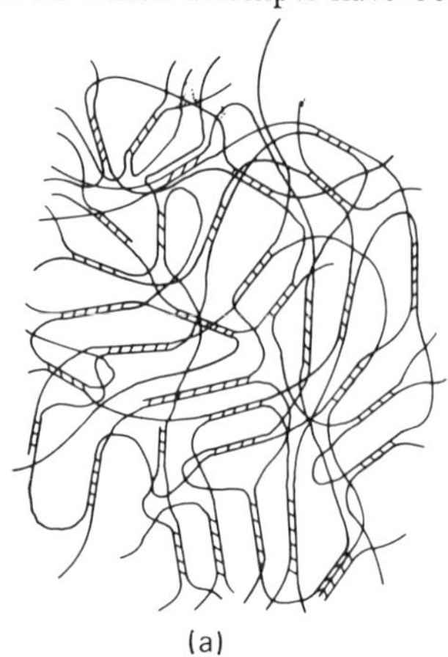

(a)

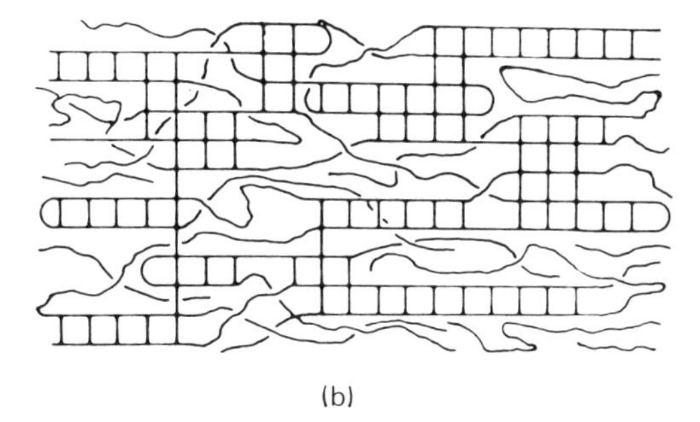

Fig. 8.12 – Models of crystalline gel structure: (a) in an unoriented structure (Hearle, *J. Appl. Pol. Sci.*: *Appl. Pol. Symp.*, **31**, 137, 1977); (b) in an oriented structure (Hearle and Greer, *Text. Prog.* **2**(4), 1970).

crystalline gel model, while Fig. 8.13 is an earlier version demonstrating how a structure with crystalline order distributed more or less uniformly without any defined crystallites might transform on annealing into a modified fringed micelle structure with a clear separation between the crystallite and the amorphous zone.

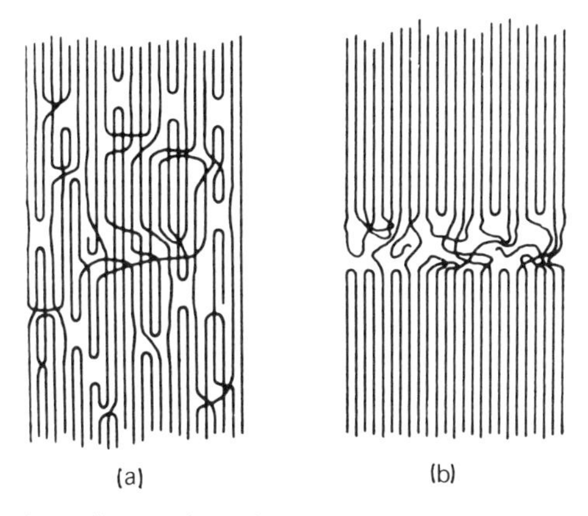

Fig. 8.13 – Annealing model (a) Initial structure. (b) After annealing. After Fischer, *et al.* (1968) (Fischer, Goddard and Schmidt, *Makromol. Chem.*, **119**, 170, 1968).

8.1.8 Perfect crystal models

In some materials, the best model may be an assembly of separate small crystals. Certainly when polymers are crystallised slowly from dilute solution, they form separate lamellar crystals which may be deposited as a mat. The lamellae will be flexible enough to adhere closely together, and provided the crystals have a reasonably high aspect ratio, the mat will have cohesion and strength. The disorder would then come from the lack of register between neighbouring crystals and the spaces between them.

Natural cellulose in plant cells is another example of material probably formed from separate crystals. The cellulose molecules are synthesised in sets of about 30 chains from enzyme complexes, and as the chain molecules form they crystallise into long fibrillar crystals, less than 100Å wide. The individual fibrils are then laid down, in particular geometric arrays, to give the solid material of the cell wall. In one sense the material is wholly crystalline, but the degree of crystallinity as measured by accessibility, density, and X-ray diffraction appears to be about 2/3. This can be accounted for by the crystalline surface, the space between crystallites, and lack of register between their lattices.

8.1.9 The choice between models

It has already been implied that any of the variety of models, described in the previous pages, may be valuable in particular circumstances, for particular materials, for particular reasons, for particular people. But we should also question whether there are any general reasons for a preference for certain forms of models.

There have been recent advances in the application of non-linear irreversible thermodynamics; but a simpler version of a similar idea may be given here.

Because of the complex entanglement of polymer chains, we recognise that it is difficult – impossible, usually – to provide conditions which lead to the chains sorting themselves into anything approaching the perfect order which would be the true minimum energy state preferred at equilibrium below the melting point. Once a certain level of order has been reached, any improved packing at one place causes a disturbance elsewhere. So let us assume that a particular material can only be organised to a given level of crystallinity, say 50%.

We now attempt to consider the free energy of different models, all with the same 50% 'crystallinity'. Let us choose a scale for order so that it gives a linear variation in entropy as the material changes (in a hypothetical zone small enough to allow it to reach a uniform structure) from a perfect crystal to a completely non-crystalline material (ignoring the difficulties of definition of an amorphous structure). This change in entropy is shown, as $(-TS)$, in Fig. 8.14. The problem now is to know how the internal energy changes. Since the interactions between neighbouring atoms are rather sharply defined, it seems likely that the structure would have to get close to perfect packing before there is a steep reduction in internal energy, and that moderate ordering will not lower internal energy very much. Such a variation in internal energy, U, is shown in Fig. 8.14. The free energy is then given by the sum $(U - TS)$, and a temperature below the melting-point must show a lower value in the perfect crystal $(U_c - TS_c)$ than in the amorphous material $(U_A - TS_A)$. At the restricted level of 50% order, a uniform structure would show the value F_1, while a two-phase combination would show the lower value F_2 and would thus be preferred.

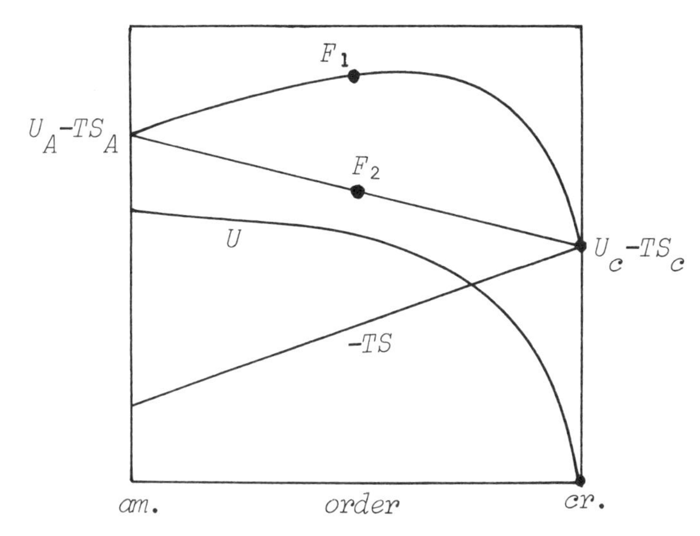

Fig. 8.14 – Possible variation of entropy, internal energy, and free energy with order ranging from amorphous (am) to crystalline (or). A uniform structure gives F_1; a two-phase mixture gives F_2.

Of course, if the lines curved the other way, the reverse would be true. However, the argument indicating that two-phase structures may be preferred to uniform structures seems plausible. Given a limited total ordering, free energy can probably be lowered by separation into regions of higher and lower order.

The kinetic situation, based on the approach to a given metastable equilibrium at a local energy minimum, works the other way. Separation into distinct regions involves more relative chain movement, more and higher energy barriers to cross. Consequently rapid quenching may lead to a more uniform partially ordered structure, while slow crystallisation or annealing leads to larger and more perfect crystallites, more widely spaced.

It must be stressed that the arguments given in this section are very crude, and that it is grossly inadequate to try and represent structure by a single order parameter. However, the reasoning does justify a preference for two-phase models.

8.1.10 A unified view: major structural parameters

The structure of a perfect crystal is characterised by a few parameters needed to define the unit cell of the lattice. The structure of a very highly disordered material may be characterised by a few statistical parameters. But a structure of intermediate order requires a vast number of parameters for its complete specification: effectively, the position of almost every atom must be specified individually, so that the number of parameters needed approaches Avogadro's number 6×10^{23}. Clearly complete specification is impossible, and this leads to the attempt to simplify the situation by means of structural models.

An alternative approach is to try and indentify a limited list of parameters in order to describe the most important characteristics of any structure. The six given in Table 8.1 seem to be particularly significant.

Table 8.1 – Major structural parameters.

Parameter	
Degree of order	
Degree of localisation of order	general character of structure
Length/width ratio of units	
Degree of orientation	
Size of units	consider separately
Molecular extent	

The **degree of order** could be defined theoretically by some statistical measure of the correlation in position of neighbouring chain segments. Operationally it could be defined by taking some experimentally measurable quantity, such as density, and using this to construct a linear scale between zero for an amorphous material and one (100%) for a perfect crystal. With a material like polyethylene terephthalate, this is relatively easy since rapid quenching gives a glassy amorphous solid whose density, at 20°C, can be measured; the crystalline density can be calculated from the crystal lattice parameters determined by X-ray diffraction; and the density of partially crystalline specimens can be measured. With other polymers, there may be greater difficulty, and there could be anomalies analagous to the fact that, unusually for a solid, the density of ice is less than that of water at 0°C. Other methods such as accessibility, X-ray diffraction, and other special techniques may also be used for an operational definition, though one measure will not necessarily vary linearly with another.

The **degree of localisation of order** would be defined theoretically in terms of the variability of the degree of order, measured over zones with a size of the order of the fine structure, namely about 10 nm. The value would be zero for a uniform structure and one (or 100%) for a two-phase structure with a complete separation into crystalline and amorphous regions. Operational definitions are more difficult, and circumstantial evidence would be used to estimate values.

The **length-width ratio** of localised units is a more obvious quantity. It would be directly measurable if the crystallites could be observed by electron microscopy or otherwise. It is most conveniently taken on a logarithmic scale, ranging from infinity for a fibril of infinite length, through 1 for a cubic micelle, to minus infinity for an infinitely wide lamella.

We can take these first three parameters together as characterising the general nature of the structure. They may be represented on a three-dimensional plot as shown in Fig. 8.15 and the various structural models discussed previously can be located on the plot. The block outlined on the figure is intended to suggest a range of commonly occurring structures of partially crystalline polymers. When dealing with any particular material, it is instructive to try and locate its position on the diagram. Thus high-density polyethylene would tend to be near the corner A, highly ordered, not very localised, lamellar; cotton would be near B, again highly ordered, but localised and fibrillar; and a 'hard-soft' block copolymer might be near C, less ordered, but clearly micellar.

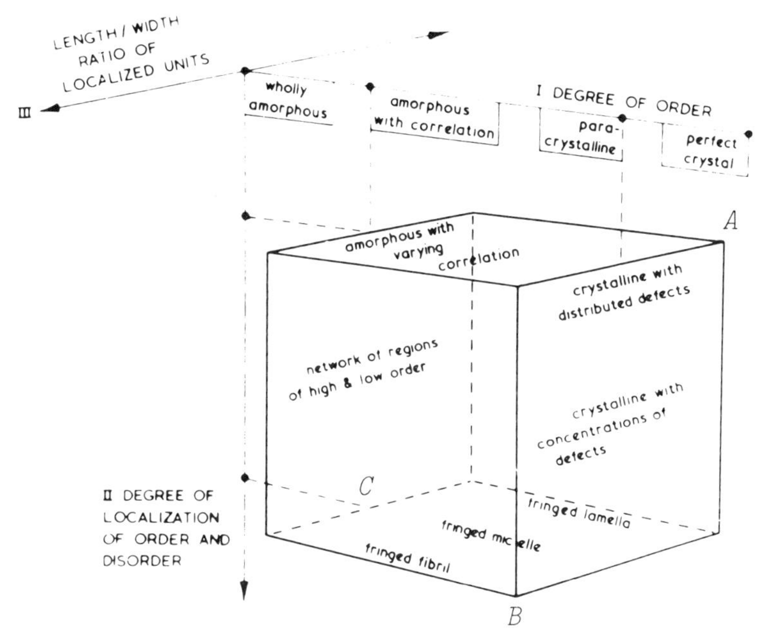

Fig. 8.15 – Parametric location of structures (Hearle, *J. Pol. Sci.*, *C*, **20**, 215, 1967).

One important lesson from this approach is that there is a continuous range of structures, and that the different structural models merely show positions where simple forms can be described. Between these positions, there are parts of the diagram where one form of structure merges gradually into another. Thus a uniform structure changes gradually from a wholly amorphous form, through some correlation and then a paracrystalline form, to the perfect crystal. At intermediate degrees of order, there is a progression from uniformity, through varying correlation and regions of high and low order, to the clear separation of a two-phase structure; or alternatively the uniform disturbance of the para-crystal may merge into localised defects, accumulations of defects, and the separation of amorphous regions. Obviously there is also a gradation from long to short fibrils, and then through long and short micelles to thick and thin lamellae; and it would be pedantic to try and define precisely when an attenuated fibril became an elongated micelle, or a wide micelle became a lamella.

Fortunately the other three parameters can be considered separately, so that we do not have to draw four-dimensional diagrams or worse. Any of the structures identified by position on Fig. 8.15 can occur over a range of values of orientation, size of units, and molecular extent.

The **degree of orientation** would be defined theoretically as an average of the orientation of molecular segments. Operationally it can be estimated from the birefringence or from other measurements. Perfect orientation occurs when all the chains are lined up in the same direction; zero orientation occurs, when over the specimen as a whole, there is no directional preference. Partial orientation is when the distribution of direction is biassed.

The **size of the localised units** would be given by the diameter of fibrils, a mean linear dimension of micelles, or the thickness of lamellae. It can vary from low values in a structure composed of small crystallites, closely spaced, to large values in a coarse structure.

Molecular extent is more difficult: it is not possible to imagine any method of direct measurement, and it is doubtful whether anyone who has drawn a structural model has considered it. But it is a quantity which may have a considerable effect on strength. The analagous quantity at a much larger scale is clearer. It is well known that the strength of a textile yarn or a non-woven fabric depends not only on fibre length but also on the extent to which the fibre is folded back on itself. It is the extent, as shown in Fig. 8.16, which is really important. Similarly molecular extent may be more important than molecular length. Chain folds are sources of weakness, just as much as chain ends.

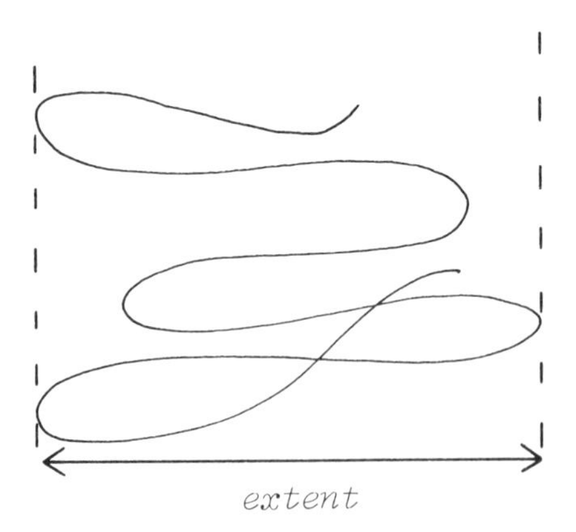

Fig. 8.16 – Extent of a long chain in a structure.

Molecular extent is a general parameter covering both crystallographic folding and a general meandering around. In a fringed micelle or lamellar structure it is related to the number of tie-molecules, since only chains which continue on to the next crystallite will contribute to extent. In a polymeric melt or solution, the molecular extent will be related to the chain dimension discussed in Chapter 3, and will depend on chain stiffness, interactions and flow; in the solid the same extent may be locked in, or it may be increased by drawing. Molecular extent can only be estimated from the circumstances of formation of the structure or from the interpretation of properties.

We may note that degree of orientation and molecular extent apply to amorphous polymers as well as to partially crystalline polymers. We must also stress that the six parameters are only a beginning, though hopefully they are by far the most important features. There are many other structural features which might be characterised: the detailed packing within crystalline or amorphous regions, the nature of defects or the form of correlation; other features of the shape of localised units; the distribution of segment orientations or unit sizes, as well as the mean values; and so on up to Avogadro's number.

Finally it is wise to emphasise the difference between three parameters that are often confused: order, orientation and extent. As shown in Fig. 8.17, they

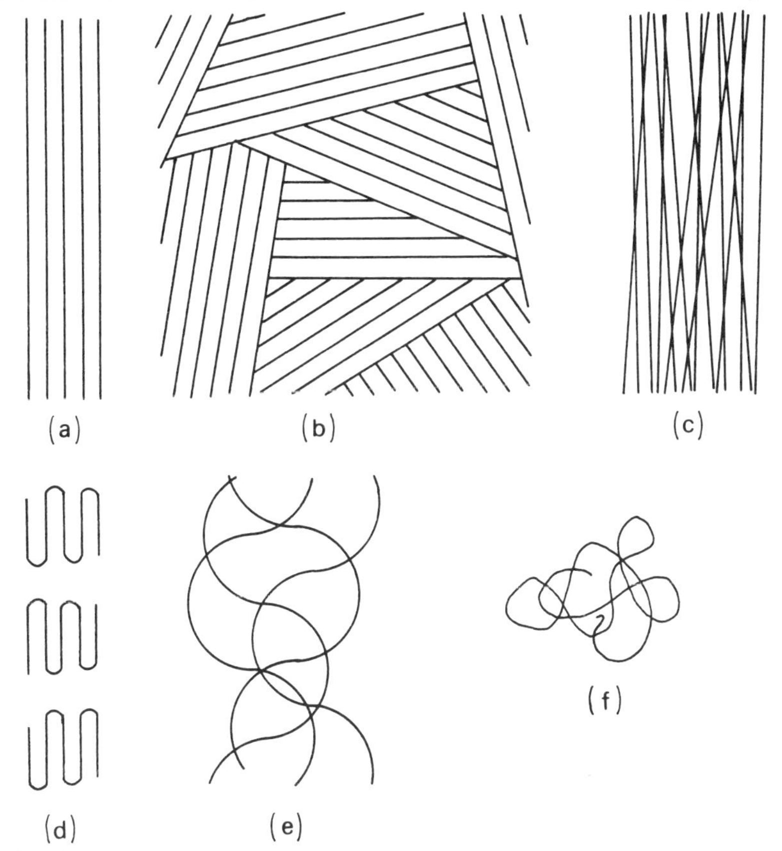

Fig. 8.17 – (a) Maximum order, orientation, and extent. (b) High order, zero orientation. (c) Low order, high orientation. (d) High orientation, low extent. (e) Zero orientation, high extent. (f) Minimum order, orientation, and extent.

do come together at the extremes: a perfectly ordered structure would have perfect orientation and maximum extent; and a structure of maximum disorder would have no preferred orientation and low extent. But in between – and we are rarely concerned with the extremes – any combinations are possible, as also shown in Fig. 8.17. A high degree of order occurs with separate perfect crystals which may be oriented in all directions; conversely, if a deviation of only a few degrees in orientation is allowed, a high degree of disorder can be achieved. A structure of low extent can be achieved by folding chains back and forth, but still keeping all the segments, except those in the fold perfectly oriented; conversely a chain following a sequence of semi-circules would have a high extent, but no preferred orientation of individual segments.

8.1.11 Manufactured materials and larger scale structures

The discussion so far in this chapter has centred on the forms of fine structure found in typical partially crystalline polymers. But some comment should also be made on the relation between structure and the circumstances of formation, particularly as the method of manufacture may lead to specific features on a larger scale than the way in which the molecules pack together at a scale of around 10 nm.

When the manufacture involves the slow cooling of a large mass of polymer, the crystallisation will proceed from a few nuclei, which are either previously present or form spontaneously, to give lamellar crystals, which then branch repeatedly and spread out rapidly to form spherulites, as discussed in section 7.5. The fine structure is probably best regarded as fringed lamellar, Fig. 8.8(b), though the lamellae are really long ribbon-like strips, and there may well be special features such as twisting of the lamellae as they radiate outwards.

The spherulitic texture can perhaps be visualised in the following way. Start with a lot of long, narrow strips of paper. Stick some of these together in an oriented bunch at one place, and then allow them to curve away so as to spread out into a circle. Now stick on other strips of paper so as to radiate outwards in the gaps between the first set. This gives an idea of a spherulite, but we must note the imperfections of the model:

(a) the centre should be a single lamella, which then branches to give the multiple ribbons; (b) the first set of strips are not special – all the branches are equivalent; (c) the branches will spread out into three dimensions, to grow into a sphere; (d) at the growing surface, there may be a more complicated intermingling, twisting, bending and varying of the lamellae, than is suggested by tidy-minded models – on the other hand, there may be some regular periodicity of twisting or waving; (e) there is a fine structure of chain molecules folding back and forth across the thickness of the ribbons, and then fringing off to link the ribbons by parts of chains (tie-molecules) which pass through disordered regions between the lamellae.

Slow and uniform cooling can be achieved in the laboratory or in the factory so that the structure formed fits into the last chapter as well as this one. But many manufacturing operations in the extrusion, moulding or casting of plastics, which are carried out as fast as possible to reduce costs, will involve both large temperature differences, associated with the patterns of heat flow, and also mechanical flow of the polymer molecules, associated with the mechanics of the system. Both of these effects will influence the pattern of crystallisation. Some principles can be suggested. Surfaces must be coolest and will crystallise first to give skins which may differ from the bulk of the material. Crystallisation will tend to follow a 'front' at a particular temperature. Flow in the crystallising material will lead to local orientation. But mainly these complexities are a question of engineering detail, rather than fundamental physical principle: the limitations must be at least understood and preferably exploited.

When we turn from large plastic objects, thick sheets, and coarse monofils to thin films and fine fibres, the situation changes. Cooling at a rate which is impossible for large masses can readily be achieved. The material can easily be taken down through the temperature of maximum rate of crystallisation, shown in Fig. 8.18, without there being time for much crystallisation to occur. The crystallisation then occurs at a lower temperature. Sometimes, as in polyester fibres, the material cools to a glass, and crystallisation only occurs on drawing or annealing.

At low temperatures, smaller nuclei will be stable, and consequently the number of available nuclei increases as temperature falls, as indicated in Fig. 8.18. As nuclei get closer, spherulites – if that is the form of crystallisation – must get smaller. In the limit, there will only be the central single lamellar crystal, or perhaps a small sheaf, and we note that the limiting form of spherulitic crystallisation, as nuclei get closer, is very similar to the modified fringed micelle structure shown in Fig. 8.5. Thus a fringed micelle structure, such as that in Fig. 8.7, may be a good working model for many fibres and films. Alternatively, there is

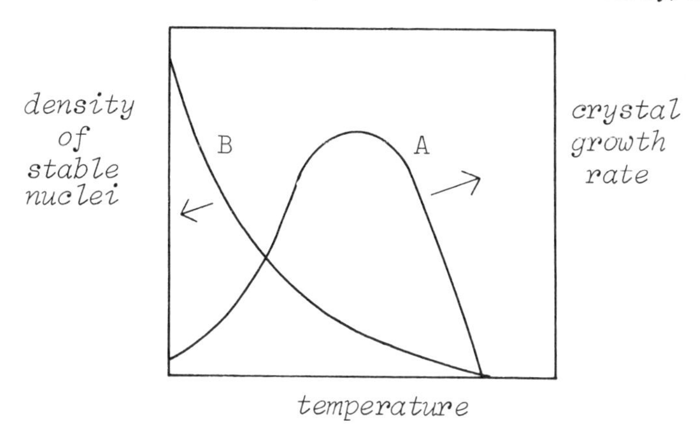

Fig. 8.18 – Variation of crystallisation rate and number of stable nuclei with temperature. At A there is very slow crystallisation on many nuclei.

evidence to suggest that something like the crystalline gel form discussed in section 8.1.7 may form first, and then transform to a micellar structure on annealing.

It is also necessary to take account of the fact that fibre and film formation is a dynamic process. The molten stream falls down under gravitational and aerodynamic forces and its temperature profile will depend on the rate of transfer of heat. The crystallisation will be related to the thermal and mechanical gradients. For example if the cooling through the temperature of maximum rate of crystallisation is delayed, spherulites will form and then either join up in complete crystallisation or remain as occluded spherulites within another structure formed by crystallisation at a lower temperature. Alternatively rapid cooling, as usually preferred, will minimise the occurrence of spherulites.

A high rate of shearing, as in the hole through which the polymer is extruded, or a high rate of attenuation (elongational flow) in the thread line or the film will lead to orientation of the polymer molecules. In competition with these effects, there will be the tendency to disorientation due to thermal vibrations. However, if the orientation can be maintained at the point of solidification, then it will be locked into the structure. This is the way in which the so-called POY (partially oriented) yarns are formed by very high-speed wind-up, which leads to elongational flow in the critical region of solidification. These POY nylon and polyester yarns require drawing by only about 1.5 or 2 times as compared to the 4 or 5 times typical of the old undrawn yarns made by lower-speed spinning. The structure of the yarns used by the textile industry is, of course, further modified by the changes which occur on drawing, though this is poorly understood. Similar changes occur in the unixial or biaxial drawing of films.

The comments made so far in this section have been related to commercial formation from the melt. In formation from solution, concentration gradients influence the structure as well as thermal and mechanical gradients. This gives added complexities. For example, a polyacrylonitrile solution will often precipitate into a spongy structure with solvent left in the voids between the solid material. On drying and stretching, the structure collapses, the voids empty, and their surfaces coalesce, but nevertheless residual effects of the void structure remain in the final acrylic fibre.

Another degree of complexity arises in the regeneration of a viscose solution into cellulose film or rayon fibres. Chemical reactions are taking place and their progress influences the course of solidification and crystallisation. By control of the sequence of events different structures can be obtained. If solutions have structure, for example if they contain liquid crystals, this gives a possibility of forming other solid structures, such as very highly crystalline and highly oriented materials. The greatest complexity comes in the natural systems where biological control determines the structure. For example, as mentioned in section 8.1.8, cellulose is synthesised as crystalline fibrils. These are then laid down in specific patterns. Fig. 8.19 shows regular sheets lying on top of one another at an angle

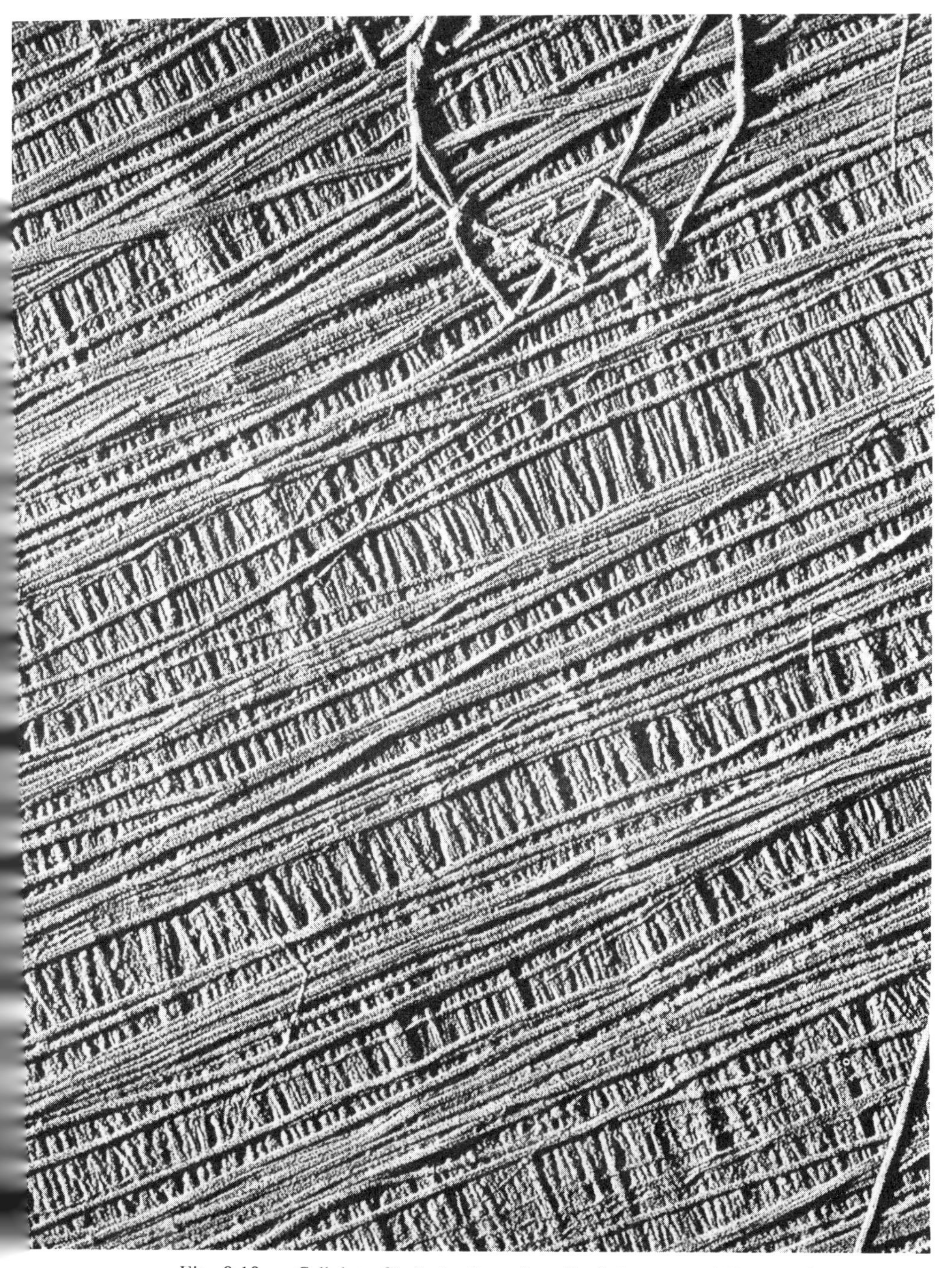

Fig. 8.19 – Cellulose fibrils in the cell wall of the seaweed Valonia (Preston, private communication).

in a seaweed cell wall. In cotton, the fibrils spiral around the cell at a characteristic angle, with frequent reversals in the direction of spiralling. Similarly the proteins in hair or skin or muscle form special patterns of structure.

Apart from the structures which arise in manufactured materials as an indirect consequence of the manufacturing process, with only limited indirect control possible, there can also be deliberately introduced large-scale structures. For example bicomponent fibres may be made by extruding more than one component through the same hole. The interaction of the responses of the different materials leads to special properties. The cellular structure of biological materials is another example of specific forms.

Above about 0.5 μm, where optical microscopy or direct visual observation is possible, structure is thought of as morphology or design. At the molecular scale, structure is chemistry. At around 10 nm, it is physical fine structure. There is however a difficult and neglected scale of structure at around 100 nm, where design merges into physics.

8.2 THERMAL RESPONSES

8.2.1 Transitions

Thermal transitions will occur within the separate components of a partially crystalline polymer in ways which have already been discussed. The simplest situation will be one in which the disordered regions have a single glass-to-rubber transition and the crystals melt. An idealised version of the transitions as found by dynamic modulus tests would then be shown in Fig. 8.20(a). The various consequences of the transitions will show up through the appropriate rule of combination in composite systems, although these rules are usually complicated and have not yet been well worked out. And obviously it is necessary to have a reasonable model of structure before any rules can be applied.

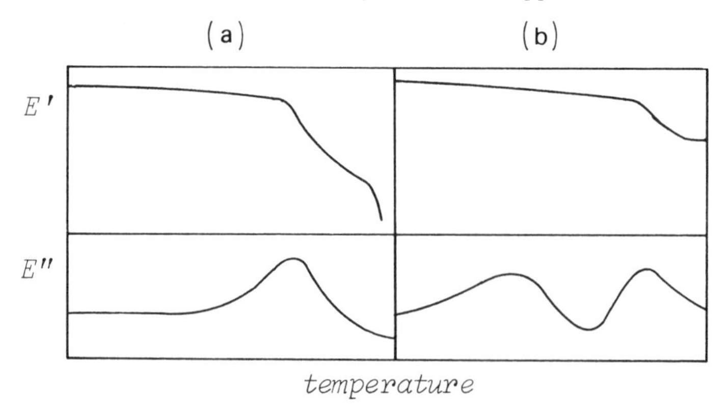

Fig. 8.20 – Thermal transition in a partially crystalline polymer; as shown by real (E') and imaginary (E'') moduli. (a) With a single glass–rubber transition, and then melting (as shown by polystyrene). (b) With split glass–rubber transition, as shown by polyethylene terephthalate.

Apart from these principal transitions, there may also be crystal lattice transitions from one pattern of packing to another, and there may be minor transitions as particular groups achieve a degree of freedom of movement either in the amorphous or the crystalline regions. These transitions will show up in various ways in mechanical, thermal and other experiments, but their effects are usually small and of more academic than practical importance.

There is however a group of commercially valuable partially crystalline polymers where the glass transition is split into two comparable parts. For example, in nylon and polyester PET an aliphatic sequence, of -CH_2- groups or -CH-O-CH_2-CH_2-O-CO-, becomes flexible well below room temperature through rotation round the main chain bonds†, but the hydrogen bonds between the -CO.NH- groups or the association between the benzene rings hold the disordered regions in a firm network until a transition above room temperature is reached. The latter transition is often referred to as the glass transition, but it is misleading to suggest the presence of a single glass transition: the development of mobility is divided.

Fig. 8.20(b) represents an idealised sequence of transitions which gives good properties in tough plastics, films and fibres. The low temperature transition allows some freedom of movement to the disordered chain segments, thus allowing the structure to readjust and provide a more uniform sharing of load between the chains. In other words local stress concentrations are minimised, thus giving toughness. Nevertheless, because of the limited nature of the transition, the material remains firm, and it is only above the higher transition that there is a marked drop in stiffness (real part of the modulus). By contrast, polyethylene shows a single low temperature transition and is too soft for many purposes; and isotactic polystyrene shows a single high temperature transition and is too brittle. Polypropylene is an uneasy compromise with a broad transition around room temperature, which does give a partial mobility but has the disadvantages associated with a high loss modulus in the working range.

At temperatures well above those of common use, melting occurs but this should be below the temperature of chemical decomposition in order to allow for processing through the melt.

8.2.2 Setting and annealing

Any transition necessarily involves the possibility of setting. The transition, as shown up by a drop in stiffness and a peak in loss modulus, shows that some part of the structure which was rigid has become mobile at the higher temperature. If the material is deformed and then cooled through the transition, it becomes rigid in the deformed state: the material has been set.

The effects at the transitions in the disordered regions are similar to those in wholly amorphous polymers. The set is a temporary set since there is no

† In nylon, the low temperature transition is further split into two parts for reasons that are not well understood.

permanent change of structure. There always exists a reference state reached by taking the material above the transition temperature free of any external restraint: it will then settle to an equilibrium state of minimum energy governed by the interconnections of rubbery chain segments (tie molecules) between the crystalline regions. Cooling from this state will give the reference state below the transition.

It is not necessary to understand the mechanism of the transition to know that setting will occur. It may be a blocking of bond rotation, or it may be, as in nylon, a formation of temporary cross-links by hydrogen bonding. With the latter mechanism, moisture also affects the transition. For example, cellulose in the form of viscose rayon shows a transition somewhere above 200°C when dry, with the peak being lost in the region of charring, and somewhere below 0°C when wet. Clearly, in this situation, setting occurs through wetting and drying at some intermediate temperature.

Permanent setting is associated with the crystalline regions, and is effectively synonymous with annealing. Heat treatment at temperatures close to the melting point, which is not in fact a well-defined temperature in polymers, leads to structural changes in the assembly of crystalline and disordered regions. If the material is in a deformed state (for example, stretched, sheared, bent, or twisted) while it is heated, the stresses will relax through the structural rearrangment and the deformed state will be stabilised or set. The set is permanent since there is no way of de-annealing and getting back to the less favourable form, except by melting and going back through the whole sequence by which the first state was reached.

One argument on annealing or setting has the following form. It is known that small or imperfect crystals melt at a temperature below the 'true' melting-point. The variation of melting point with size or perfection is shown schematically in Fig. 8.21(a). If the specimen has been rapidly crystallised, for example

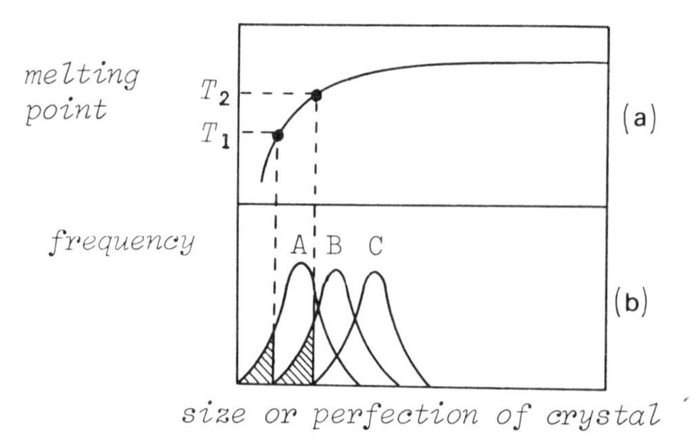

Fig. 8.21 – (a) Variation of melting point with size and perfection of crystal. (b) Distribution of crystallite size, (A) after rapid cooling, (B) after annealing at temperature T_1, (C) after annealing at higher temperate T_2.

in some commercial process, it is likely to contain many small imperfect crystals as shown by the distribution A in Fig. 8.21(b). Heating to T_1 would melt many of these crystals, but allow larger and better crystals to form giving the distribution B. During this rearrangement, segments will move so as to relieve stresses. The process could be repeated again at T_2, T_3 and so on, giving successively higher distributions of size and perfection.

This simple argument suggests that permanent heat setting at a given temperature will be effective against subsequent treatments at lower temperatures, but can be overcome by treatments at higher temperatures. For some materials, such as nylon 66, this simple rule does seem to hold provided account is also taken of the influence of moisture and stress on melting. But in other materials the behaviour is different: for example, it appears that polyester fibres can be repeatedly re-set at the same or even lower temperatures. Generally, there is a lack of good quantitative data which shows the effect of various sequences of heat setting treatments. Surprisingly the best source of information at present is often the behaviour in technological processes. In nylon 66 and polyester PET with melting-points over 250°C, commercial heat-setting operations are carried out between about 180 and 240°C.

Annealing may also be explained, without postulating localised melting, by means of the movement of defects out of crystals, and this is very likely to be a mechanism in polyethylene with its simple repeat, though probably not in more complicated polymers. It may also be possible for crystal dimensions to change without actual melting. Fig. 7.42 shows that the fold length in single crystals can change. It must however be remembered that growth of a polymer crystal must usually involve either secondary nucleation or considerable rearrangement of chains inside and outside the crystal. It is not just a question of adding on more units.

Finally one must note that melting is less simple than is suggested by the simple argument, and that the multiple melting effects discussed in the next section will interact with setting and annealing.

8.2.3 Melting

The softening-point of polyethylene is often quoted as about 115°C, but Fig. 7.68 showed that single crystals could melt anywhere between 120 and 135°C and that the extrapolated value for large crystals was about 145°C. Clearly melting of polymers is nothing like the sharp change found in simple crystals, with melting-points determined to within a fraction of a degree, as a means of identification, by simple visual observation of the change from solid to liquid, and easily detected by the latent heat or the volume change.

Thermodynamically, the melting-point is the temperature at which there is no difference in free energy between the crystal and the melt. Consequently:

$$\Delta F = \Delta U - T_m \Delta S = 0 \tag{8.1}$$

or

$$T_m = \Delta U / \Delta S \tag{8.2}$$

where Δ referes to the difference between the states, U is the internal energy, S is the entropy, and T_m is the melting point.

The thick line in Fig. 8.22 shows the situation graphically, with the melting-point occurring sharply at the intersection. However when we are dealing with a bulk sample of partially crystalline polymer, there are other features which must be considered:

(a) Small crystals have a higher internal energy, due to their surface energy, by which we really mean that the segments on the surface are not able to lower the internal energy by regular bonding on all sides with their neighbours in a crystal lattice.

(b) Internal imperfections or defects also raise internal energy, by denying the correct bonding interaction.

(c) The reduction in bonding may also increase the entropy, thus lowering the free energy, though this effect is less than the internal energy effect, and, if it was dominant, would lead to complete melting.

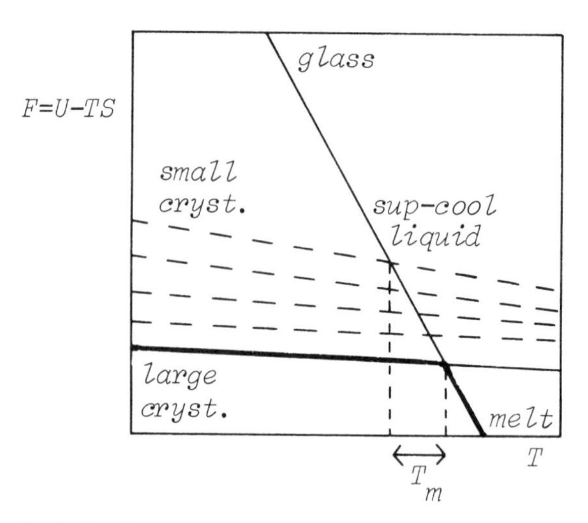

Fig. 8.22 – Free energy relations in melting of a simple substance.

These effects are well-known with simple substances, and the only special feature of polymers is the common occurrence of small, imperfect crystals, which can grow or perfect on annealing. However, there are other features which are peculiar to polymers.

(d) There will be a difference in free energy depending on whether there are chain folds or not at the surface of the crystal. The chain fold itself is an unfavourable form with high internal energy, but it leads the chain back into the low energy state within the crystal, and so interacts with the whole pattern of crystallisation.

(e) The entropy, and perhaps to a lesser extent the internal energy, of the disordered material depends on the interconnections between the crystals. The reason is illustrated schematically in Fig 8.23. If a chain segment melts, the total entropy change ΔS_T has two parts:

$$\Delta S_T = \Delta S_S + \Delta S_A \tag{8.3}$$

The first term ΔS_s relates to the segment which actually moves from the crystalline to the disordered phase, and may vary to some extent insofar as the entropy of the amorphous material is influenced by the restraint of the crystallites to which it is connected. The second term ΔS_A comes in because the entropy of a chain will depend on the length between the fixed points: a greater free length gives more freedom. (In discussing rubber elasticity, the argument was put the other way: the entropy decreases as the length increases for a fixed number of repeats between network points). We thus have the unusual effect in polymers, that the entropy change on melting is a result not only of the change associated with the material which changes phase, but also of the resulting change in the material which is already in the second phase. A corrollary of this is that melting-point will be influenced by stress, since the change will be more marked if the connecting segment is relatively extended than if it is slack.

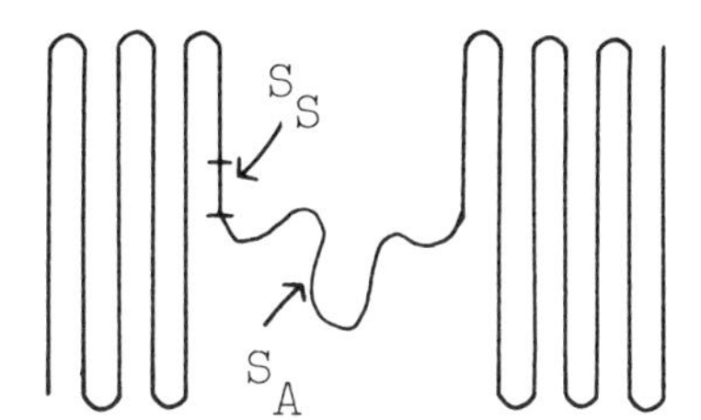

Fig. 8.23 – Change on melting of part of a tie molecule. S_S is the melting segment. S_A is the associated tie molecule segment.

The overall result of these several effects is that the full line of Fig. 8.22 will spread into bands, shown by dotted lines, and a range of melting-points is possible dependent on the particular circumstances.

A simple-minded view of the situation would suggest that rapid quenching would lead to the smallest, least perfect crystals and that annealing would lead to a rising melting-point. A similar behaviour would be expected in polymers which can be quenched to a glassy state, but crystallise at some intermediate temperature. The reality in many polymers is different.

Fig. 8.24 shows a common response. When the rapidly quenched and crystallised material is heated in a differential scanning calorimeter (DSC) it shows a melting endotherm at a temperature T_1. Annealing at a lower temperature T_a, followed by testing in the DSC, leads to a partial endotherm at a temperature T_2, lower than T_1. Thus annealing has apparently generated a less stable form. Annealing for a longer time, or at a higher temperature causes an increase in T_2 and an increasing dominance of the second endotherm. Eventually there is only the second (T_2) endotherm, and its temperature goes above T_1.

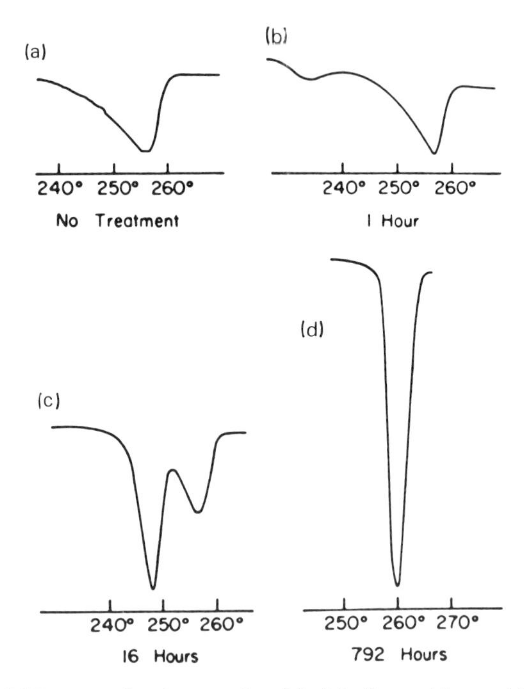

Fig. 8.24 – DSC traces of undrawn nylon 6.6: (a) after rapid quenching; (b) after annealing at 220°C for 1 hour; (c) for 16 hours; (d) for 792 hours. After Bell, Slade and Dumbleton (1968).

The most straightforward interpretation of these results is that rapid quenching leads to crystallisation in a form I which is then progressively converted to a form II. There are however, alternative views, such as an explanation in terms of recrystallisation during heating in the DSC†. It is however certain that the rapidly quenched material is different from the annealed material, and that this is not showing up merely as a rising melting-point. There are also differences in mechanical properties between the two forms: for example, form II is brittle and fractures before yielding; but form I will yield and draw at room temperature ('cold-draw') with a conversion to form II.

Three other experimental results may be noted. Treatment of form II at a temperature between T_2 and T_1 will convert it back to form I. Either form may be found in oriented materials, drawn fibres, as well as in unoriented, undrawn

† It should be noted that the ensuing discussion reflects the personal view of the author.

specimens. The melting-point of form I shows very little change in a variety of circumstances, such as rates of cooling and rates of heating: it is, in fact, probably the most constant temperature for any polymer material.

If one takes the view that there are two forms, the changes can be explained in terms of the free energy diagram shown in Fig. 8.25. It is proposed that the classical view shown in Fig. 8.22 corresponds to liquid and Form II. In Fig. 8.25, another line, which is intermediate in level (U) and slope (S), represents form I. The other requirement is that there would be little hindrance to the conversion from the melt to form I over low energy barriers; but the conversion to form II should be more difficult, with a greater movement of the molecules and higher energy barriers.

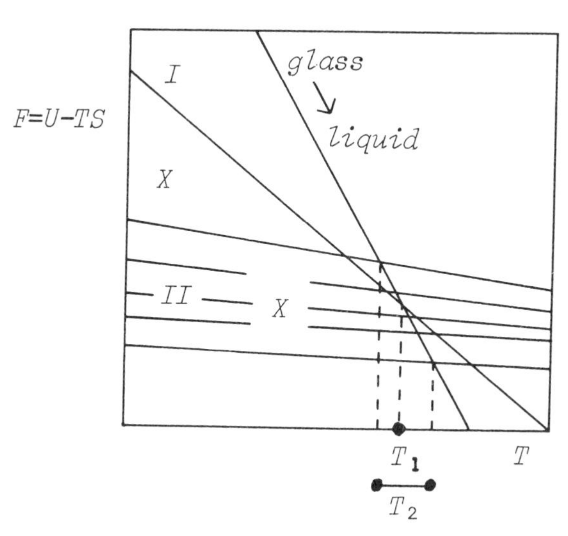

Fig. 8.25 – Free energy relations for form I, form II (at different levels of perfection and size), and the glass to melt. X indicates location of energy barriers.

When the melt is cooled, even rapidly, it would easily drop to the lower energy form I. In some polymers the barriers would be high enough to allow the melt to cool to a glass, but at a temperature T_c this would crystallise with the free energy drop to form I. However the crystallisation occurred, form I would melt at the temperature T_1. Annealing at T_2 would lead to a slow drop to the lower energy form II, with a free energy which would cross the liquid line at T_2, higher than T_1. Further annealing would lead to larger and better crystals with lower free energy and a progressively higher melting-point.

The postulation of another thermodynamic state, intermediate in internal energy and entropy, appears to explain the experimental results.

It is suggested that form II is a two-phase structure with crystalline and amorphous regions, such as the micellar model shown in Fig. 8.5, with the size and perfection of the crystallites being dependent on the extent of annealing. On the other hand, form I may be the dynamic crystalline gel described in section 8.1.7. This is a more-or-less continuous single-phase structure, which would have a single melting-point. It will be easily formed since it merely involves closely neighbouring segments locking into crystalline register, whereas the formation of micelles involves growth on nuclei by means of appreciable movement of chains. One can also imagine that small form II crystals would 'melt' to form I by the chains shuffling relative to one another, increasing their entropy without much unfavourable increase in internal energy.

Table 8.2 gives a comparison of free energy in the various forms, and indicates that it is reasonable to regard the dynamic crystalline gel as intermediate in internal energy and entropy. It is therefore a plausible structure for the rapidly quenched and crystalline material.

Table 8.2 – Internal energy and entropy of different forms.

Form	Internal energy	Entropy
Partially crystalline micellar, II	LOW, because of high level of intermolecular bonding in crystallites (*decreasing* strongly with improved size and perfection of crystallites)	LOW, because of great restriction on movement of chains in crystallites (slight decrease with size and perfection)
Dynamic crystalline gel, I	MEDIUM, because intermolecular bonding is reduced	MEDIUM, because segments are breaking and re-forming, giving more disorder with interchange between the states, subject to no relative movement of whole molecules
Liquid (melt)	HIGH, because of reduced bonding	HIGH, because more local mobility *and* movement of whole molecules; a disordered structure.

8.3 MECHANICAL RESPONSES

8.3.1 Modes of deformation and a common response

The complexity of structure of partially crystalline polymers, described in the previous sections, is matched by a variety of modes of deformation, which are summarised in Table 8.3. Many of these occur more simply in wholly crystalline or wholly amorphous materials, have been discussed in previous chapters, and only require a brief reference now. But there are other effects which depend on the interaction of the structural components.

Fig. 8.26(a) shows a common form of stress-strain response of an unoriented, partially crystalline polymer.

The curve is typified by an initial elastic deformation, followed by yield and then fracture. Sometimes there is a maximum in the curve at the yield point. This may be approximated by the simpler idealised form which was suggested in Appendix B.3.4. Orientation will lead to the change of stress-strain curve shown in Fig. 8.26(b).

In some partially crystalline polymers, special molecular effects, such as a crystal lattice transition, will lead to different forms, as for example in wool. In other materials, larger scale features of structure will have an important influence.

Table 8.3 – Modes of deformation of partially crystalline polymers.

Recoverable, Elastic or Viscoelastic, Structural Strain		
	Energy dependent	Entropy-dependent
Small strain	crystal lattice (7.2.1)†	
	glassy chain (6.2.1)	
	cross-linked net (6.3)	
Large strain	crystal transition (7.2.5)	rubber elasticity (4.2.5)
Plastic Change of Structure		
	Crystal	Amorphous
	recrystallisation	glassy chain conformation
	unfolding, refolding	chain slippage
	defects moving	chain breakage
		cross-links breaking and reforming
Loss of Continuity		
	cavitation	
	transverse crack initiation and propagation	
	axial splitting	

† The numbers in brackets refer to the relevant sections.

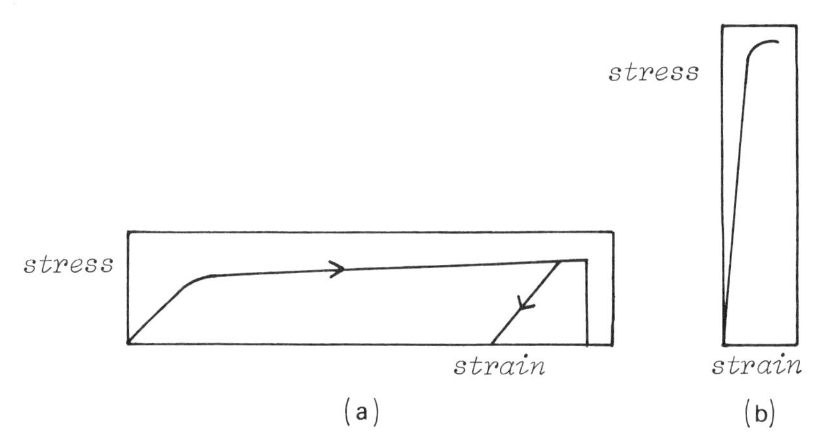

Fig. 8.26 – (a) Typical stress strain curves of an unoriented crystalline polymer. (b) Curve for an oriented polymer.

8.3.2 Elastic and viscoelastic deformation

Whenever a material is under stress it must suffer elastic strain which will lead to some recovery on removal of stress. The short-range elastic deformation involves the change of bond angles or bond spacing in ways that have already been discussed in sections 2.3.1 and 7.2. Such deformations are associated with vibrations at around 10^{13} Hz, and so, in any practical sense, the deformation is independent of time. The large-strain deformations, due to lattice transitions (section 7.2.5) or rubber elasticity (Chapter 4) have also been dealt with in earlier chapters. Since they involve comparatively long-range movements of chain segments past one another, there will be a viscous drag giving rise to viscoelasticity in the form of primary (recoverable) creep.

Often it is possible to deal with the elastic or viscoelastic deformation on the basis of a combination of known properties of the separate crystalline and amorphous components. The problem is then an exercise in the mechanics of composites and it must be remembered that the modulus parallel to the direction of fibrillar orientation or along lamellae will be close to that of the stiff component, while the modulus in the cross direction will be close to that of the soft component. The components themselves, particularly crystalline lattices, will be anisotropic; and if the structure is not simply fibrillar or lamellar, but is micellar for example, the mechanics is even more difficult to analyse. The mechanical effects of interaction, such as the consequences of different Poisson's ratio in the components of the structure or of localised strain differences, must also be considered.

In some partially crystalline polymers with a very fine structure, it may be impossible to use a macroscopic two-phase model of components with defined mechanical properties. It may be necessary to go back to first principles, discussed in Chapter 2, and consider the behaviour of individual irregular chain segments linking individual crystalline regions.

One thing is clear. No simple treatment can be generally applicable. The characteristics of each form of structure must be determined, and appropriate methods of analysis and explanation adopted.

8.3.3 Yield and drawing

Yield is a plastic deformation with a change of structure which is irreversible; or, to be more explicit, the deformation is not recovered on removal of load. Sometimes the original structure is destroyed and cannot again be produced (except by repeating the whole manufacturing history). In other instances, the original structure, or something close to it, can be retrieved by introducing mobility into the plastically deformed structure by heat or by the absorption of water or some other plasticiser. A material which has suffered such a temporary plastic deformation is thus likely to suffer dimensional changes in subsequent use. The structure may also sometimes be restored by applying a stress in the opposite direction. For example, kink bands, which are found in compression, may be pulled out by tension.

Since the yield involves large deformation, over energy barriers, it is always time and temperature dependent in the practical range of times, and yield stress increases with rate of deformation and decreases with temperature. Under constant load, the yield continues as secondary (irrecoverable) creep.

Yield can occur separately in the individual components. Some crystal defects may move or crystals may suffer a plastic deformation, but the strong co-operative effects within the crystal will usually make this more difficult than yield in the amorphous material. And yielding in one part of a composite tends to relieve stresses on other parts.

If the amorphous material is a pure elastic rubber, then yield will not occur. If it is glassy, then yield can occur, as discussed in section 6.2.4 because the energy barriers, which prevent motions within or between the polymer molecules, will be overcome under stress.

More interesting are the polymers without a simple single glass transition. For example, in nylon, the aliphatic sequences are rubbery at room temperature, but the -CO.NH- groups hold the chains together in a network unitl a transition temperature (often miscalled the glass transition) is reached. Under stress these associations will rupture, chain segments will move, and fresh associations will form. This often gives rise to a minor lower yield point in these materials. A more dominant effect occurs in cellulose where there is a very high degree of hydrogen-bonding between chains, and a marked yield occurs when these start to break under stress and re-form in new positions. Other plastic changes will occur when chain ends move through the structure or when chains break, without leading to complete rupture.

In many partially crystalline polymers, yield can occur through an overall change of structure. This occurs in polyethylene and in the upper yield point of nylon. The initial extension occurs mainly in non-crystalline regions, but as the stresses on the chains rise they eventually cause the crystalline regions to break up and reform in new ways.

The mechanism at the moelcular level can be envisaged in various ways. It may be that there is a local melting, partly due to a change of melting-point with stress and partly due to the heat generated once the plastic deformation is progressing: the melting would allow chain elongation, relief of stress, flow away of heat, and recrystallisation. Alternatively, there may be a major plastic deformation with loss of crystallinity. For example, one could imagine a continuous unfolding within the crystal. Between these two extremes there are many other possibilities with varying dregrees of structural disruption.

At a rather larger scale of structure, it is proposed that the drawing of polyethylene leads to a change from a lamellar crystalline morphology to a fibrillar morphology, as illustrated in Fig. 8.27.

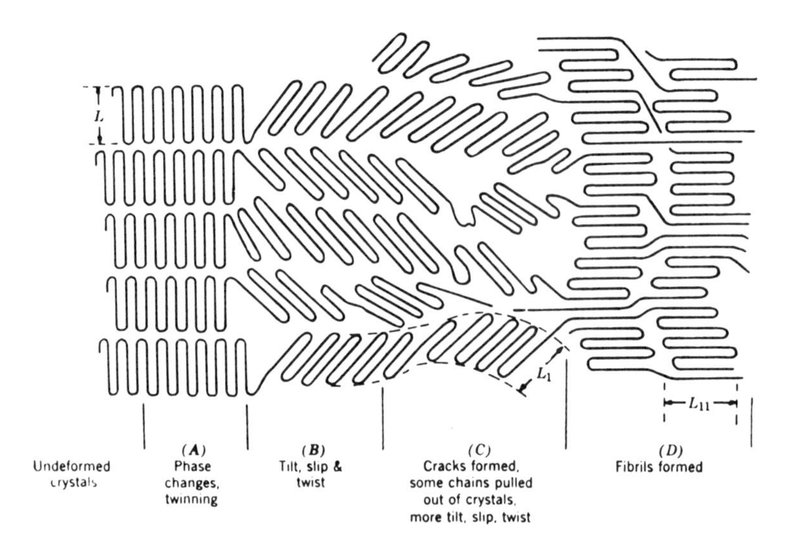

Fig. 8.27 – Suggested form of change from lamellar to fibrillar structure. After Peterlin (1965) (Peterlin, *J. Pol. Sci.*, *C*, 9, 61, 1965).

8.3.4 External and internal inhomogeneity in yield

Yield can take place more or less uniformly throughout the material, but it may also be concentrated locally. On a gross scale, necking is a consequence of the instability of a falling stress-strain curve as discussed in Appendix B.6.4. The drop in stress after yield may be due to structural causes, the fact that once disruption starts it becomes easier to deform neighbouring units, or to the heating which results from the plastic deformation.

At a finer scale, yield may also concentrate into particular zones, either within the neck or throughout a specimen which is yielding uniformly on a gross scale. These effects are most easily shown up by the yield in compression of an oriented polymer such as a drawn fibre. Fig. 8.28(a) illustrates the kink bands which occur on the inside of a bent polyester fibre. The compressive stress is forcing the material to shorten, and, when a yield stress is reached, this is easily achieved by internal buckling under the influence of the resolved shear stress. The yielding may be on any scale. It could be a slip in a crystal lattice, but it is more likely to be yielding over a larger scale by means of the molecular mechanisms already discussed.

The material on the outside of the local band will have to elongate more (just as the soldiers on the outside of a turning colum have to march faster) and may also yield. Fig. 8.28(b) shows a picture of kink-bands on the outside of a bend in an undrawn unoriented, or only slightly oriented, fibre. The local mechanisms of yielding are not as easy to imagine, but the effect may occur through yielding under the influence of shear stress.

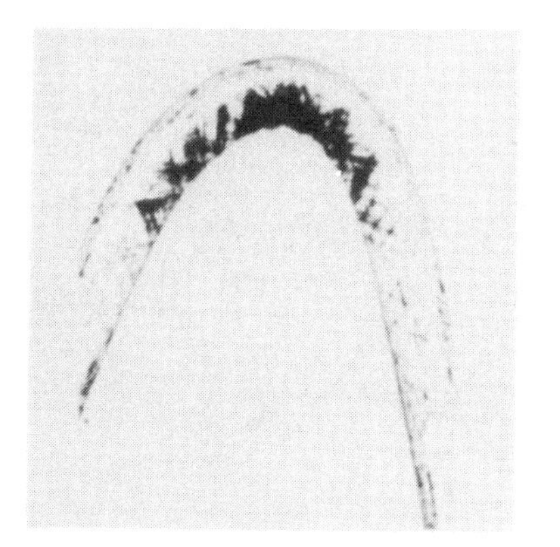

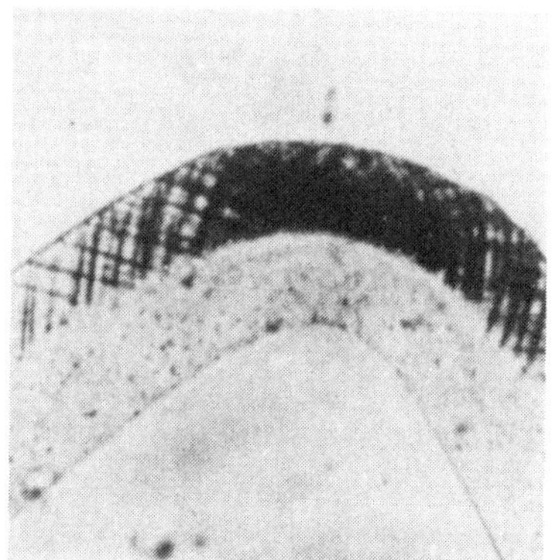

Fig. 8.28 – Kink-bands as viewed between crossed polars in a bent fibre. (a) Drawn PET, with yield in compression on the inside of the bend. (b) Undrawn PET with yield in tension on the outside. After Jariwala (1974) (Jariwala, Ph.D. thesis, Manchester).

When yielding occurs through elongation, though not when it occurs through uniform shear, there will be tendency to volume increase. In necking, this is avoided by the decrease in thickness at the neck. But in local yielding, it may only be possible to accommodate the volume change by cavitation, which is often associated with kink bands.

8.3.5 Orientation, size of units, and molecular extent.

Fig. 8.29 shows a typical family of stress-strain curves for partially crystalline polymers. As orientation increases, there is an increase in modulus, yield stress and strength and a decrease in breaking extension. There are several reasons for this. Firstly, the non-crystalline material may be oriented and less easy to deform by pulling into oriented alignments. Secondly, if crystallites are disoriented, they can help the elongation by rotating into alignment. Thirdly, the localised crystalline and amorphous units will be anisotropic in response. However, the details of these effects can only usefully be discussed in detail in relation to particular materials.

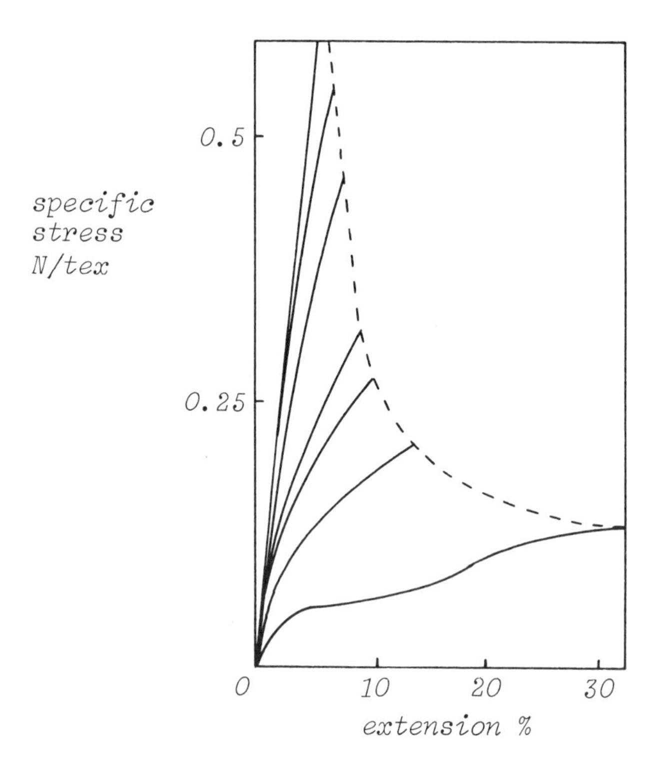

Fig. 8.29 – Stress-strain curves of a series of cellulose fibres (regenerated from cellulose acetate). After Work (1949).

The size of crystalline units generally has more influence on strength, than on the earlier part of the stress-strain curve. Thus finally textured regenerated cellulose fibres are much stronger than those with coarse texture: in the 1950s the strength of rayon fibres for tyre cords was doubled, by changing the manufacuring conditions in order to cause the formation of a large number of very small crystals. One reason is that stress concentrations are less in a fine structure. The other is that there will be a lower proportion of free chain ends between crystals when the gaps are small, than there will be when the gaps are large: in the fine texture the crystallites, which are strong units in the structure, will be more effectively linked by the molecules.

For the reasons already mentioned, molecular extent is an even more difficult parameter to study, but it has a considerable influence on mechanical properties.

8.3.6 Fracture and fatigue

Ultimately, the material will break. At the level of fine structure, there are three causes operating: chain breaking; chain slippage; and chain separation. But the local separation will depend on the stress distribution in the material, and particularly on stress concentrations at flaws.

Experimental work is now showing up a number of characteristic forms, but the analysis by fracture mechanics has not yet progressed very far. Apart from rupture under monotonically increasing stress, there is also fatigue failure under repetitive stresses.

The subject of failure will be taken up in another volume.

8.3.7 Comparative summary

The varied patterns of crystallisation, linked with disorder, and the different molecular responses of different polymers lead to a wide variety of mechanical responses, which will be described in more detail and related to structure in Volume 2. However the behaviour of some typical examples will be briefly summarised here.

Fig. 8.30(a) shows a typical stress-strain curve for low-density polyethylene, and is characteristic of many unoriented crystalline polymers. The initial steep,

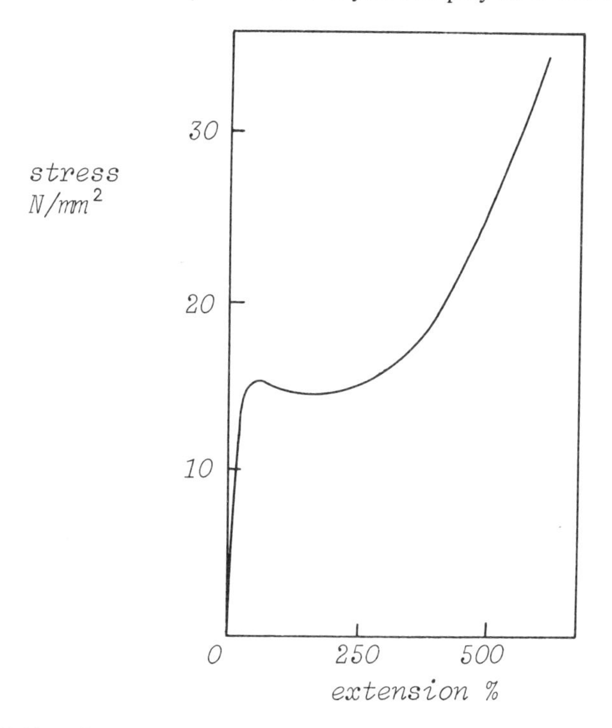

Fig. 8.30(a) – Stress-strain curves in extension to break of partially crystalline polymers. Unoriented low-density polyethylene. After Boenig (1966).

and linear, portion is an elastic deformation of the composite crystalline/amorphous network. But eventually the forces exerted on the crystallites become large enough to disrupt them, and the material yields with a progressive reorganisation of the whole crystallisation pattern, involving large-scale relative movement of segments in both crystalline and amorphous regions. Various detailed mechanisms may be proposed ranging from localised melting and recrystallisation to plastic yield within crystals, accompanied by some interchange of segments between positions of order and disorder. The long-range effects due to continuity along polymer chains will have an important influence. There may be changes in macroscopic geometry, such as necking, and in microscopic fine structure, such as a switch from lamellar to fibrillar morphology. At higher strains, the reorganisation and increase of orientation becomes more difficult, and consequently the resistance to elongation increases. Eventually the structure jams and high stress concentrations lead to fracture, with localised or distributed chain breakage. Beyond the yield point, there is poor recovery, as indicated in Fig. 8.31(a).

Another example of drawing is shown in Fig. 8.32, though here, in polyethylene terephthalate, there is the difference that the unoriented material is amorphous, but crystallises on drawing. The yield point – or draw force – decreases with temperature: in rapid drawing, which is effectively adiabatic, this leads to instability, and the material shows a characteristic draw ratio, with partial drawing leading to an alternation of drawn and undrawn material.

Material which has been oriented by drawing in this way can be regarded as a new processed material with increased stiffness and strength and reduced breaking extension: apart from any more subtle structure effects, the direct geometric reasons for this are shown in Appendix B.6.4. In the limiting case, typical of commercial fibres, the material is drawn to the maximum practicable

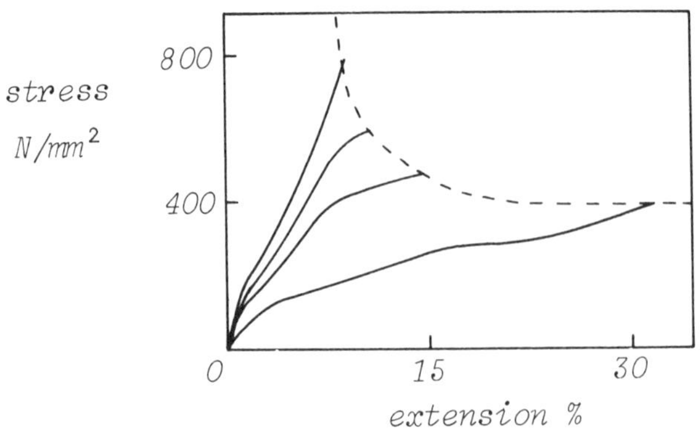

Fig. 8.30(b) – Stress-strain curves in extension to break of partially crystalline polymers. Polyethylene terephthalate fibres, oriented to varying extent by drawing. The dotted line is the locus of breaking points. After Marshall and Whinfield (1953).

extent so that the post-yield region is reduced to the minimum compatible with avoidance of breakage during production. However, the particular conditions of drawing, namely rate and temperature, and subsequent thermo-mechanical treatments, as well as differences in type of polymer, do cause big differences in properties as illustrated in Fig. 8.30(b) and (c).

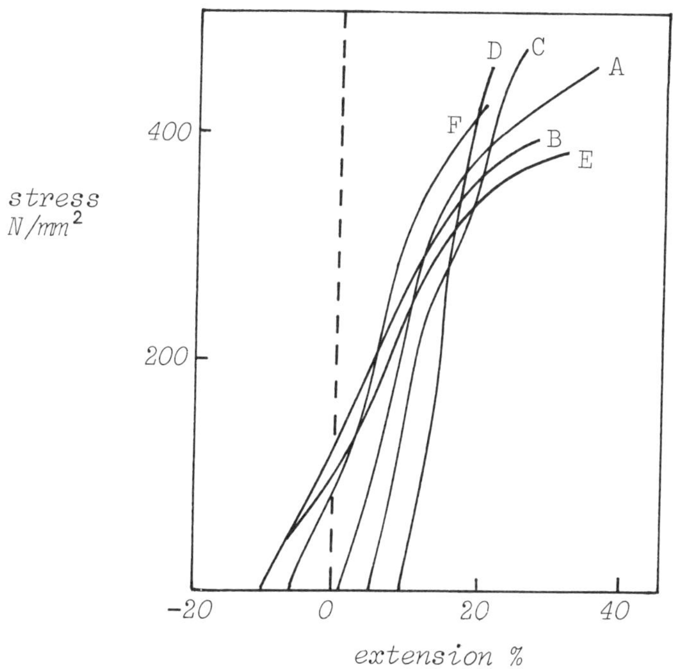

Fig. 8.30(c) – Stress-strain curves in extension to break of partially crystalline polymers. Nylon 66 yarn (78 dtex) subject to various thermal treatments: (A) as received; (B) 200°C, zero tension; (C) 200°C, 30 gf; (D) 200°C, 75 gf; (E) 200°C, zero tension + boiling water, zero tension; (F) 200°C, 10 gf + 160°C, 50 gf + boiling water, zero tension; (F) 200?C, 10 gf + 160?C, 50 gf + boiling water, zero tension. The lengths are related to the original length as received. After Hearle, Sen Gupta and Matthews (1971).

In stretched films, it is possible to demonstrate the difference between the stress-strain curves parallel and perpendicular to the direction of orientation as shown in Fig. 8.30(d).

The behaviour described above is simple and widespread and can be interpreted, at least qualitatively, as the deformation of a composite system of crystalline regions embedded in and interlinked with amorphous material. The crystals will be stiff and elastic, with a modulus and yield stress dependent on the chain conformation, high in extended chains and lower in helical chains, and on the molecular interactions. The amorphous material may be rubbery or

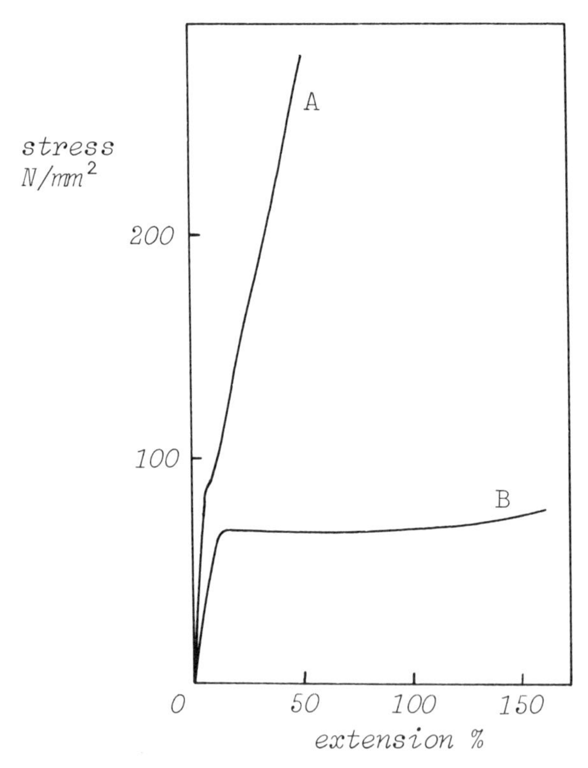

Fig. 8.30(d) – Stress-strain curves in extension to break of partially crystalline polymers. Polycarbonate film which has been uniaxially stretched: (A) parallel, (B) perpendicular to orientation. After Schnell (1964).

glassy, and will be influenced by cross-linking and plasticisation as well as time and temperature. For the reasons given in Appendix B.6.5 there will be a dependence on whether the composite assembly tends to be lamellar of fibrillar. The curves for rayon in Fig. 8.30(e) show the difference between ordinary rayon with a micellar texture and newer rayons (high wet modulus) with a fibrillar texture: the dotted lines are from simple treatment of a composite system, given in Appendix B.6.5, Fig. B.23. There will also be an influence of the nature of the interconnection between the components. The degree of orientation of each phase will also play a major role. But although there are big differences, the framework of explanation remains consistent. This is not so with some more specialised structures or patterns of response.

The stress-strain curves of the regenerated cellulose fibres (rayon) shown as Fig. 8.30(e) follow the behaviour already described. But the natural plant fibres are different, as shown in Fig. 8.30(f) A, B: their response, which shows no yield point, is that of an assembly of crystalline fibrils helically wound around cell walls, as a consequence of the mechanism of biosynthesis. Another material

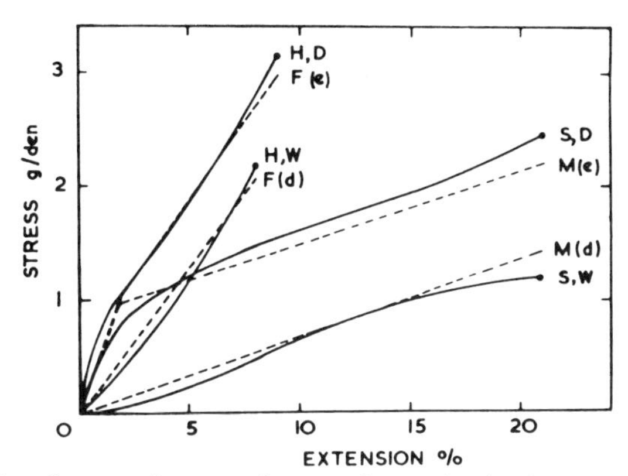

Fig. 8.30(e) – Stress-strain curves in extension to break of partially crystalline polymers. Rayon: S = standard (micellar); H = high-wet-modulus (fibrillar); D = Dry (65% r.h.); W = wet. The dottoed lines are the theoretical predictions from Fig. B.23.

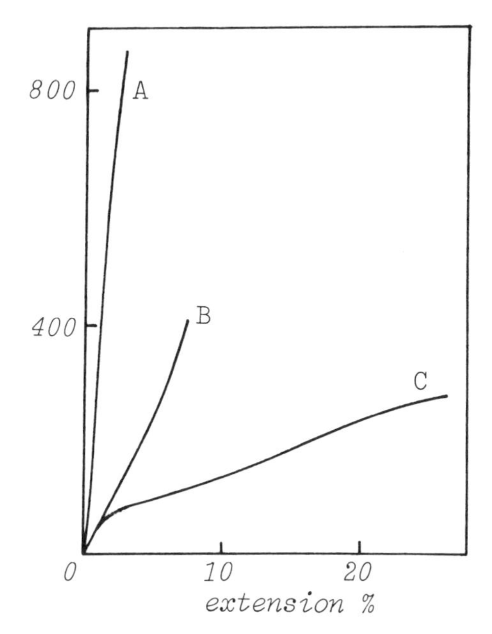

Fig. 8.30(f) – Stress-strain curves in extension to break of partially crystalline polymers. Cellulose fibres: (A) flax; (B) cotton; (C) rayon. After Meredith (1958).

which is dominated by elastic deformation of the crystal lattice is the highly oriented, highly crystalline, aromatic polyamide fibre, Kevlar, with the stress-strain curves shown in Fig. 8.30(g).

The typical range of reasonable elastic recovery in crystalline polymers is from 2% to 20%, and usually the yield point is the limit of elasticity as illustrated

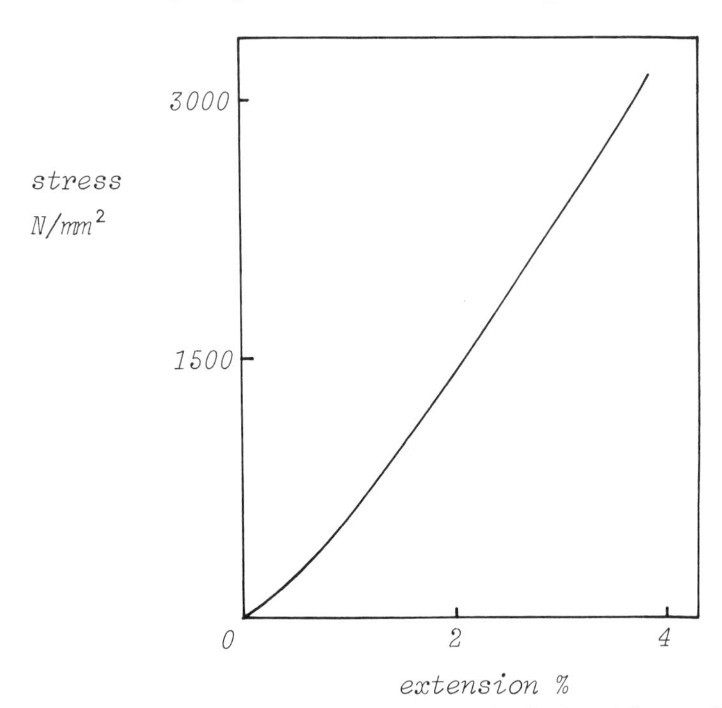

Fig. 8.30(g) – Stress-strain curves in extension to break of partially crystalline polymers. Kevlar fibre. After Konopasek (1975).

for rayon in Fig. 8.31(b); but some materials are different. Although they contain crystalline regions, the high-extension Spandex fibres, Fig. 8.30(h) are

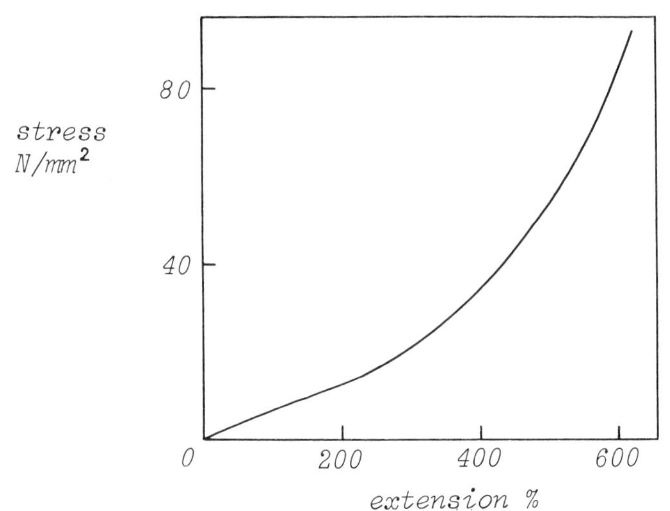

Fig. 8.30(h) – Stress-strain curves in extension to break of partially crystalline polymers. Spandex, segmented polyurethane fibre. After Wilson (1967).

best regarded as amorphous elastomers: they are block copolymers dominated by a 'soft' rubbery component but with 'hard' segments forming crystallites, which crosslink the elastic network. Fig. 8.31(c) shows their good recovery from large strains, although the complexity of response is shown by the difference between the first loading curve and the behaviour after repeated cycling.

Wool and hair are more interesting. The stress-strain curves Fig. 8.30(i) are

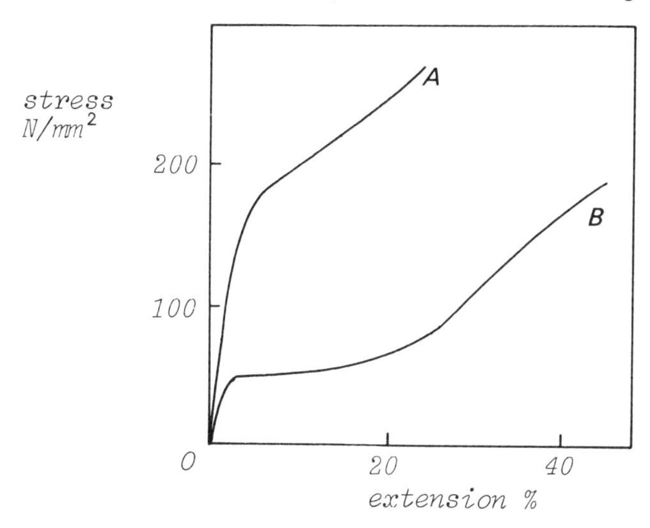

Fig. 8.30(i) – Stress-strain curves in extension to break of partially crystalline polymers. Wool: (A) 0% r.h.; (b) 100% r.h. After Speakman (1927).

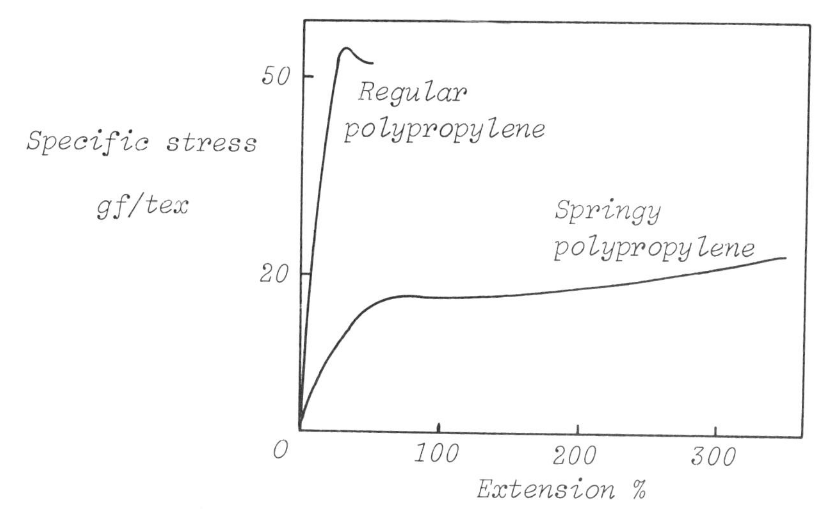

Fig. 8.30(j) – Stress strain curves in extension to break of partially crystalline polymers. Regular orientated (A) and Springy (B) polypropylene. After Konopasek (1975).

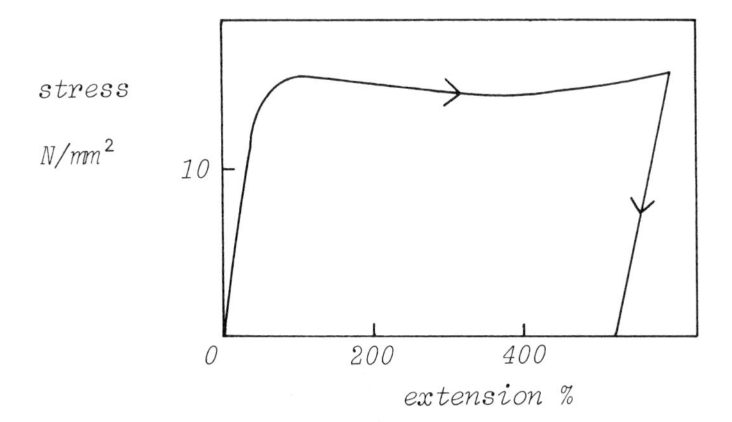

Fig. 8.31(a) – Recovery behaviour of partially crystalline polymers. Low-density polyethylene. After Boenig (1966).

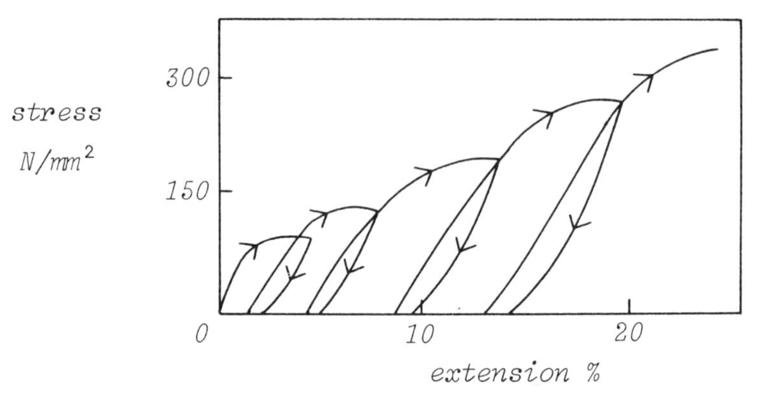

Fig. 8.31(b) – Recovery behaviour of partially crystalline polymers. Rayon, regenerated cellulose fibre. After Meredith (1945).

of the usual shape, but illustrate the importance of water as a plasticiser. What is unusual is that the recovery is good beyond the yield point, as shown in Fig. 8.31(d) remaining complete up to 30% extension and good up to a breaking elongation of about 50%. However, the path of the return curve is different to the extension. This particular behaviour can be explained by the detailed mechanical response of a system of crystalline fibrils, with a phase transition from α- to β-forms (see section 7.1.5) embedded in a rubbery matrix. The structure itself is a consequence of the manufacture in the cell wall of a mixture of particular proteins, under the natural control of the genes.

However unusual structures can also be made artificially. For example, by appropriate conditions of crystallisation and treatment, it is possible to obtain

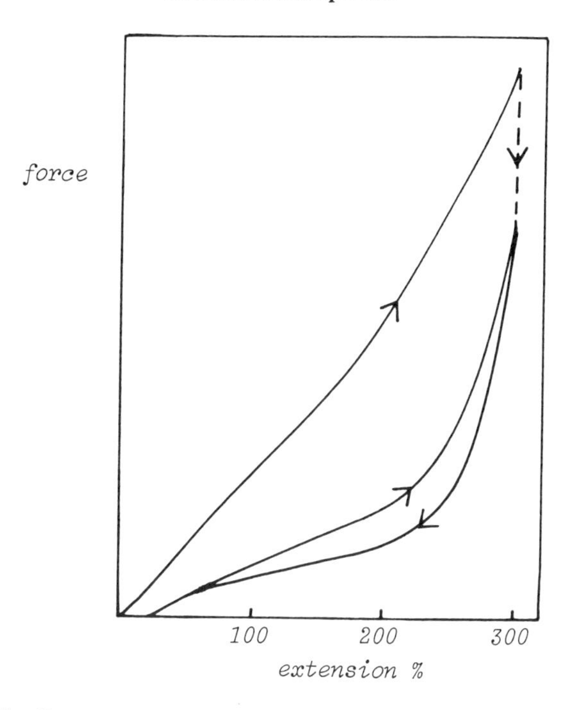

Fig. 8.31(c) – Recovery behaviour of partially crystalline polymers. Spandex: (A) first loading; (B) after repeated cycling. After Hughes and McIntyre (1976).

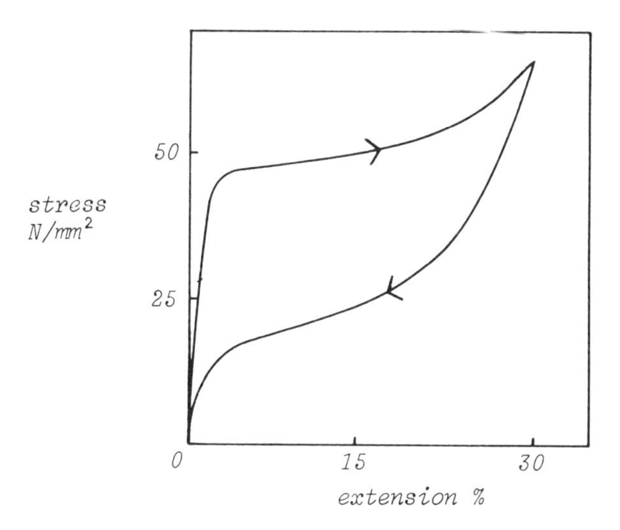

Fig. 8.31(d) – Recovery behaviour of partially crystalline polymers. Wet wool. After Hearle (1975).

polypropylene with a lamellar structure, which can open up like a set of leaf springs. The extensibility of this springy polypropylene is illustrated in Fig. 8.31(i), in contrast to that of regular oriented polypropylene. Once again there is good recovery from large extensions, with a different path in recovery, as shown in Fig. 8.30(e).

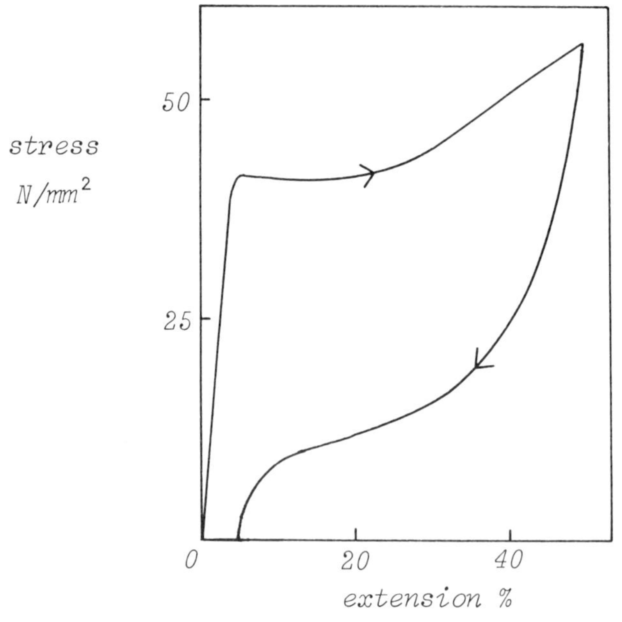

Fig. 3.11(e) – Recovery behaviour of partially crystalline polymers. Springy polypropylene. After Spraque (1974).

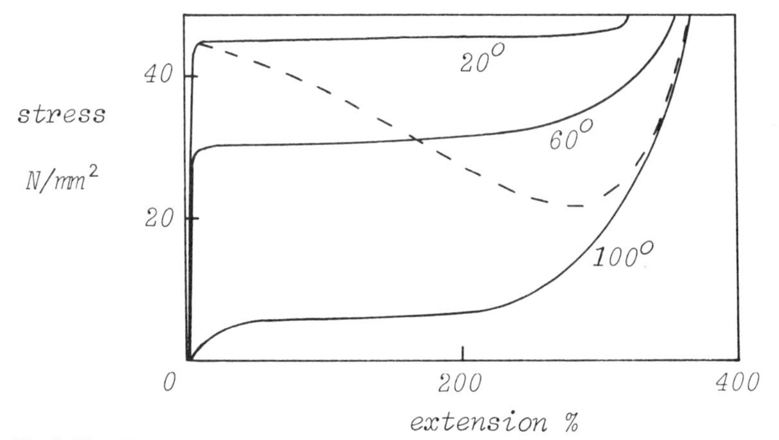

Fig. 8.32 – Drawing behaviour of polyethylene terephthalate at 20°C, 60°C and 80°C. The dotted curve is for rapid, adiobatic elongation. After Marshall and Thompson (1954).

Inevitably, the stress-strain curves included in the preceding figures have been drawn to different scales. In order to show the wide range of properties which can be found in crystalline polymer materials, their breaking points are listed in Table 8.4, and demonstrate the hundred-fold range in strength and breaking extension. One may also note that although the modulus of ordinary commercial polyethylene samples may vary by at least a ten-fold range from about 0.1 to 1 GPa, highly oriented laboratory-made samples reach moduli of 70 GPa or more, not far short of the calculated crystal modulus of about 250 GPa. Even so, only a small sample of the possible properties found in crystalline polymers are shown by these values.

Table 8.4.

Material	Breaking point	
	Stress N/mm^2	Extension %
Polyethylene, unoriented	3.5	600
Polycarbonate, $\perp$ orientation	80	160
Spandex	90	600
Wet wool	200	45
Dry wool	250	25
Springy polypropylene	260	350
Polycarbonate, $\parallel$ orientation	280	45
Rayon	350	25
Cotton	400	7
Polypropylene	500	80
Nylon	400 to 500	40 to 15
Polyethylene terephthalate	400 to 800	30 to 10
Flax	900	5
Kevlar	3200	4

8.4 CONCLUSION

In constrast to the detailed analysis and specific information in earlier chapters, this Chapter has been general and discursive. This is inevitable for two reasons. Firstly, the behaviour of partially crystalline polymers is very much dependent on the nature of the particular polymer and, even more, on the particular form of fine structure. Some of the effects have been mentioned here, but others, such as the way in which a particular form of lamellar crystallisation leads to springy polypropylene, have not. Secondly, the theoretical framework of explanation has not yet been adequately built, except in some isolated instances.

There is much more to be done in order to explain the behaviour of assemblies of long-chain polymer molecules which are neither simple disordered nor fully ordered, but are in some state of intermediate order, not unlike that of a parliamentary assembly.

APPENDIX A

A Discursion on some General Physics

> The main actors in the drama of the kinetic theory are rapidly-moving molecules; the principle events in their lives are collisions with one another and with the boundary of a containing vessel.
>
> Sir James Jeans, *Kinetic Theory of Gases*

A.1 WHY THE DISCURSION?

There are certain branches of physics which are of great relevance to our study of the physics of polymers. Some are well known, but it is useful to recapitulate their main features, so that we can bring out the analagous aspects in the treatment of polymer problems. Readers may even welcome the opportunity of revision! Some other aspects are often neglected in conventional physics courses and so it is even more important to review them here.

We shall start with an important, but neglected, piece of thermodynamics; and go on to reminders about the treatment of ideal gases, and the way in which the equations of statistical mechanics can be applied to condensate materials (liquids and solids).

It is important to note that this Appendix and the following one are not polymer physics: they consist of relevant general physics and mathematics. Where, as often happens, the two get presented together there is a danger that the general topics may be regarded as peculiar to polymers. It is for this reason that they are dealt with separately in this book.

A.2 THE THERMODYNAMICS OF EXTENSION

A.2.1 Energy and entropy

For any reversible, isothermal change in a system, it follows from the first and second laws of thermodynamics that:

$$\mathrm{d}W = \mathrm{d}U + \mathrm{d}Q = \mathrm{d}U - T\mathrm{d}S \tag{A.1}$$

where
$\mathrm{d}W$ = total work done on the system
$\mathrm{d}U$ = change in internal energy,
$\mathrm{d}Q$ = change in entropy,
T = absolute temperature.

We now take, as our system, any specimen of length l in simple extension under a force F, as in Fig. A.1. A slight increase in force will lead to an extension $\mathrm{d}l$, and the work done will be $F\mathrm{d}l$ (neglecting the second order term in $\mathrm{d}F\mathrm{d}l$). We shall assume that this is the only source of work: in reality, there may be others, but they are usually small and can be ignored. (Examples of these other sources are work done under lateral contraction by atmospheric or other pressure, or energy changes associated with evaporation or absorption of moisture.)

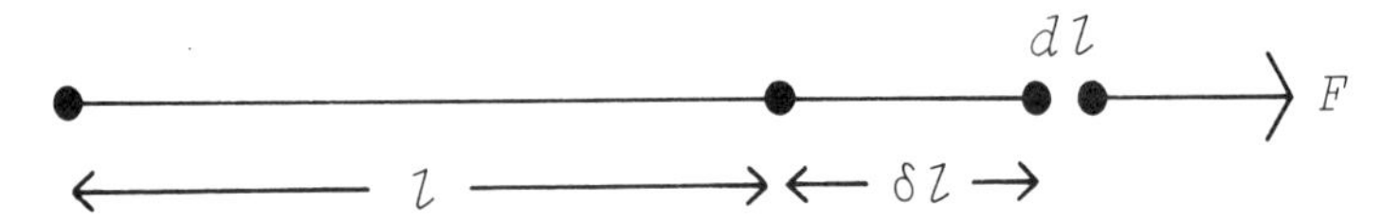

Fig. A.1 – Linear specimen in extension.

We thus get:

$$F\mathrm{d}l = \mathrm{d}U - T\mathrm{d}S \tag{A.2}$$

$$F = (\partial U/\partial l)_T - T(\partial S/\partial l)_T \tag{A.3}$$

Equation (A.3) shows that the tension in an extended specimen is made up of two terms: one depending on internal energy changes and one depending on entropy changes. In crystalline solids, it is the energy term which predominates: there is no appreciable disordering on straining a crystal lattice, but a lot of work is done in pulling the atoms out of their lowest energy arrangement. In many polymer systems we shall find that the entropy term plays an important part.

A.2.2 Separation of effects

It is useful to see how the values of the energy and entropy terms can be estimated separately by experiment. For this purpose we introduce the Helmholtz free energy A as an auxiliary quantity. This is defined as:

$$A = U - TS \tag{A.4}$$

Hence from (2.3):

$$\left(\frac{\partial A}{\partial l}\right)_T = \left(\frac{\partial U}{\partial l}\right)_T - T\left(\frac{\partial S}{\partial l}\right)_T = F \tag{A.5}$$

Thus the equilibrium length, when $F = 0$, is the position of minimum Helmholtz free energy, involving a balance between internal energy and entropy effects.

For a general change, we have:

$$\mathrm{d}A = \mathrm{d}U - T\mathrm{d}S - S\mathrm{d}T \tag{A.6}$$

$$= F\mathrm{d}l - S\mathrm{d}T \quad \text{from (A.2)} \tag{A.7}$$

Hence:

$$\left(\frac{\partial A}{\partial T}\right)_l = -S \tag{A.7}$$

But it is a general theorem that:

$$\frac{\partial}{\partial T}\left(\frac{\partial A}{\partial l}\right)_T \equiv \frac{\partial}{\partial l}\left(\frac{\partial A}{\partial T}\right)_l \tag{A.8}$$

and so, by substitution from (A.5) and (A.7):

$$\left(\frac{\partial F}{\partial T}\right)_l = -\left(\frac{\partial S}{\partial l}\right)_T \tag{A.9}$$

The entropy contribution to the force F, given by the second term in (A.3), is thus equal to $T(\partial F/\partial T)_l$. The partial differential coefficient, $(\partial F/\partial T)_l$, is the rate of change of tension with temperature for a specimen held at constant length: this can be measured experimentally. Hence we get a measure of the entropy contribution, and, by subtraction from the total force F, a measure of the internal energy contribution.

A.2.3 Other relations

We can also derive a relation for the coefficient of linear expansion, namely the fractional change in length with temperature at constant (usually zero) force. Using another general theorem, and then substituting from (A.9), we have:

$$\begin{aligned}\frac{1}{l}\left(\frac{\partial l}{\partial T}\right)_F &\equiv -\frac{1}{l}\left(\frac{\partial l}{\partial F}\right)_T\left(\frac{\partial F}{\partial T}\right)_l \\ &= \frac{1}{l}\left(\frac{\partial l}{\partial F}\right)_T\left(\frac{\partial S}{\partial l}\right)_T \\ &= \frac{1}{l}\left(\frac{\partial S}{\partial F}\right)_T \end{aligned} \tag{A.10}$$

This means that expansion is positive when stretching the system leads to an increase of entropy – a greater degree of disorder – and is negative, when the reverse is true.

Another relation of some interest is the change in temperature during adiabatic extension. This is most conveniently given, with substitution from (A.9), by:

$$\left(\frac{\partial T}{\partial F}\right)_S = -\left(\frac{\partial T}{\partial S}\right)_F \left(\frac{\partial S}{\partial F}\right)_T$$

$$= -\left(\frac{\partial T}{\partial S}\right)_F \left(\frac{\partial S}{\partial l}\right)_T \left(\frac{\partial l}{\partial F}\right)_T$$

$$= \left(\frac{\partial T}{\partial S}\right)_F \left(\frac{\partial F}{\partial T}\right)_l \left(\frac{\partial l}{\partial F}\right)_T$$

$$= -\left(\frac{\partial T}{\partial S}\right)_F \left(\frac{\partial l}{\partial T}\right)_F \tag{A.11}$$

Since the change in heat $dQ = TdS$, we see that $(1/T)(\partial S/\partial T)_F = C_F$, the thermal capacity at constant force. Hence:

$$\left(\frac{\partial T}{\partial F}\right)_S = \frac{T}{C_F}\left(\frac{\partial F}{\partial T}\right)_l \left(\frac{\partial l}{\partial F}\right)_T$$

$$= -\frac{T}{C_F}\left(\frac{\partial l}{\partial T}\right)_F \tag{A.12}$$

A.3 THE TREATMENT OF AN IDEAL GAS

A.3.1 Definition and kinetic analysis

An ideal gas is defined as a random collection of elastic, point particles (molecules) of mass m moving about with all possible velocitites, with no intermolecular forces between the particles. The relation between pressure and volume is then derived by calculating the change in momentum when a particle hits the wall and rebounds with its component of velocity perpendicular to the wall reversed in sign. The sum, of the momentum changes for all collisions in unit time gives the force on the walls. For a given number of particles, the collisions will be less frequent in a large container, since the time taken to cross the container will be greater, and so pressure p falls as volume v increases. Conversely, for a given volume of container, the pressure increases with the number of particles N. The actual expression derived from such a kinetic calculation is:

$$p = \frac{1}{3}\frac{N}{v} m\overline{u^2} \tag{A. 13}$$

where $\overline{u^2}$ is the mean square velocity.

For one gram of a gas of molecular weight M_0 the number of particles is N_0/M_0, where N_0 is Avogadro's number. Thus:

$$pv = \frac{1}{3}(N_0/M_0)m\overline{u^2} \tag{A.14}$$

Alternatively, the gas equation may be proposed by idealisation of the experimentally determined Boyle's and Charles' laws, combined with the definition of an absolute temperature T, as:

$$p = (1/M_0)RT/v = (N_0kT/M_0)(1/v) \tag{A.15}$$

where R = gas constant (per gram molecule)
k = Boltzmann's constant.

Usually, the gas equation is written as $pv = RT$, for a gram-molecule of gas; but, for the comparisons to be made later, it is preferable to express it in the above form, valid for 1 gram of gas.

Comparing (A.14) and (A.15) we see that they are compatible if:

$$m\overline{u^2} = 3kT \tag{A.16}$$

Since the mean kinetic energy of a particle will be $\frac{1}{2}m\overline{u^2}$, we note that (A.16) could also be obtained by using the law of equipartition of energy and ascribing energy $(\frac{1}{2}kT)$ to each of the three degrees of freedom of the point particle.

Further development of the kinetic theory of gases leads to calculations of the distribution of velocities, to other properties of ideal gases, and to modifications which enable the theory to be corrected for real gases, and indeed to some extent, to be applied to liquids and solids made up of small molecules.

A.3.2 Derivation by thermodynamics and statistics

Instead of using the kinetic approach, with its easily understood mechanistic model, the same relations can be derived from thermodynamics and the statistical theory of matter. The classical statistics of Boltzmann, which are valid in this problem, lead to entropy S^* being defined as:

$$S^* = k \log_e W \tag{A.17}$$

The thermodynamic probability W is a measure of the degree of disorder of the system. For a perfectly ordered system, there is only *one* form of arrangement (everything must be in its place!), and, if we put $W = 1$, we get $S^* = 0$. For a disordered system, there are a large number of arrangements which are all substantially equivalent: we can thus define W as the number of ways of arranging the system, while still satisfying the defined external conditions.

Fortunately, we do not usually need to know the absolute value of the entropy, and nor is this defined in classical thermodynamics. We are only concerned with the differences in the entropy of two states:

$$\Delta S^* = S_2^* - S_1^* = k\,(\log_e W_2 - \log_e W_1)$$
$$= k \log_e(W_2/W_1) \tag{A.18}$$

Consequently, we also never need to define the actual number of arrangements W: it is sufficient to be able to produce arguments showing that a certain change doubles the number of states, or changes them by 10%, or by whatever factor is applicable.

If we decide to measure entropy S from some arbitrarily defined zero by pugging $S = S^* - S_1^* = 0$ at $W = W_1$, we get:

$$S = k \log_e(W/W_1) \tag{A.19}$$

(We note that S^* refers to an absolute value of entropy, which is explicitly defined when we go over to the limited number of states permitted by quantum mechanics, while S refers to entropy defined in relation to an arbitrarily chosen zero, as is done in practice).

Returning to the ideal gas, we see that it is by definition a random collection of particles distributed in a large volume. We can imagine this volume v as divided into a large number of small elements of volume v_1. The number of ways of distributing a single particle among these elements is equal to the total number of elements, namely to v/v_1. For two particles – since we are not excluding the possibility of more than one particle being in the same element – the number of ways is $(v/v_1)^2$, and in general, for N particles, it is $(v/v_1)^N$. Hence W must be proportional to $(v/v_1)^N$. We can now choose to measure entropy from a state (S_1^*, W_1) in which $(v/v_1)^N = 1$, namely when $v = v_1$, and so, in general:

$$W/W_1 = (v/v_1)^{N_0/M_0} \tag{A.20}$$

If we consider a gram of gas with N_0/M_0 particles, (A.19) thus gives:

$$S = k \log_e (v/v_1)^{N_0/M_0} \tag{A.21}$$
$$= (kN_0/M_0)(\log_e v - \log_e v_1)$$

Going back to classical thermodynamics, we note that for an isothermal volume change in an ideal gas, the work done is $(-p\mathrm{d}v)$, and so (A.1) becomes:

$$-p\mathrm{d}v = \mathrm{d}U - T\mathrm{d}S \tag{A.22}$$

But by definition there are no intermolecular forces in an ideal gas, and so there can be no changes in internal energy U. Hence:

$$p = T(\partial S/\partial v)_T \tag{A.23}$$

Differentiating (A.21), we then get:

$$p = (N_0 kT/M_0)(1/v) \tag{A.24}$$

This is, once again, the ideal gas equation, given before as (A.15). We now see that it can be obtained in three ways: by a kinetic argument; empirically, by idealisation from experiment. and by an argument using statistical thermodynamics.

Although more abstract and difficult in its initial concepts, the third method is simplest in the mathematical details of the analysis. When we deal with polymer molecules, we find the same three possibilities; but the kinetic argument is too difficult to use except with an unrealistically simplified model, and the behaviour of real materials in experiments is complicated. The statistical thermodynamic approach must be used to obtain the fundamental equations, and the discussion has been introduced particularly as a reminder of its basis.

A.3.3 Real gases

The classical treatment of ideal gases, as outlined here, is not quite perfect: it must be modified slightly to follow quantum mechanics, and in some special instances this leads to important consequences. Furthermore, even for common real gases, corrections must be introduced to allow for the finite size of the molecules, for the possible occurrence of rotation and vibration, and for the inter-molecular forces of attraction. This leads, for example, to van der Waals' equation of state:

$$p + \frac{a}{v^2} \; (v - b) = R'T \tag{A.25}$$

where a, b and R' are constants.

As the temperature falls and the molecules become less energetic, the attractive forces finally overcome the tendency of the molecules to shoot about in space: the system condenses to a liquid, with freedom of movement restricted to a limited volume, or to a solid, on which the molecules do no more than vibrate about fixed locations. Near the conditions for condensation, there are considerable complications and deviations from the behaviour of either ideal gases, or liquids, or solids.

A.3.4 A reminder about entropy and disorder

Equation (A.17) is a relation between entropy and disorder. However, it must be remembered that disorder must be dynamic to contribute to entropy: the equivalent states, or at least a reasonable random selection of them, must occur at random over the time under consideration. A disordered structure which is 'frozen' into immobility will have a low entropy just like a crystal which is truly frozen in an ordered lattice. An amorphous material in a liquid state will have a low entropy but a similar amorphous material in a liquid state will have a high entropy – in the glass, the structure is fixed in a particular state of disorder, but in the liquid the states of disorder are continually changing as a consequence of thermal vibration.

A.4 THE DEFORMATION OF SOLIDS AND FLOW OF LIQUIDS

A.4.1 A regular, one-dimensional ideal solid

In the ideal gas, the kinetic energy is dominant and the potential energy of inter-molecular forces is negligible. In the ideal solid the reverse is true.

For simplicity let us define a regular, one-dimensional solid, as illustrated in Fig. A.2(a) consisting of a string of point particles (atoms or molecules) spaced at intervals x_0. Under strain to a spacing x, the internal energy, U, for convenience taken per particle, must then be of the form shown in Fig. A.2(b). If the origin is located at the point of minimum energy, we must have a relation of the form:

$$U = \tfrac{1}{2}a(x - x_0)^2 + \tfrac{1}{3}b(x - x_0)^3 + \ldots \qquad \text{(A.26)}$$

We note the implicit acceptance that the equilibrium position is the position of minimum energy, and hence the first two terms in the series disappear with the choice of origin.

Under a force F, the spacing will increase as shown in Fig. A.2(c). The

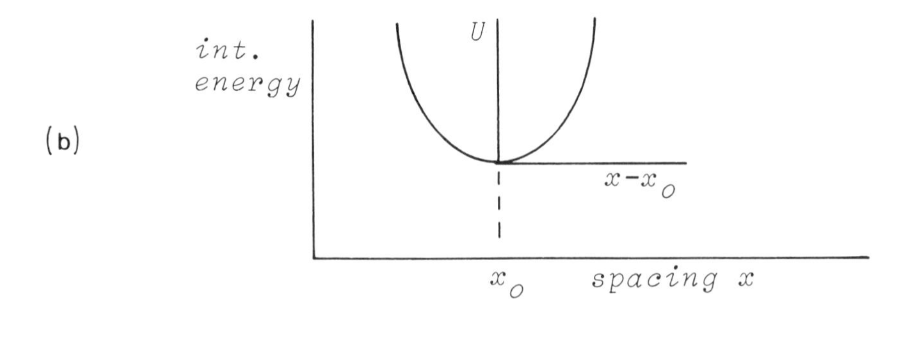

Fig. A.2 – (a) Idealised structure of one-dimensional solid. (b) Local variation of internal energy with spacing. (c) Structure under load.

relation between force and deformation will be:

$$F = \mathrm{d}U/\mathrm{d}x = a(x - x_0) + b(x - x_0)^2 + \ldots \tag{A.27}$$

When $(x - x_0)$ is small, this becomes Hooke's Law:

$$F = a(x - x_0) \tag{A.28}$$

If the effective cross-sectional area occupied by a particle is A, (A.28) can be transformed into the more usual relation between stress, f, and strain, ϵ, in terms of the modulus E:

$$f = F/A = (ax_0/A)\,[(x - x_0)/x_0]$$

$$= E\epsilon \tag{A.29}$$

It is important to note that Hooke's Law is an inevitable simplification, and would only cease to be valid at small enough strains if the first term in (A.26) was absent. Non-linearity develops at larger strains when the higher order terms have to be taken into account.

At very large separations, it is plausible to postulate that the internal energy curve has the form shown in Fig. A.3(a), and consequently the force varies as in Fig. A.3(b). A region of decreasing force is inherently unstable, since the separation of one pair of particles in the sequence can increase indefinitely while the others retract to closer positions, as illustrated in Fig. A.3(c). Consequently

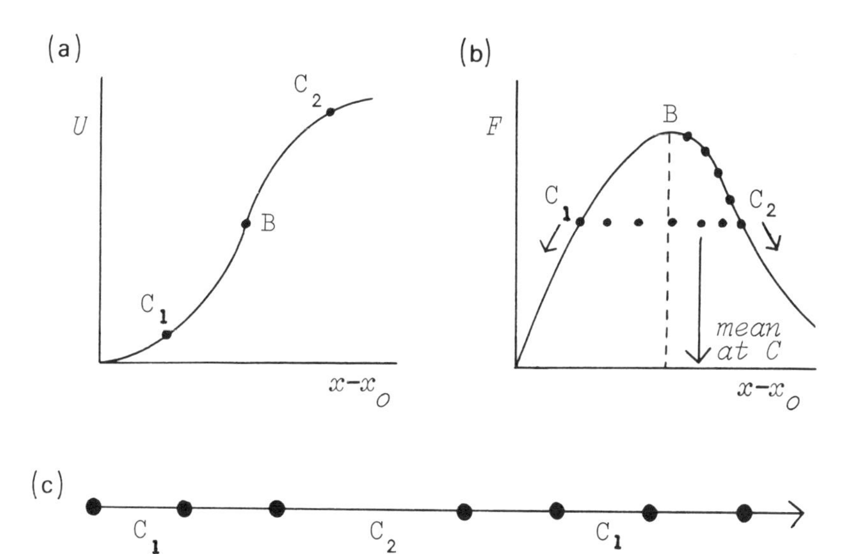

Fig. A.3 – (a) Variation of internal energy with spacing over a larger distance. (b) Variation of force with spacing. (c) Situation in region of decreasing force-elongation relation: the corresponding positions are marked as C on the other diagrams.

the maximum B in the force curve, corresponding to the point of inflection in the energy curve, is the breaking point of the ideal crystalline solid. One interesting aspect of this argument is that the most important features of the response can be predicted if the form of the potential energy function is known up to the point of inflection: the more distant, and less certain, part can be ignored.

The use of (A.26) and (A.27) provides the best basis for further mathematical development of the subject, but there is a more direct physical representation which is instructive. Consider the force F as due to a hanging weight as shown in Fig. A.4(a). Since an increase in spacing of a pair of particles from x_0 to x_1 will lower the weight by $(x - x_0)$, we can associate a potential energy

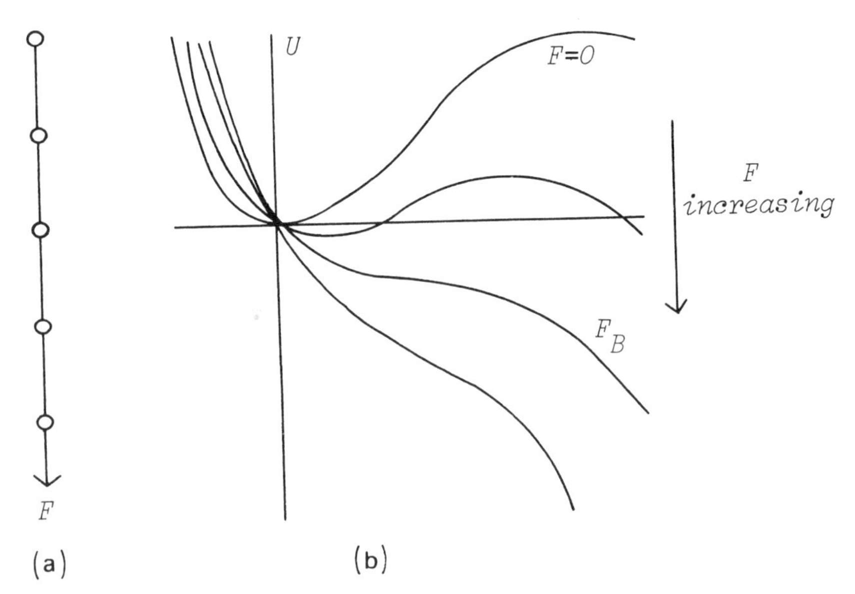

Fig. A.4 – (a) Representation of structure under a gravitational load. (b) Energy relations under increasing force, with instability beyond F_B.

$(-F(x - x_0))$ with each particle. Consequently, the total energy $[U - F(x - x_0)]$ will vary with increasing value of F in the way shown in Fig. A.4(b). Equilibrium at the position of minimum energy at the bottom of trough will occur at increasing values of x. Eventually F will reach a value F_b when there is no trough and the situation is unstable, so that fracture occurs. A little thought about the geometry of the system shows that this must occur at the position corresponding to the point of inflection: the more distant, and less certain, part can be ignored.

A.4.2 Real elastic solids

The previous section clearly relates to a highly idealised situation. In real elastic solids, the following features must be taken into account.

(a) Real crystals are three-dimensional. The displacements of the point particles and the force on the system are vector quantities, and consequently the single equations of the previous section must be generalised into a set of equations to take account of the directional effects. The energy of interaction between particles must be characterised in three dimensions, and if there is appreciable directionality this will lead to anisotropy of behaviour. However, because of the regularity of crystal lattices, the treatment can still be applied to a single unit cell, which replaces the pair of particles in the previous section. At the macroscopic level, the anisotropy is taken into account in the generalised form of Hooke's Law discussed in B.3.6.

(b) Glasses are irregular assemblies of particles, and within crystals, there are often defects which depart from the perfect regularity. The same principles still apply, but it becomes necessary to predict the average response of a set of differing interactions between neighbouring particles – a statement which is simple in its fundamental physics, but difficult in practice because of the complexity of the geometry and the vast number of interactions.

(c) In some circumstances, there may be quantum mechanical effects, and the classical mechanics will not be valid.

(d) More important, as the temperature rises it becomes necessary to take account of the influence of thermal vibrations. This is the converse of the situation in gases where the interaction energy is introduced as a small correction of the thermal motion: in solids, the thermal effects are a small correction of the ideal theory based on interaction energy. When the thermal vibrations become large enough, they disrupt the lattice completely and the material melts or sublimes. Once again, the behaviour is more complex in this temperature range when the kinetic and potential energies are comparable.

At lower temperatures, the thermal vibrations have two consequences which should be noted here. At equilibrium, the lattice spacing will be fluctuating and due to the asymmetry of the energy variation with position, this will lead to a shift in the mean spacing, which shows up as thermal expansion, and to a possible change in modulus.

More significantly, if we look at Fig. A.4(b), we see that when the force is close to the critical value, which gives instability and fracture, the residual energy barrier has become very small. Consequently a thermal vibration may take the separation over the hump and into the region of instability: fracture is thus inherently time and temperature dependent, since the frequency of jumping over a residual barrier of a given height will increase with increasing temperature.

A.4.3 Liquid flow: ideal viscosity

In the ideal liquid, we postulate:

(a) a system with an infinite set of equivalent minimum-energy locations for the positions of the point particles, which represent the atoms or molecules, as shown in Fig. A.5 (a);

(b) a thermal energy which leads to a rapid approach to the equilibrium distribution. A liquid thus flows to its macroscopic state of minimum energy, in the simplest situation to the bottom of a jug, where the gravitational pull is balanced by the upthrust of the container. In the extreme case, this will happen instantaneously (hydrodynamics without viscosity), but usually the visous resistance slows down the response.

When the forces are not balanced, there will be flow since the force will bias the direction of jumps over energy barriers, as illustrated in Fig. A.5 (b). It is convenient to use the form $U - F(x - x_0)$, introduced in Fig. A.4, although x_0 is now a co-ordinate of the initial position of a particle and not a spacing.

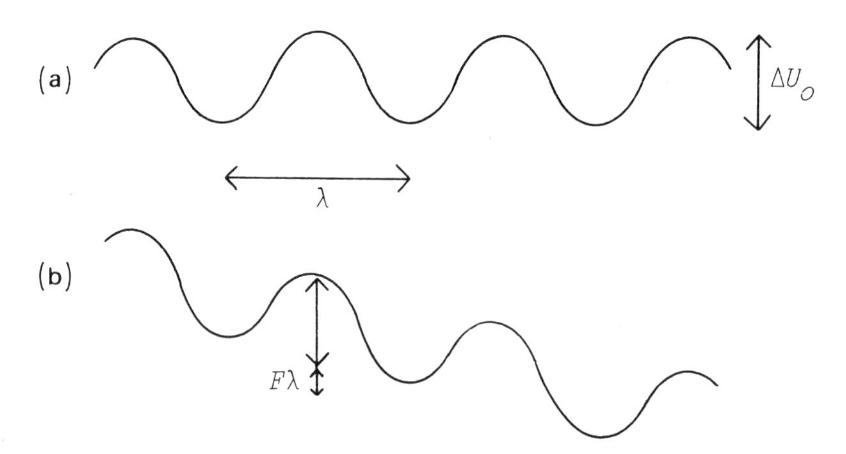

Fig. A.5 – (a) Energy levels for positions of molecules in a liquid. (b) Energy levels under an applied force.

Statistical mechanics can be applied to this problem by means of the basic equation for the number of jumps n per unit time over an energy barrier of height ΔU:

$$n = \nu \exp(-\Delta U/kT) \tag{A.30}$$

where ν is the natural frequency of vibration of the particles, typically of the order of 10^{13} Hz for individual atoms.

In the absence of force, the barriers will be symmetrical, as in Fig. A.5 (a), and the number of forward jumps will equal the backward jumps. But in the presence of force, as in Fig. A.5 (b), the barrier will be asymmetrical, and will be given by:

$$\Delta U = \Delta U_0 \pm F\lambda \tag{A.31}$$

where ΔU_0 is the undisturbed barrier
λ is half the spacing between positions
and $\pm$ relate to backward and forward moves.

Consequently, the net number of forward moves n_f is given by:

$$\begin{aligned} n_f &= \nu \exp[-(\Delta U_0 - F\lambda)/kT] - \nu \exp[-(\Delta U_0 + F\lambda)/kT] \\ &= \nu \exp(-\Delta U_0/kT)\,[\exp(F\lambda/kT) - \exp(-F\lambda/kT)] \\ &= \nu \exp(-\Delta U_0/kT) \sinh(F\lambda/kT) \end{aligned} \tag{A.32}$$

A more complicated treatment of the geometry is needed to relate this an applicable equation for viscous flow, but it is unnecessary to go into that detail here.

The important point is that n_f if a measure of the rate of flow of the particles making up the liquid, and thus the predicted viscosity law is:

rate of flow $\propto$ sinh (force)

The first term in the series expansion of the hyperbolic sine function gives Newton's law:

rate of flow $\propto$ force.

A.4.4 Creep, yield, and flow of solids

We must now combine the features of the previous sections, by considering what happens when there is a sequence of available locations, but the thermal vibrations do not lead to a rapid attainment of equilibrium.

The simplest situation is illustrated in Fig. A.6(a) (A), with a set of equivalent positions: this would apply, for example, if there were successive displacements between the sheets of a crystal lattice from one position of register to the next. In more realistic situations, there might be a variety of barrier heights depending on local differences in packing of atoms in amorphous glasses or on the conformations of crystal defects, which will move more easily than whole sheets.

In other circumstances, the energy levels may rise, as indicated for a pair of neighbouring positions in Fig. A.6(b) (A): this might be a regular effect associated with a change of crystal lattice, or an irregular effect due, for example, to an increasing resistance to defect movement.

If a force F is applied, the energy levels will shift as shown in the sets of curves in Fig. A.6. A number of predictions can now be made.

The energy barrier disappears when the forces reaches a value F_y, and so the material must yield, either continuously in the situation of Fig. A.6(a) or to the new minimum energy position in Fig. A.6(b). There is nothing in the argument so far to prevent this flow being instantaneously complete, but, in reality, the inertial effects of wave propagation would limit the rate.

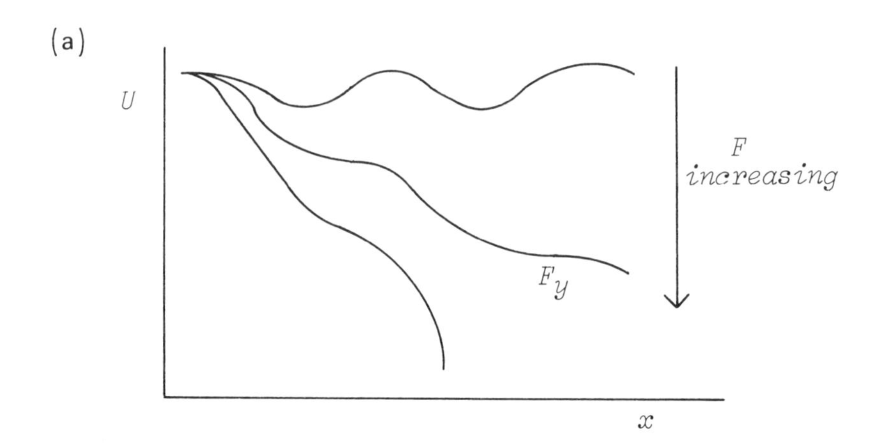

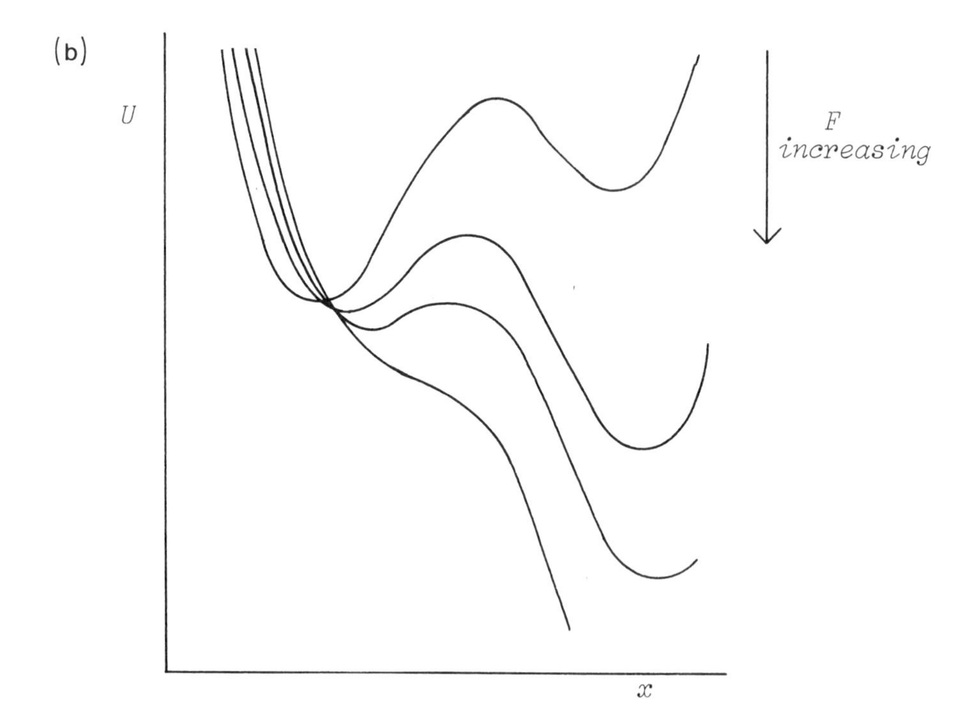

Fig. A.6 – (a) Set of equivalent energy levels in a solid. (b) Set of increasing energy levels.

In the absence of thermal vibrations, there would be no recovery mechanism, operating when the force is removed so that the deformation would be permanent – or at least would remain until it was reversed by an opposite force, or released by increased mobility. This is a mechanism of plastic yield and the resulting force-deformation diagrams would be those shown in Fig. A.7.

If there is some thermal vibration in the material, jumps over the residual energy barriers will occur at stresses less then F_y. However if the thermal vibrations are small, there will be no jumps under zero load, no spontaneous recovery mechanism, and so the deformation will be plastic. Except for the fact that the yield stress will be time and temperature dependent and the magnitude of deformation will depend on the flow rate and the time available, the diagrams of Fig. A.7 will still apply.

It will be seen that the curves in Fig. A.7(a) are very similar to those in Fig. A.5. Consequently the plastic flow can be regarded as a viscous mechanism. In another terminology, it is secondary creep: a continuing time-dependent deformation under stress, which is not spontaneously recoverable on removal of stress.

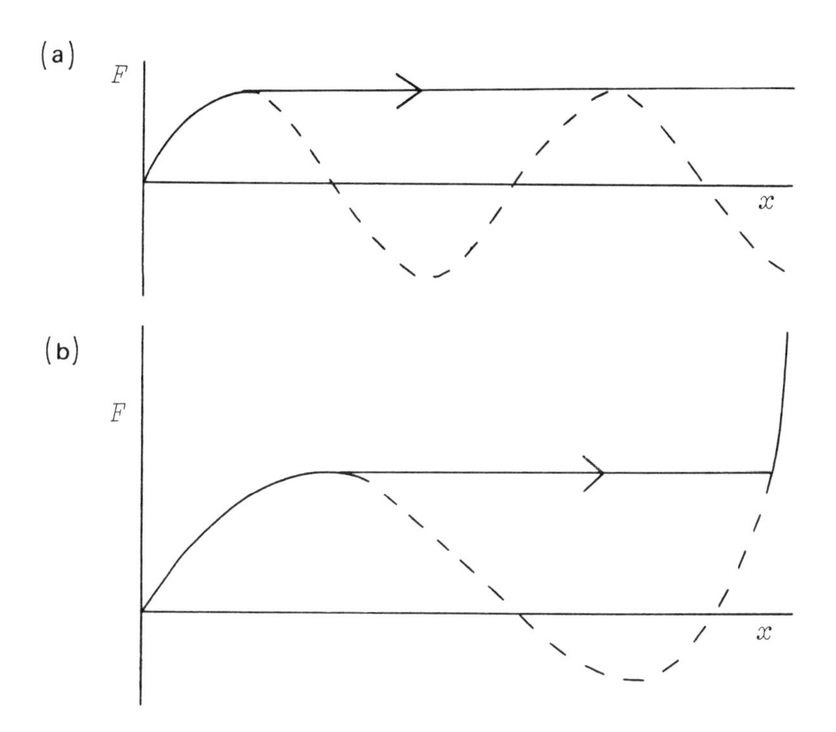

Fig. A.7 – Force-deformation relations. The dotted lines are a differentiation of Fig. A.6 in an unstable region. The full lines show actual response. (a) From Fig. A.6(a). (b) From Fig. A.6(b).

If the thermal vibrations are larger, in relation to the height of the energy barriers, there may be, in the situation of increasing energy levels, shown in Fig. A.6(b), a slow recovery under zero stress due to jumps over the barrier from the high level to the low level. In this situation, any stress, even a very small one, will cause a time-dependent deformation, as the relative energy levels change due to the bias applied by the force, but there will also be time-dependent recovery. This is a mechanism of primary creep. We note two requirements: the deformation must be to a less favourable state, so that there is an inherent tendency to return; and the temperature must be high enough to activate the recovery.

One final point should be made. The derivation of the hyperbolic sine law of flow, given in the previous section, on the basis of Fig. A.5(b), is only valid for small forces which do not appreciably alter the shape of the energy barriers. It is then reasonable to use the form $(\Delta U_0 \pm F\lambda)$ to give the bias between forward and backward jumps. When forces are larger, the complete shape of the barrier must be considered; and, indeed, in the limit, it is the point of inflection and not the height of the barrier which determines whether or not yield occurs. Close to the ultimate yield stress, it is the shape in the region of the point of inflection which determines the magnitude of the residual energy barrier.

Any further analysis requires an assumption about the shape of the potential energy function for interaction between the atoms in the material. With parabolically shaped barriers it can be shown that the effective yield force F_t, with a time t available, is given by:

$$F_t/F_y = 1 - [kT/\Delta U \log_e \nu t]^{\frac{1}{2}} \qquad \text{(A.33)}$$

The time and temperature dependence of yield are predicted by this equation.

A.4.5 The significance of co-operation

Except for an occasional surfacing of disturbing facts, the preceding discussion has been based on an extension of the ideas used in the statistical mechanics of ideal gases, namely that we can deal with an assembly of independent small molecules, represented as point particles, able to take up a variety of locations at different energy levels. There has been the tacitly assumed picture that the energy levels are a separate framework defined in space. In reality, condensed systems are highly co-operative; and the energy levels for a particular atom are generated by all the other atoms in the system, which are themselves subject to movement.

One consequence of this co-operation is that the increase of thermal vibrations in a crystal not only increases the likelihood of jumping a barrier, but also lowers the barriers. At a certain temperature, this effect suddenly becomes cumulative, and so there is a sharply defined temperature at which the crystal melts. In considering any thermal transition, we need to consider whether it is essentially a co-operative phenomenon, in the sense that a crystal can only be defined by specifying the positions of many molecules, or one where the molecules can act independently, as in the dissociation of diatomic molecules into monatomic: the former will give sharp transitions, while the latter willl show progressive change over a temperature range.

Another basic feature of co-operation is related to the frequency of jumps over energy barriers, as given in (A.30). The thermal energy available to a given unit is kT, which is a value independent of the size of the unit; but the energy barrier is the absolute value ΔU, not a normalised value of energy per unit mass, and will thus increase in proportion to the number of units interacting together. If the nature of the deformation mechanism within the structure requires many units to move in unison, this will reduce the time and temperature dependence; but of individual small moleules or atoms can move singly then time and temperature effects will be more important.

A.4.6 Summary of effects

It is clear from the above discussions that the deformation and flow of condensed materials is determined by the pattern of energy levels and the magnitude of thermal vibrations. In any particular material, it is necessary to consider the structural geometry in order to see how forces are transmitted and what modes of deformation are possible. But whatever the detail, the basic framework of statistical mechanics remains the same: it is this which leads to the rather facile statement that all materials, for example polymers and metals, behave in the same way. The similarlity of features of response does not imply a similarity of mechanisms.

Table A.1 summarises the different responses of materials, in terms of the relation between ΔU and kT.

Table A.1 – Summary of mechanical behaviour of material.

$\Delta U/kT$	Material	Response	Determining factors
very low	gas	separation to fill available volume	high temperature, weak interactions between atoms and molecules, little co-operation
low	liquid	maintains condensed volume: but flows freely	
intermediate	liquid *or*† solid	slow viscous flow primary (recoverable) creep	
high	solid	elastic + secondary creep	
very high	solid	elastic and yield or fracture	low temperature, strong interactions, high co-operation

† Depending on whether there is a particular low energy reference state to which the system recovers.

APPENDIX B

Notes on Formal Aspects of Elasticity and Rheology

'. . . . The theorems of geometry and mechanics are certainly useful to a person who wishes to apply these disciplines, since they present alternate forms, characteristics, and classifications which are often more immediately helpful than the definitions and axions.

To sum up, finally, when a physicist or experimenter asks, "What shall I do with this theory?", he should be answered with the brevity of the ancients: "Learn it." . . .'

C. Truesdell, *Rational Mechanics of Deformation and Flow,*
Proceedings of the Fourth International Congress on Rheology

B.1 INTRODUCTION

B.1.1. The relation between stress and strain

The previous appendix was concerned with the physics of deformation, and so used the laws of thermodynamics and statistical mechanics in relation to the atomic and molecular structure of materials. This second appendix considers what can be done by mathematical representation of the macroscopic behaviour of materials, usually treated as a structural continuum.

If we are given any piece of material, we can study changes in its state of strain either simply, by measuring its external dimensions, or, to get more detailed knowledge, by attempting to follow the internal deformations. In general, the state of strain will depend in a complicated way on the whole time sequence of forces applied to the material, and on its previous history (or on a full – often impossibly full – statement of its exact internal structure and state at the start of the study). Theoretical studies of the mechanical behaviour of materials have largely been concerned with ways of reducing this generality, and applying simpler relations. But it is important to remember that it is the generality and complexity which are real, and that it is the theoretical laws which are artificial and approximate (or worse) but essential to any development of the subject beyond experimental empiricism.

For solids, the simplest approximation – which is called **Hooke's Law** – is to assume that deformation is proportional to applied force, or, in other terms, that stress is proportional to strain. The whole history of the material is disregarded, and it is postulated that the displacement at any instant can be calculated from simple proportionality to the load at that instant. In practice, this is a reasonable approximation for many materials. Furthermore for the simplest solids, which are isotropic, only two constants of proportionality (moduli of elasticity) are needed, since any local state of strain can be expressed as a combination of three principal shear strains, related to shear stresses by a shear modulus, and volume strain, related to hydrostatic stress by a bulk modulus. The great theory of linear elasticity of isotropic materials is then concerned with the mathematical analysis of the consequences of this simple statement in two ways: firstly, in relation to other statements of local stress and strain, and other elastic constants such as Young's modulus and Poisson's ratio; secondly, in relation to the distribution of stress and strain when particular forces or surface displacements are applied to particular shapes of specimen, with the additional restriction that the strains are small.

For liquids, different simple assumptions are made. The bulk properties are assumed to be linearly elastic as for solids: volume strain is proportional to hydrostatic stress. But the other approximation – **Newton's law** – is that the local shear stress is simply proportional to the local rate of shear strain, with the constant of proportionality being the coefficient of viscosity. Both the previous history and the instantaneous state of strain are regarded as having no influence: only the first time-derivative of strain plays a part. For many materials this is a reasonable approximation to the behaviour above the melting-point. The theory of hydrodynamics is then concerned with the details of the flow of liquids in particular circumstances, as governed by internal and viscous forces – the latter being given by Newton's law. Where viscous forces alone are concerned, it follows from the similar form of the basic relations that results from the theory of elasticity can be applied also to the flow of liquids, by replacing strain by rate of strain in the equations.

In the discussion, we have tacitly assumed that it is possible to subdivide the material into local zones which are small enough to be regarded as elastic or viscous continua without any effective inhomogeneity of structure. Where these two requirements are incompatible, the situation is more complicated, although one can assume that the elastic or viscous constants define a continuum on as small a scale as is needed.

We thus have the two simple idealisations which are introduced in elementary physics as elasticity and viscosity, and elaborated in edifices of applied mathematics. The fundamental equations are:

$$p = K\epsilon_{\mathrm{v}} \quad \text{for simple solids and liquids,} \tag{B.1}$$

$$\sigma = G\epsilon_{\mathrm{s}} \quad \text{for simple Hookean solids,} \tag{B.2}$$

$$\sigma = \eta \frac{\mathrm{d}\epsilon_{\mathrm{s}}}{\mathrm{d}t} \quad \text{for simple Newtonian liquids,} \tag{B.3}$$

where p = hydrostatic stress (positive in tension),
K = bulk modulus,
$\epsilon_v = (dv/v)$ = volume strain,
σ = shear stress,
G = shear modulus,
ϵ_s = shear strain,
t = time,
η = coefficient of viscosity.

Unfortunately – or fortunately, if one is looking for challenging subjects of study – these idealisations are inadequate for the study of polymer materials. They are usually only useful as very rough approximations, though there is a limited range of conditions in which they are valid to a reasonably high degree of accuracy. In this chapter, we shall look at ways of introducing less restrictive assumptions. Of course, there is no value in going to complete generality and putting:

$$\sigma = f(\mathcal{H}, \epsilon_s, d\epsilon_s/dt, d^2\epsilon_s/dt^2, d^3\epsilon_s/dt^3, \ldots) \quad (B.4)$$

where $f(\ )$ is a general function of the variables, which should also be generalised to include all directions, and $\mathcal{H}$ is a statement of the instantaneous state of the material (or of the complete integrated past history).

The topics which are introduced in this appendix, are formal treatments which are not specific to polymers, but are necessary to an understanding of polymer behaviour.

B.1.2 The development of the subject

Before going on to detailed discussion, it will be useful to list the ways in which (B.1), (B.2) and (B.3) can be generalised, or (B.4) simplified.

(a) *Non-linear elasticity*

Stress is regarded as a general single-valued function of strain alone. Equations (B.1) and (B.2) are generalised:

$$p = f(\epsilon_v) \quad (B.5)$$

$$\sigma = f(\epsilon_s) \quad (B.6)$$

Equation (A.27) used a power series for the function.

(b) *Anisotropic linear elasticity*

Equations (B.1) and (B.2) are generalised into a set of linear equations, which take account of the directions of stress and strain.

(c) *Anisotropic non-linear elasticity*

A set of non-linear functions is needed.

(d) *Non-linear viscosity*

Stress is regarded as a general function of rate of strain alone:

$$\sigma = f(\mathrm{d}\epsilon_s/\mathrm{d}t) \tag{B.7}$$

The hyperbolic sine function, (A.33) is one example.

(e) *Linear visco-elasticity*

The system is regarded as a combination of linear elastic and linear viscous elements. In the simplest form, (B.2) and (B.3) are added:

$$\sigma = G\epsilon_s + \eta(\mathrm{d}\epsilon_s/\mathrm{d}t) \tag{B.8}$$

In the full version, a bigger combination of linear terms is needed.

(f) *Anisotropic linear visco-elasticity*

A directionally dependent set of linear visco-elastic equations is used. (B.9)

(g) *General rheology*

This final heading covers all other attempts to represent the general stress and time-dependent deformation of materials. For example, at its simplest it would cover the replacement of the two terms in (B.8) by non-linear functions of ϵ_s and $(\mathrm{d}\epsilon_s/\mathrm{d}t)$, but it might also include other aspects of past history or the effect of higher time derivatives of strain. It would certainly include anisotropy, and could also bring in time-dependent volume stress-strain relations.

(h) *Longer-range influences*

Implicit in all the above generalisations is the assumption that response can be related to the situation at a point in the material. But in some circumstances longer-range effects will play a part. Equation (B.4) then becomes generalised to include variables on the right-hand side at locations over a zone in the vicinity of the point at which the stress is being evaluated.

B.2 THE DEFINITIONS OF STRESS AND STRAIN

B.2.1 Stress: mass or volume basis

Conventionally stress is defined in physics and engineering as force per unit area. But suppose we want to talk about the stress in a single molecular chain. What is its area of cross-section? This is not well-defined, since a molecule does not have a sharp boundary and further we ought to associate some inter-molecular space with each molecule; nor is it easy to measure; nor is it usually the relevant and important descriptive parameter. The quantity which is well defined is the mass per unit length (or linear density) of the chain: this can be calculated precisely from a knowledge of atomic weights and of the number of repeating units in a given length. It thus seems convenient to define stress as:

$$\text{specific stress} = \frac{\text{force}}{\text{mass per unit length}} \tag{B.10}$$

Modulus and strength (tenacity) values will follow the same usage.

The absolute unit of specific stress in the SI system will be Nkg^{-1}m.

The dimensions are $(\text{MLT}^{-2})/(\text{ML}^{-1}) = \text{L}^2\text{T}^{-2}$, and this suggests the identity between specific modulus and $(\text{wave velocity})^2$ in elastic materials.

There is also a technological justification for the use of specific stress. It is the relevant quantity whenever we want to compare materials on the basis of equal weights rather than of equal volumes. Technically this is often appropriate in design calculations, and commercially materials are commonly sold on a weight basis. Engineers often 'correct' values of stiffness or strength, based on conventional stress, by dividing by the density and thus obtaining the specific stress equivalent.

With homogeneous solids, either the conventional stress or the specific stress can be satisfactorily defined and measured. But where areas of cross-sections are ill-defined or difficult to measure, the use of specific stress is necessary. This has long been recognised in dealing with textile fibres, yarns and fabrics. The recognised international textile unit of specific stress in N/tex: 'tex' is a unit of linear density, suitable for textile materials, and is equal to g/Km.† It is however still common to find g/tex. Another common unit is g/den, denier being g/9000 m.

The use of specific stress corresponds to the energy per unit mass, in J/kg in the SI system, where conventional stress is related to energy per unit volume, in J/m^3.

The specific stress f is related to the conventional stress σ through the density ρ by the equation:

$$\sigma = \rho f \tag{B.11}$$

Equation (A.26) is also correct if f is expressed in N/tex, σ in kN/mm^2, and ρ in g/cm^3.

It may also be noted that the value of the specific stress is numerically equal to the length, in appropriate units, of a specimen whose weight in a standard gravitational field equals the force on the specimen. This is why units of length have often been used for what is really a specific stress; in particular the specific stress at break has been referred to as the breaking length. The unit Km or Km wt is numerically equal to gf/tex.

The various choices of definition and of systems of units lead to a bewildering variety of units of stress, since we must not only include the dimensional equivalence of specific stress, $(\text{velocity})^2$, stress/density, specific energy, and length-force, but also the choice between various metric and imperial units and between inertial and gravitational measures of force. There is a similar set of choices for stress units. At one extreme we find Pascals as the units of stress, at the other we find inches, or even $\text{psi}/(\text{g/cm}^3)$ as units of specific stress. Some useful conversions are given in Table B.1.

† This is 10^{-6} times the absolute SI unit, namely $\text{Kg}\,\text{m}^{-1}$.

Table B.1 – Unit Conversions.

	Specific stress		Stress – density in g/cm^{-3} times
1 ——	N/tex, kJ/g, GN $m^{-2}/g\ cm^{-3}$, $(km/s)^2$	1 ——	GN/m^2, J/mm^3
—	10.2 gf/dtex, 11.3 gf/den	—	
—	102 gf/tex, kmf, kgf $mm^{-2}/g\ cm^{-3}$ 239 cal/g, 430 Btu/lb	—	102 kg/mm^2 145 ksi
10^3 ——	mN/tex, J/g, MPa/g cm^{-3}	10^3 ——	N/mm^2, MPa
—		—	10^4 bar, 9869 atm, $1.02 \times 10^4\ kg/cm^2$
—	145,000 psi/g cm^{-3}	—	145,000 psi, lbf/in^2
*10^6 ——	N/kg m^{-1}, J/kg, Pa/k gm^{-3}, m^2s^{-2}	10^6 ——	
—	3.94×10^6 inchf, psi/(lb/cu.in)	—	7.5×10^6 mm Hg
—		—	
10^9 ——		*10^9 ——	Pa, N/m^2, J/m^3, kg $m^{-1}s^{-2}$
—	10^{10} dyn/g cm^{-1}, erg/g	—	$10^{10}\ dyn/cm^2$

* Strict SI.

Notes: (a) Other multiples also used.
(b) Nm^{-2} = Pa.
(c) Gravitational force units are often written as g, g-wt or pond; lb or lb-wt; km, km-wt or Rkm.

B.2.2 Strain

There is no problem about the definition of small tensile strains. If a spacing x increases by dx, as in Fig. B.1(a), then strain ϵ_{xx}, is given by:

$$\epsilon_{xx} = \mathrm{d}x/x \tag{B.12}$$

At large strains, there is a real choice. At its simplest, we can see this as a difference depending on whether the denominator x is the spacing before or after deformation. If x is taken to be the deformed length and x_0 the original length, the commonly used definitions are:

'common' strain, as used in this book, ϵ

$$\epsilon = (x - x_0)/x_0 = \delta x/x_0$$

or

$$(1 + \epsilon) = x/x_0 \tag{B.13}$$

Green's strain, ϵ'

$$(1 + 2\epsilon')^{\frac{1}{2}} = x/x_0 \tag{B.14}$$

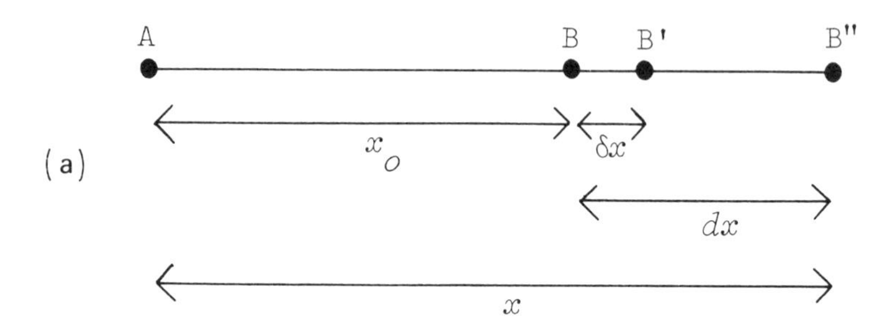

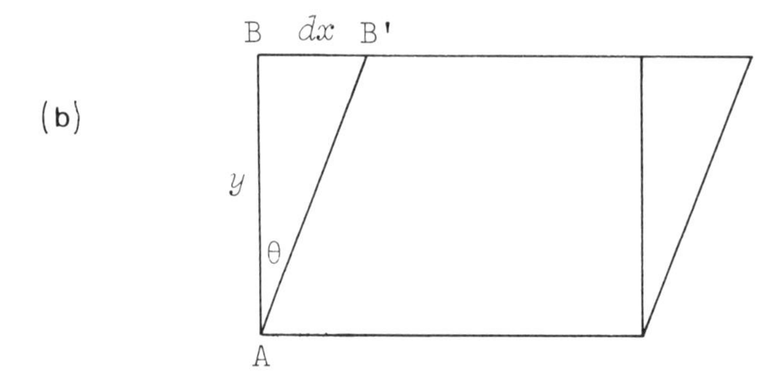

Fig. B.1 – (a) Tensile strain: a line AB, between two points in the material, deforms to AB' in small strain or AB'' in large strain. (b) Small shear strain.

Natural or logarithmic strain, e

This is given by:

$$\mathrm{d}e = \mathrm{d}x/x \tag{B.15}$$

and, thus on integration:

$$x/x_0 = \exp e \tag{B.16}$$

or

$$e = \ln(x/x_0) \tag{B.17}$$

All these quantities become identical for small strains.

Alternatively, we may use: extension ratio, λ

$$\lambda = x/x_0 \tag{B.18}$$

The various measures are related by the equations:

$$\lambda = 1 + \epsilon = \mathrm{d}\lambda/\mathrm{d}e = \mathrm{d}\epsilon'/\mathrm{d}\lambda \tag{B.19}$$

$$e = \ln\lambda = \ln(1+\epsilon) \tag{B.20}$$

Small shear strains ϵ_{xy} are commonly defined, from the construction in Fig. B.1(b), by an equation similar to (B.12):

$$\epsilon_{xy} = \mathrm{d}x/y = \tan\theta \tag{B.21}$$

There are more complications when there are large strains, and the mathematically most useful relations give forms which reduce at small strains to half the above value. Care must therefore be taken to see which definitions are in use.

The brief account here only touches on the questions of definitions of strain. The subject develops through mathematical analysis of the displacement of points in the material by appropriate partial differential equation statements, recognition of the three-dimensional effects which make strain a second rank tensor, and derivation of a variety of useful relations, invariants, and simplifications arising from the choice of axes. For a more detailed account, the reader is referred to standard texts such as Williams' *Stress Analysis of Polymers* (Horwood 1980).

B.2.3 Forms of stress

At large strains a similar problem arises in the definition of stress, depending on whether this is related to the starting or the actual cross-section. We thus have:

σ_0 = nominal stress = force/original area

σ_t = true stress = force/actual deformed area

f_0 = nominal specific stress = force/original linear density

f_t = true specific stress = force/actual deformed linear density.

Since the mass M is invariant it follows that, with a force F:

$$f_0 = F \div (M/x_0) \quad \text{(B.22)}$$

$$f_t = F \div (M/x) \quad \text{(B.22)}$$

and thus:

$$f_t = f_0(x/x_0) = (1+\epsilon)f_0 \quad \text{(B.23)}$$

For the conventional stresses, it will be necessary to account for the volume changes (the Poisson's ratio effect in simple extension). We may note that this problem does not arise when one is considering the scalar quantities energy per unit mass, or, except as a minor correction, for energy per unit volume.

Stress, like strain, exists in three dimensions and is a second rank tensor. Fig. B.2 illustrates the three easily recognised tensile stresses acting along three perpendicular axes, and three shear stresses, associated with couples about these axes. We note that deformation of the material, rather than movement of the whole specimen, must be due to balanced pairs of tensile forces or shear moments. We can also abstract a uniform hydrostatic stress acting equally in all directions and related to volume strain.

As with strain there is a useful field of mathematical treatments of stress.

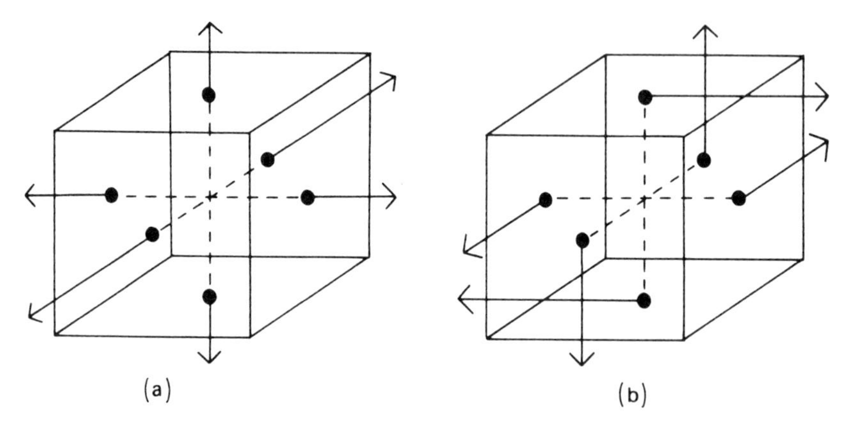

Fig. B.2 – Stresses given by forces on the face of a unit cube. (a) Three directions of tensile stress. (b) Three directions of shear stress.

B.2.4 Modulus, compliance, and other elastic constants

The simple definitions are:

$$\text{modulus} = \text{stress/strain} = \sigma/\epsilon = E$$

$$\text{compliance} = \text{strain/stress} = \epsilon/\sigma = C$$

Apart from the use of different definitions of stress and strain, we then have a choice for non-linear stress-strain relations between:

$$\text{secant modulus} = \sigma/\epsilon$$

$$\text{tangent modulus} = \mathrm{d}\sigma/\mathrm{d}\epsilon$$

The modulus and compliance give the relation between stresses and strains in the same direction thus defining a tensile (or Young's) modulus, a shear modulus, or a bulk modulus. But a given stress will also cause strain in other directions. The commonly used quantity for isotropic materials is:

$$\text{Poisson's ratio} = -(\epsilon_{yy}/\epsilon_{xx})$$
$$\text{for a uniaxial stress } \sigma_{xx}$$

There are many other quantities which could, in principle, be defined and would be constants in a Hookean material at small strains.

B.2.5 Strain energy

The work done in deforming a material is given by integrating the product of force and displacement. Consequently when this is related to a unit cube, as in Fig. B.3, we have:

$$\text{strain energy per unit volume} = W = \Sigma\int\sigma\,\mathrm{d}\epsilon \tag{B.24}$$

The summation is included to cover all six directions of possible application of stress.

For small-strains and linear elasticity this reduces to

$$\begin{aligned} W &= \Sigma\int E\epsilon\,\mathrm{d}\epsilon \\ &= \Sigma\tfrac{1}{2}E\epsilon^2 = \Sigma\tfrac{1}{2}C\sigma^2 \end{aligned} \tag{B.25}$$

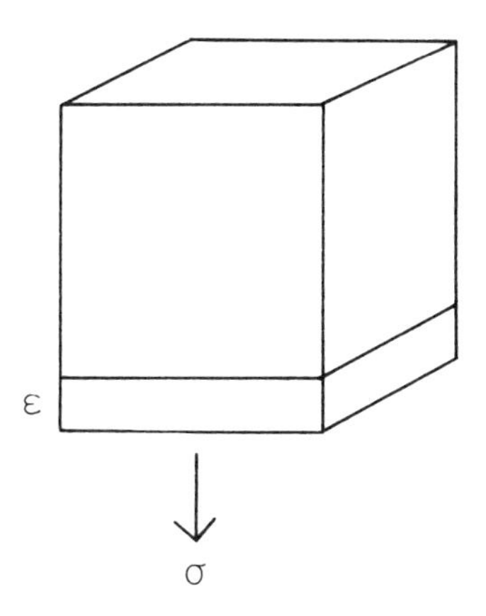

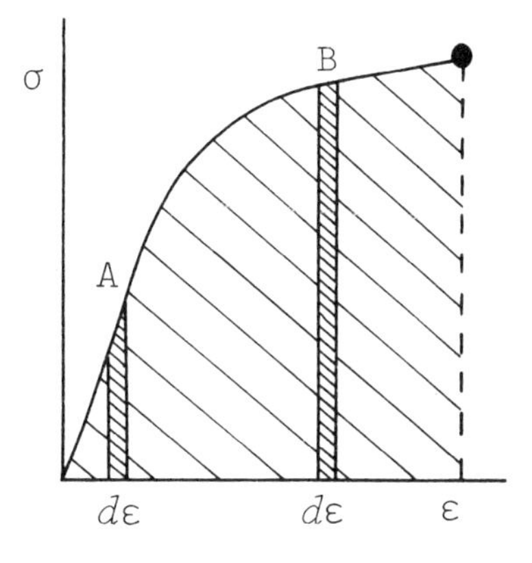

Fig. B.3 – (a) Tensile stress in a unit cube. (b) Linear (A) and non-linear (B) regions of stress-strain curves, showing closely shaded area as strain energy increments $\mathrm{d}W = \sigma\mathrm{d}\epsilon$, with W as total energy, given by open shaded areas on integration.

Equation (B.25) shows the relation between the moduli and compliances and the strain energy. If specific stresses and moduli are used the relation is to strain energy per unit mass.

B.2.6 Easy experimental modes of deformation

The definitions of stress and strain relate to the most useful ways of characterising a uniform deformation of the material. Experimental simplicity leads to another set of quantities related to a test specimen. The three easily applied modes of deformation of a test specimen, as shown in Fig. B.4(a) are extension, twisting and bending.

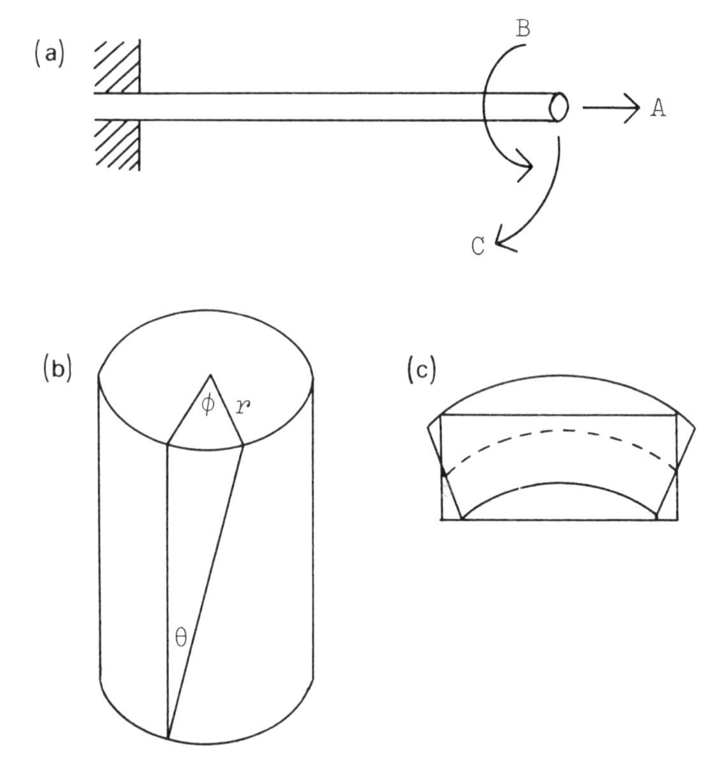

Fig. B.4 – (a) Three easy modes of deformation of a rod assumed to be fixed at one end: (A) extension (B) twisting or torsion; (C) bending or flexing. (b) Relation of twisting to shear. (c) Relation of bending to extension and comparison.

Extension does give a uniform deformation of the material, and so there are simple relations:

tensile force = tensile stress × area

or specific stress × linear density

spring constant = modulus × area

But twisting and bending give a non-uniform deformation, with zero strain at the 'centre' of a specimen and a maximum at the surface, and a consequent faster increase of specimen stiffness in proportion to the square of the area of cross-section, and not to the first power as in extension. The detailed treatment in relation to different shapes of specimen and other complications, arising with large strains, combined deformations, and non-linearity, will be found in other texts, but the basic relations will be given here.

Twisting is related to shear. If, as in Fig. B.4(b), we define twist as the angle of rotation ϕ per unit length of specimen, then, in the simple case:

$$\text{shear strain at radius } r = r\phi \tag{B.26}$$

$$\text{torque} = \int \sigma r \mathrm{d}A = \int n r^2 \phi \mathrm{d}A \tag{B.27}$$

$$\begin{aligned} \text{torsional rigidity} &= \text{torque per unit twist} \\ &= k_1 G A^2 \end{aligned} \tag{B.28}$$

where k_1 is a shape factor (1 for a circle)
G is shear modulus
A is area.

Bending gives extension on the outside and compression on the inside with a natural plane of zero strain in the centre, as shown in Fig. B.4(c). In the simple case, for a specimen with a radius of curvature R or curvature $c\,(= R^{-1})$, we have:

$$\text{strain at distance } r \text{ from neutral plane} = r/R = rc \tag{B.29}$$

$$\text{bending moment} = \int \sigma r \mathrm{d}A = \int E r^2 c \mathrm{d}A \tag{B.30}$$

$$\begin{aligned} \text{flexural rigidity} &= \text{moment per unit curvature} \\ &= k_2^2 E A^2 / 4\pi \\ &= EI \end{aligned} \tag{B.31}$$

where k_2 is a shape factor (1 for a circle)
E is tensile modulus
I is 'moment of inertia' of cross-section.

The value of r/R at the surface of a specimen is often referred to as the bending strain.

It is important to note that if the material response is non-linear with a difference between tension and compression, the neutral plane will shift in position. Thus if, as often happens in oriented polymers, there is easy yield in compression, the neutral plane will move towards the outside, in order to give the minimum energy of deformation.

B.3 ELASTICITY

B.3.1 General relations

In generalising the analysis of deformation of solids, one important route of development retains the simple assumption that stress is a single valued function of strain, but discards linearity. Equation (B.2) thus becomes:†

$$\sigma = f(\epsilon) \tag{B.32}$$

Graphically such a relation is shown in Fig. B.3(b). It is a direct consequence of the above statement that there can be no hysteresis – elastic recovery is perfect. And it is an obvious indirect consequence that all the work done in deformation is returned during recovery and so can be regarded as stored elastic energy: the system is said to be conservative.

In the macroscopic situation, which we are considering, it is immaterial whether the energy is stored as internal energy of deformation or as heat in the surroundings, provided it can be recovered. Clearly the introduction of the general function will complicate the application of elasticity theory, but the details of this lie outside the scope of the present book.

It is a part of polymer physics to determine – experimentally or theoretically – the forms of the functional relations between stress and strain. In some applications numerical values of the function may be found and used for computation. Alternatively, as in rubber elasticity, a particular non-linear function is theoretically derived or empirically fitted. Or an arbitrary polynomial may be used to relate stress σ to strain ϵ:

$$\sigma = a_1\epsilon + a_2\epsilon^2 + a_3\epsilon^3 + a_4\epsilon^4 + \quad \ldots\ldots\ldots \tag{B.33}$$

It should be noted that, provided the first term exists, Hooke's law will be valid as an approximation at small enough strains, with an elastic modulus a_1.

B.2.2 Strain energy function

In many theoretical studies it is easier to calculate energies than forces. If W is the elastic energy per unit volume‡ which can be expressed as a function of ϵ, we then get:

$$\sigma = \mathrm{d}W/\mathrm{d}\epsilon \tag{B.36}$$

The situation is shown graphically in Fig. B.3(b).

In thermodynamic terms, or in terms of molecular calculations, W will be the free energy. There are some complications and differences of precise definition, depending on whether deformation is adiabatic or isothermal, at constant volume *or* constant pressure.

† For the sake of generality we drop the suffix which restricts the equation to shear behaviour. In general, relations will apply to other forms of stress and strain as well.

‡ Or per unit mass, if specific stress is used (see B.2.1).

Often calculations lead more directly to W in terms of a length between two units rather than a strain. For example, as discussed in Appendix A, in the deformation of a crystal, when entropy changes are negligible, W is given by internal energy U and can be expressed as a function of the distance x between atoms in the lattice.

It will be seen that the point of inflection in the energy diagram corresponds to a peak in the force diagram: this is very important because it represents a point of instability, when an increasing deformation is opposed by a decreasing force. There are various empirical or theoretical potential functions, which may be used to express the variation of U with x.

At equilibrium let $W = W_0$, $x = x_0$: the strain $\epsilon = (x - x_0)/x_0$ will be zero. From the definition of equilibrium, we know that this is a point of minimum free energy, where stress $= \mathrm{d}W/\mathrm{d}\epsilon = 0$. Hence we must be able to express the free energy as a potential with the second term missing:

$$W = W_0 + b_2\epsilon^2 + b_3\epsilon^3 + b_4\epsilon^4 + \ldots \tag{B.35}$$

and

$$\sigma = \mathrm{d}A/\epsilon = 2b_2\epsilon + 3b_3\epsilon^2 + 4b_4\epsilon^3 + \ldots \tag{B.36}$$

Comparing coefficients with (B.33), we see that:

$$a_1 = 2b_2$$

$$a_2 = 3b_3 \quad \text{etc.}$$

Hence:

$$W = \tfrac{1}{2}a_1\epsilon^2 + \tfrac{1}{3}a_2\epsilon^3 + \tfrac{1}{4}a_3\epsilon^4 + \ldots \tag{B.37}$$

The first term is the usual expression for elastic energy in a Hookean material, and we note that any material where deformation is dependent on an elastic displacement from a position of minimum free energy will show this behaviour. If the bottom of the trough is parabolic in form the material will obey Hooke's law to a first approximation.

The existence of a strain energy function leads to important mathemetical consequences in the theory of elasticity.

B.3.3 Large deformations: invariants

The theory of linear elasticity is limited by the two restrictions:

(a) the strains are small;

(b) stress is proportional to strain.

As we have seen in the last section, the first almost invariably implies the second, though the converse of this is not true. As we know from simple experiments on rubbers, plastics and fibres, these restrictions are not valid for polymers. The development of the theory of elasticity after removing these restrictions has been extensively studied by Rivlin and others. There are two sorts of consequence. Firstly, there are the geometrical consequences of large deformations: this is a particular application of applied mathematics and, important as it is in practice, lies mainly outside the scope of this book on the fundamentals of polymer materials. Secondly, there is the form of the relation between stress and strain: here there are some theoretical arguments of fundamental importance.

Since we shall be dealing with large deformations, it is convenient to express the deformation in terms of the extension ratios of a unit cube, as illustrated in Fig. B.5.

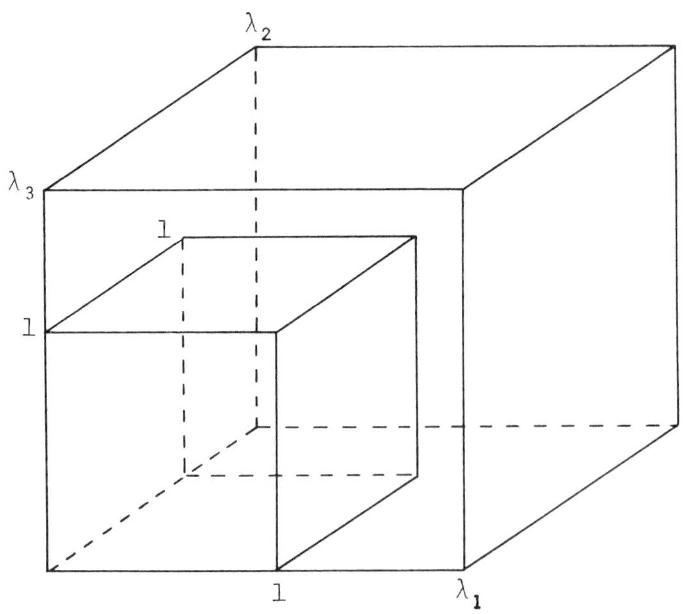

Fig. B.5 – Extension ratios of a unit cube.

Without further thought, it would seem that, in general, the strain energy W could be expressed as any function of the extension ratios. But this is not so, for reasons of symmetry. The particular restrictions will depend on the nature of the anisotropy of the material.

For an isotropic material, it is obvious that the choice of axes for λ_1, λ_2 and λ_3 is arbitrary, and so W must be a symmetrical function of the three extension ratios. Rivlin argued that W must be an even-powered function of the extension ratios.† With this limitation it is possible to represent all possible forms of W as functions of three invariants:

$$I_1 = \lambda_1^2 + \lambda_2^2 + \lambda_3^2 \tag{B.38}$$

$$I_2 = \lambda_1^2\lambda_2^2 + \lambda_2^2\lambda_3^2 + \lambda_3^2\lambda_1^2 \tag{B.39}$$

$$I_3 = \lambda_1^2\lambda_2^2\lambda_3^2 \tag{B.40}$$

† For a discussion of this point, see Treloar, *Physics of Rubber Elasticity*, 3rd edition, page 215. Oxford University Press.

If we also make the assumption (which often turns out to be reasonable) that the solid is incompressible, and its volume $\lambda_1 \lambda_2 \lambda_3$ always equals 1, we have:

$$I_3 = \lambda_1^2 \lambda_2^2 \lambda_3^2 = 1 \quad \text{(B.41)}$$

So I_3 is then not a function of strain, and the state of strain may be expressed as a function two variables I_1 and I_2. Substitution from (B.41) gives:

$$I_1 = \lambda_1^2 + \lambda_2^2 + \lambda_3^2 = \lambda_1^2 + \lambda_2^2 + 1/\lambda_1^2 \lambda_2^2 \quad \text{(B.42)}$$

$$I_2 = \frac{1}{\lambda_1^2} + \frac{1}{\lambda_2^2} + \frac{1}{\lambda_3^2} = \frac{1}{\lambda_1^2} + \frac{1}{\lambda_2^2} + \lambda_1^2 \lambda_2^2 \quad \text{(B.43)}$$

Since I_1 and I_2 completely define the state of strain, the strain energy W must be capable of expression as a function solely of I_1 and I_2. One form for the most general function is in terms of powers of $(I_1 - 3)$ and $(I_2 - 3)$:

$$W = \sum_{i=0}^{\infty} \sum_{j=0}^{\infty} C_{ij} (I_1 - 3)^i (I_2 - 3)^j \quad \text{(B.44)}$$

The quantities $(I_1 - 3)$ and $(I_2 - 3)$ are chosen in order that W, whose zero may be arbitrarily selected, becomes zero in the unstrained specimen with $\lambda_1 = \lambda_2 = \lambda_3 = 1$. In addition, we have to put $C_{00} = 0$.

All this shows is the nature of the restrictions on the form of the stress-strain function, for an incompressible, isotropic solid. It is not possible to go any further, on purely mathematical grounds, towards the choice of the particular terms in the power series. Some real structure is needed for this.

What (B.44) does do is to suggest which simple froms might be tried in an attempt to fit experimental results to an empirical relation. The two simplest forms of (B.44) are obtained by putting all the coefficients, except one of the first-order terms equal to zero:

$$W = C_{10} (I_1 - 3) = C_{10} (\lambda_1^2 + \lambda_2^2 + \lambda_3^2 - 3) \quad \text{(B.45)}$$

$$W = C_{01} (I_2 - 3) = C_{01} (1/\lambda_1^2 + 1/\lambda_2^2 + 1/\lambda_3^2 - 3) \quad \text{(B.46)}$$

The first of these equations is used in the study of rubber elasticity. The second has not been shown to be related to any real material. However, the equation given by including both first-order terms has proved useful. It was originally derived by Mooney and may be written:

$$W = C_{10} (I_1 - 3) + C_{01} (I_2 - 3) \quad \text{(B.47)}$$

B.3.4 Empirical stress-strain relations

Another approach to the problem of deviations from Hooke's Law is to make some other simple approximation concerning the form of the stress-strain relation. Thus in the theory of plasticity, which has been extensively developed and applied to metals, the stress-strain is assumed to have the form shown in Fig. B.6(a): Hooke's law is followed up to the yield stress and then plastic flow occurs at constant stress. This approach is useful for metals, but not usually for polymers. An alternative approximation, which is reasonable for many polymer materials, is to assume two linear regions of different slope as in Fig. B.6(b).

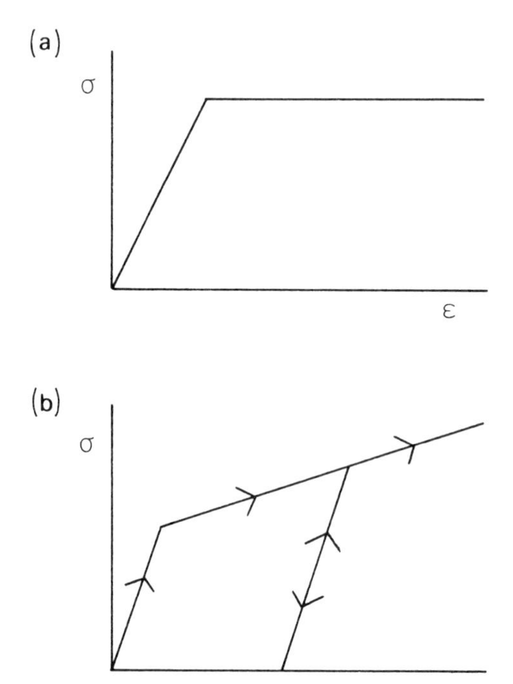

Fig. B.6 – (a) Idealised elastic-plastic behaviour. (b) Four-parameter model.

Both of these approaches depart from the concept of perfect elasticity, since recovery will be on different lines. However, so long as the deformation is continuing in one direction (and under specified time conditions) the stress may be regarded as a single-valued function of strain. It is sometimes useful to elaborate the second model by postualting that recovery takes place along lines with a slope equal to the initial slope, as is illustrated in Fig. B.6(b). It may be noted that the relation in Fig. B.6(b) is characterised by four parameters: two slopes and the strains (or stresses) at the yield and breaking points.

B.3.5 The use of models

Another convenient way of representing behaviour is by means of models. For the ideal elastic solid, the model is a Hookean spring, as in Fig. B.7(a). For plastic deformation, the model can be a frictional element under normal load, as in Fig. B.7(b). Combining these two, we get the Bingham model of Fig. B.7(c): this represents what is assumed in an ideally elastic-plastic material. The bilinear form of Fig. B.6(b) would be represented by Fig. B.7(d).

Instead of Hookean springs, alternative non-linear laws may be assumed to represent the spring behaviour. In the most general way, the differential spring constants may be taken as functions of strain.

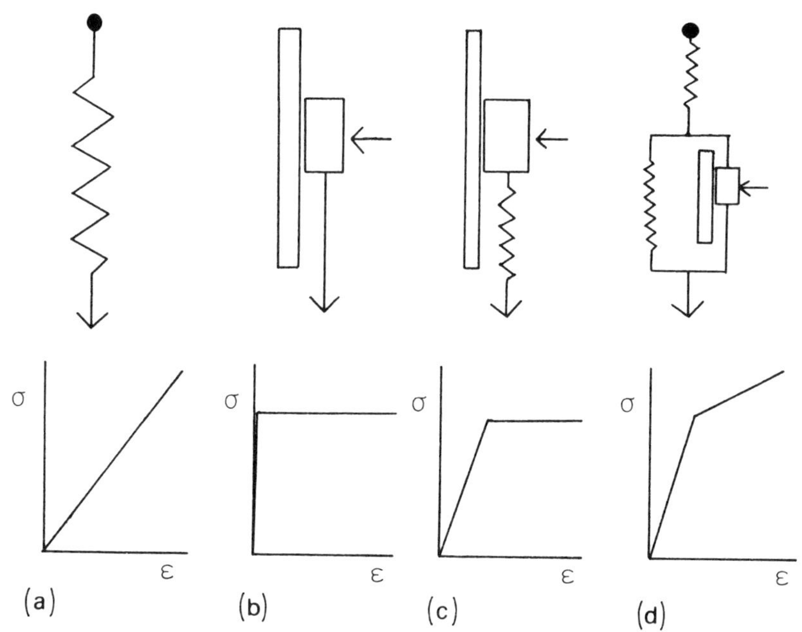

Fig. B.7 – (a) Ideal spring representing Hookean elasticity. (b) Frictional element representing plasticity. (c) Bingham model: elastic-plasticity. (d) Model for Fig. B.6(b).

B.3.6 Anisotropy in linear elasticity

We now turn to the complication of anisotropy in mechanical properties. Again there are various mathematical ways of handling the problems, and a choice of notations. The more powerful mathematical tools, such as the relation between the second rank stress and strain tensors through a fourth rank modulus or compliance tensor, give unnecessary complication in the basic approach to the problem. We shall give a brief account of a simple approach here and refer the reader to other texts for more detail.

We should first note that the simplifications which can be developed in the mathematical theory of isotropic materials, because of the freely available choice of axes, do not arise in anisotropic materials where there are axes inherent in the structure of the material. Save in special cases, axes which are convenient in relation to the imposed deformation will be inconvenient in relation to the structure, and vice versa.

We therefore consider the deformation of a unit cube of the material as illustrated in Fig. B.8. Associated with each direction there are three tensile strains, ϵ_1, ϵ_2, ϵ_3 and three shear strains ϵ_4, ϵ_5, ϵ_6. Similarity there are three tensile stresses, σ_1, σ_2, σ_3 and three shear stresses σ_4, σ_5, σ_6 though the latter appear as moments which must be appropriately balanced by couples on the system as a whole (or split into opposed couples on neighbouring faces, giving the same shear).

For the most general anisotropy of structure, with no simplifying symmetry, each strain may result in a stress in all six directions, and vice versa. In the linear, Hookean situation, the stresses are independent, as derived from different strains, and additive. We can therefore write:

$$\sigma_1 = a_{11}\epsilon_1 + a_{12}\epsilon_2 + a_{13}\epsilon_3 + a_{14}\epsilon_4 + a_{15}\epsilon_5 + a_{16}\epsilon_6 \qquad \text{etc.}$$

or, in the matrix notation:

$$\begin{bmatrix} \sigma_1 \\ \sigma_2 \\ \sigma_3 \\ \sigma_4 \\ \sigma_5 \\ \sigma_6 \end{bmatrix} = \begin{bmatrix} a_{11} & a_{12} & a_{13} & a_{14} & a_{15} & a_{16} \\ & a_{22} & a_{23} & a_{24} & a_{25} & a_{26} \\ & & a_{33} & a_{34} & a_{35} & a_{36} \\ & & & a_{44} & a_{45} & a_{46} \\ & & & & a_{55} & a_{56} \\ & & & & & a_{66} \end{bmatrix} \begin{bmatrix} \epsilon_1 \\ \epsilon_2 \\ \epsilon_3 \\ \epsilon_4 \\ \epsilon_5 \\ \epsilon_6 \end{bmatrix} \qquad \text{(B.48)}$$

The lower left portion of the matrix is omitted because it can be shown, from the existence of a single strain energy function in linear elasticity, that the matrix must be symmetrical and the same coefficients, a_{12}, a_{13} etc appear.

As an alternative to the modulus matrix of (B.48), there will be a reciprocal compliance matrix defined by:

$$\begin{bmatrix} \epsilon_1 \\ \epsilon_2 \\ \epsilon_3 \\ \epsilon_4 \\ \epsilon_5 \\ \epsilon_6 \end{bmatrix} = \begin{bmatrix} C_{11} & C_{12} & C_{13} & C_{14} & C_{15} & C_{16} \\ & C_{22} & C_{23} & C_{24} & C_{25} & C_{26} \\ & & C_{33} & C_{34} & C_{35} & C_{36} \\ & & & C_{44} & C_{45} & C_{46} \\ & & & & C_{55} & C_{56} \\ & & & & & C_{66} \end{bmatrix} \begin{bmatrix} \sigma_1 \\ \sigma_2 \\ \sigma_3 \\ \sigma_4 \\ \sigma_5 \\ \sigma_6 \end{bmatrix} \qquad \text{(B.49)}$$

This is a more natural, or useful, form when stress is regarded as the independent, imposed, variable, whereas the modulus matrix is more convenient when the deformation is imposed and independent.

In these matrices, the terms a_{11}, a_{22}, and a_{33} are tensile (Young's) moduli; a_{33}, a_{44}, a_{45} are shear moduli; the terms a_{12}, a_{13} and a_{23} are related to Poisson's ratios, which are most easily seen in the form $(-C_{12}/C_{11})$; a_{45}, a_{46} and a_{56} involve

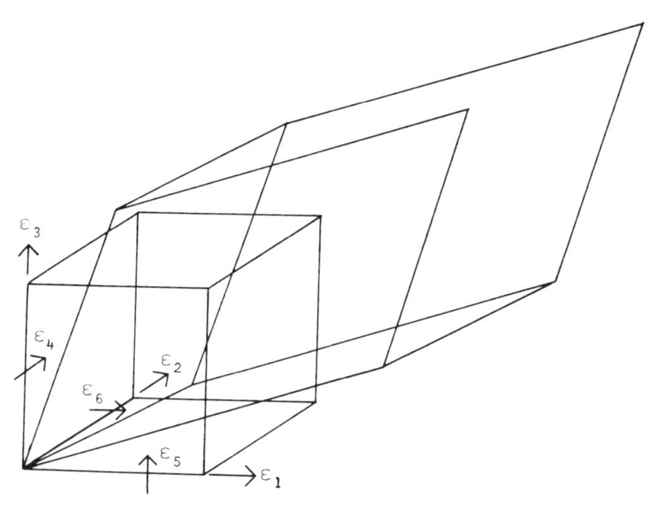

Fig. B.8 – General deformation of a unit cube, related to axes OX, OY, OZ embedded in the structure.

analagous cross-effects in shear; and the nine terms in the top right corner involve cross-effects between shear and extension. The relation to a bulk modulus can also be derived.

We see that the most general anisotropy leads to the 21 elastic constants in the matrices of (B.48) and (B.49). A similar relation may be expressed in energy terms through the 21 coefficients of the second order terms (ϵ_1^2, ϵ_2^2 ... $\epsilon_1\epsilon_2$, $\epsilon_1\epsilon_3$ ... etc.) or (σ_1^2, σ_2^2 ... $\sigma_1\sigma_2$, $\sigma_1\sigma_3$ ... etc.).

If the material has symmetry, the number of independent constants reduces considerably. It is well-known in simple physics, that in an isotropic material the number of independent constants drops to 2. The four commonly defined quantities, Young's modulus, shear modulus, Poisson's ratio and bulk modulus, are related by two equations.

In simplifying the matrices, the two factors to take into account are: (a) the equivalence of directions, which leads to identity of some constants; (b) the influence of symmetry in eliminating cross-terms. Thus in the isotropic case, where all three directions are equivalent, there can be no cross-terms in shear, and Poisson's ratio is related to shear, the equation reduces effectively to:

$$\begin{bmatrix}\sigma_1\\ \sigma_2\\ \sigma_3\\ \sigma_4\\ \sigma_5\\ \sigma_6\end{bmatrix} = \begin{bmatrix} a_{11} & a_{12} & \bullet & & & \\ \bullet & \bullet & \bullet & & 0 & \\ \bullet & \bullet & \bullet & & & \\ & & & \frac{1}{2}(a_{11}-a_{12}) & & \\ & 0 & & & \bullet & \\ & & & & & \bullet \end{bmatrix} \begin{bmatrix}\epsilon_1\\ \epsilon_2\\ \epsilon_3\\ \epsilon_4\\ \epsilon_5\\ \epsilon_6\end{bmatrix} \tag{B.50}$$

There are two special cases of limited anisotropy of particular interest.

The first is uniaxial orientation with transverse isotropy (hexagonal symmetry): one particular direction in the material is a principal axis and all directions perpendicular to it are equivalent. This often happens in fibres, and leads to 7 common parameters, namely two tensile moduli, E_L and E_T, two shear moduli, G_{TT} and G_{LT} (of which G_{TT} is related to torsion) and three Poisson's ratios, ν_{TL}, ν_{LT} and ν_{TT} as illustrated in Fig. B.9. The number of independent constants is reduced to five by the equations:

$$G_{TT} = E_T/2(1 + \nu_{TT}) \tag{B.51}$$

$$\frac{\nu_{LT}}{E_L} = \frac{\nu_{TL}}{E_T} \tag{B.52}$$

In the matrix notation, we have:

$$\begin{bmatrix} \sigma_1 \\ \sigma_2 \\ \sigma_3 \\ \sigma_4 \\ \sigma_5 \\ \sigma_6 \end{bmatrix} = \begin{bmatrix} a_{11} & a_{12} & \bullet & & & \\ \bullet & a_{22} & a_{23} & & 0 & \\ \bullet & \bullet & \bullet & & & \\ & & & \frac{1}{2}(a_{22} - a_{33}) & & \\ & 0 & & & a_{55} & \\ & & & & & \bullet \end{bmatrix} \begin{bmatrix} \epsilon_1 \\ \epsilon_2 \\ \epsilon_3 \\ \epsilon_4 \\ \epsilon_5 \\ \epsilon_6 \end{bmatrix} \tag{B.53}$$

The other common situation is biaxial orientation (orthorhombic symmetry) with three mutually perpendicular principal axes. The three directions are

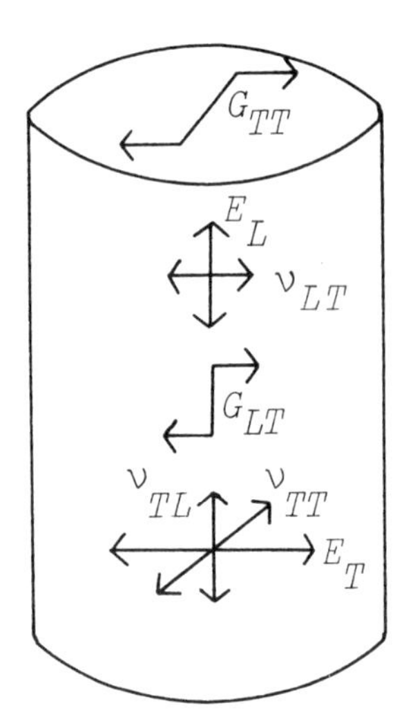

Fig. B.9 – Elastic parameters in a material with uniaxial symmetry.

different but, because they are orthogonal, many cross-terms disappear, and there are nine independent constants.

The matrix becomes:

$$\begin{bmatrix} \sigma_1 \\ \sigma_2 \\ \sigma_3 \\ \sigma_4 \\ \sigma_5 \\ \sigma_6 \end{bmatrix} = \begin{bmatrix} a_{11} & a_{12} & a_{13} & & & \\ \bullet & a_{22} & a_{23} & & 0 & \\ \bullet & \bullet & a_{33} & & & \\ & & & a_{44} & & \\ & 0 & & & a_{55} & \\ & & & & & a_{66} \end{bmatrix} \begin{bmatrix} \epsilon_1 \\ \epsilon_2 \\ \epsilon_3 \\ \epsilon_4 \\ \epsilon_5 \\ \epsilon_6 \end{bmatrix} \quad \text{(B.54)}$$

B.3.7 Anisotropic, non-linear elasticity

The theory of non-linear anisotropic elasticity still presents major difficulties in the establishment of useful formulation. With large strains and non-linearity the additivity of strains (or stresses), namely the principle of superposition, is lost. The coefficients in the modulus matrix might be replaced by non-linear functions, but many other interactions would also need to be considered.

The most useful representation, and the most economical way of recording experimental data, and introducing theoretical simplifications, may be through the strain-energy functions which for an elastic material must be capable of expression as a function of the six strains (or stresses):

$$W = f(\epsilon_1, \epsilon_2, \epsilon_3, \epsilon_4, \epsilon_5, \epsilon_6) \quad \text{(B.55)}$$

The stresses would then be given by equations such as:

$$\sigma_1 = (\partial W / \partial \epsilon_1)_{\epsilon_2, \epsilon_3, \epsilon_4, \epsilon_5, \epsilon_6} \quad \text{(B.56)}$$

Although the strain energy would have to be plotted in 6 dimensions (subject to reduction if there is symmetry), this is better than plotting 6 stresses or 21 elastic constants as functions of strain in 6 dimensions.

B.4 VISCOSITY

B.4.1 Linear and non-linear viscosity

The equation for Newtonian viscosity has been given as (B.3), and the model of such an ideal liquid is a Newtonian dashpot, as illustrated in Fig. B.10. In the same

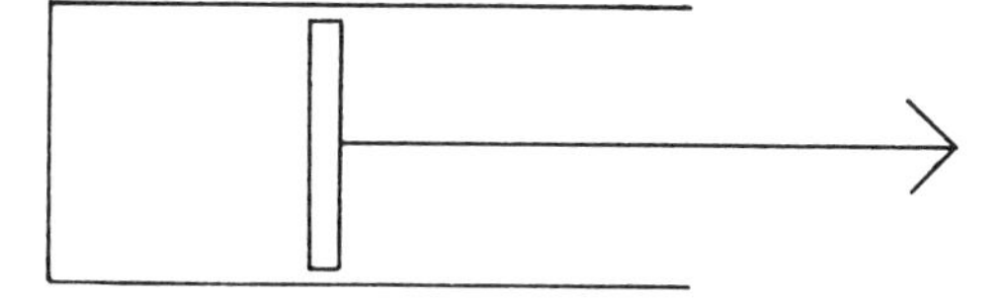

Fig. B.10 – Idealised viscous dashpot.

way as for elasticity, this behaviour may most easily be modified by maintaining the assumption that stress is a single-valued function of rate of strain; but replacing Newton's law by a more general functional relation, of which the hyperbolic sine law of flow is one example.

B.4.2 Anisotropy

In a simple material, viscosity is necessarily isotropic, since the liquid must relax to a random state. However, when there is flow, or in the influence of another external field, there may be an alignment of elements of structure so that the viscosity becomes anisotropic. The formation will be similar to anisotropic elasticity.

B.4.3 Forms of flow

It is sometimes necessary to take account of the differences between the forms of flow, shear flow and elongational flow shown in Fig. B.11. In the simple case it can be shown that the same coefficient of viscosity applies, so that we can write

$$\sigma_1 = \eta \mathrm{d}\epsilon_1/\mathrm{d}t \tag{B.57}$$

where σ_1 and ϵ_1 and the tensile stress and strain.

However in materials where there are longer-range influences, as along the length of polymer molecules, the responses may be different. This is a statement of a complication which has not yet been mentioned: that the response of the material may not be given solely by the state at a point, but may be influenced by effects over a larger element. Such effects will be different again in more complicated flow patterns.

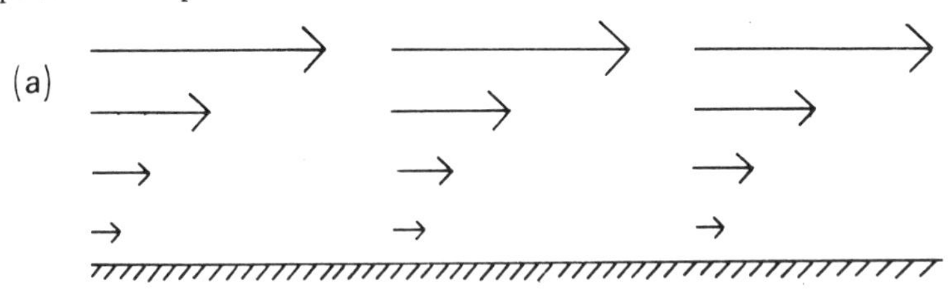

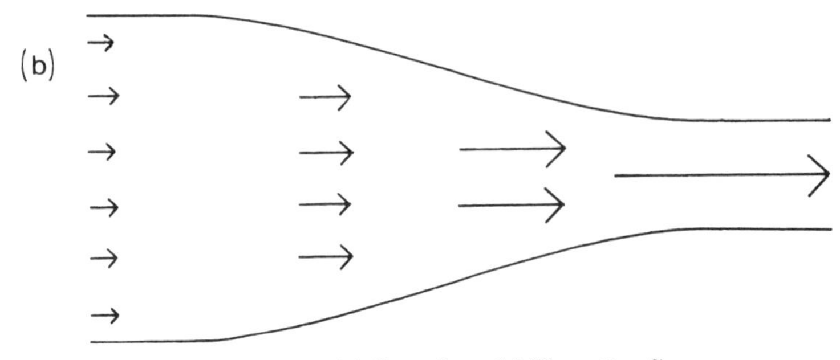

Fig. B.11 – (a) Shear flow. (b) Elongation flow.

B.5 RHEOLOGY

B.4.1 The nature of rheological experiments

Entering the general subject of rheology, one leaves the convenience of the one-to-one relations of elasticity (stress = f(strain)) and viscosity (stress = f(rate of strain)), and instead takes up the interrelations between three quantities – stress σ, strain ϵ and time t. And, because of the nature of time as a variable, this means that the whole past history and current change becomes of importance. It is only when we define some particular pattern of change, that we can accept time as a single, simple variable.

So the first problem in rheology is to describe the sort of experiments which can usefully be done to determine the rheological behaviour of materials. The second problem is to determine relations between different types of experiment. The third and fourth problems are to develop ways of analysing more complex sequences of deformation, and to study the properties of particular real and idealised materials. A great deal of work has been done in all these fields, and in this appendix we shall only discuss those parts which are needed for a study of the fundamentals of polymer physics. The wealth of detailed application can be followed up elsewhere.

There are four simple forms of rheological experiment. For convenience, they will be described in terms of tensile stress and length changes, but analagous experiments can be carried out in shear, torsion, bending or any other form of deformation.

Creep

In the elementray laboratory experiment to measure Young's modulus, weights are successively added to a long wire, and the length changes noted. In an elastic material, the time for which any given weight is left hanging is immaterial. But in a general material, further extension would take place as long as the load was applied. This time-dependent deformation under load is known as **creep**.

In a creep test, therefore, a load is applied instantaneously and remains as a constant. The general response, illustrated in Fig. B.12(a), is an instantaneous deformation followed by the creep. The term **instantaneous**, as used here, has two limitations. In practice it may just mean any time less than the shortest time in which one can apply a load or measure the length. But as experimental technique is refined and these times are made shorter, there is another limit given by the times at which the inertia of the system or the material delays the response – the former may be avoided by changing the system (for example, going from dead-weight loading to spring loading), but the latter is an absolute limitation (for typical values of the masses of atoms and the force constants between them it is of the order of 10^{-13}s).

Having defined a creep test in this way, the response may be described by a general function $\epsilon = f(\sigma, t_c)$, where t_c is the time under a constant load σ. Graphically this can be represented by a series of curves or by a three dimensional plot. Because of the behaviour of many materials, it is common to plot extension against time on a logarithmic scale as in Fig. B.12(b). Often such a plot is linear over many decades of time; and even when there is some curvature apparent on the logarithmic plot, it is still a better first approximation to assume linearity with log (time) than with time.

When the load is removed, as in Fig. B.12(a), there is generally an instantaneous recovery (often, but not necessarily, equal to the instantaneous deformation) followed by a further recovery in time. The original creep may thus be regarded as composed of two parts: primary creep, which is recoverable and equal to (– creep recovery); and secondary creep, which is non-recoverable and is the remainder.

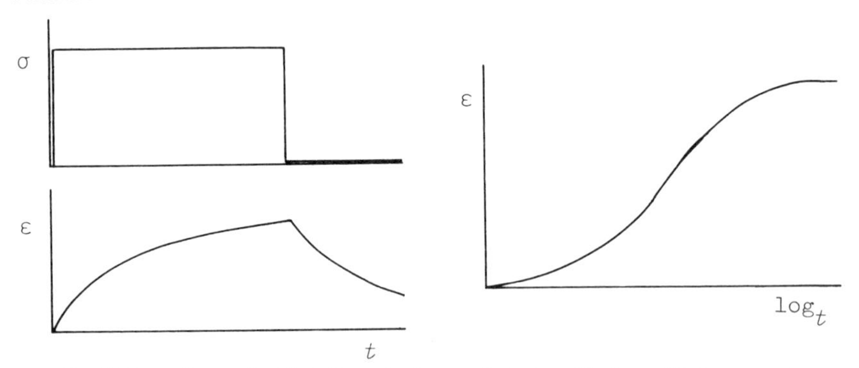

Fig. B.12 – Creep. (a) Variation of stress and strain with time. (b) A common response on a logarithmic time scale.

Stress relaxation

The converse of creep is stress relaxation: the change in stress with time under constant strain, as illustrated in Fig. B.13. Experimentally, the specimen is extended as rapidly as possible, and then the tension changes are followed by

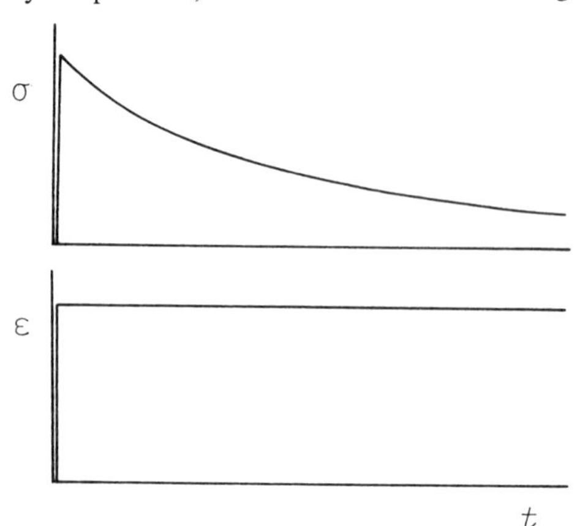

Fig. B.13 – Stress relaxation.

means of a load cell (force transducer) and recorder. The results can be represented by a general function $\sigma = f(\epsilon, t_r)$, where t_r is time at constant-strain ϵ. The stress may decay ultimately to zero or to some constant value. As with creep, it is often more convenient to plot against log (time) since this gives a closer approximation to linearity. If a specimen is released after a stress relaxation, it will generally show only a partial instantaneous recovery, followed by more recovery with time, but still leaving some permanent deformation.

It may be noted that if a complex sequence of length changes is imposed, the direction of stress relaxation may show a reversal: in other words the stress may increase with time and then fall again. Analagous effects can occur in creep, when a specimen may contract under a tensile load or extend under a compressive load, due to a complicated previous history.

Load-elongation

Time dependence may also be studied by varying the rate at which a simple load-elongation test is carried out. It is necessary to specify the rate in some particular way – most simply, as constant rate of elongation or constant rate of loading. The pattern of events is illustrated in Fig. B.14. The general functions are $\sigma = f(\epsilon, \mathrm{d}\epsilon/\mathrm{d}t)$ or $\sigma = f(\epsilon, \mathrm{d}\sigma/\mathrm{d}t)$.

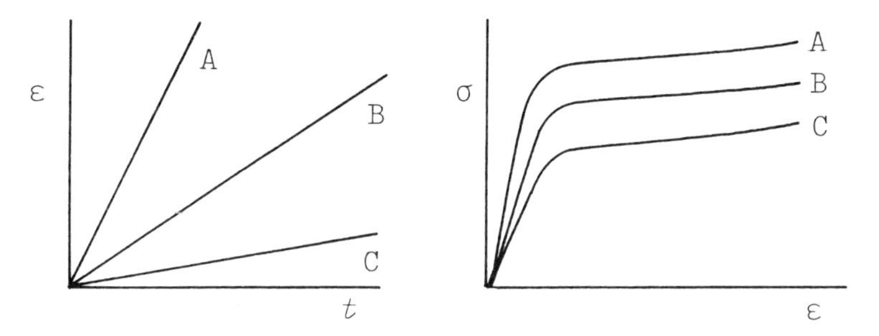

Fig. B.14 – Stress-strain curves at different rates of extension, from fast A to slow C.

Dynamic tests

Another very useful technique is to subject the material to repeated cycles of stress or strain. In the simplest form of dynamic test these will be imposed contant-amplitude sinusoidal variations of either stress or strain. In general, in a non-linear material, a sinusoidal variation of one will not lead to a sinusoidal variation of the other, and the amplitude of variation of the dependent variable may also change with time, so that the general relation would be as in Fig. B.15(a). But since the amplitudes are usually small (though often superimposed on a fixed deformation), it is commonly possible to regard both stress and strain as varying sinusoidally with an angular frequency ω, though out of phase by an angle δ; as in Fig. B.15(b):

$$\epsilon = \hat{\epsilon} \sin \omega t \tag{B.58}$$

$$\sigma = \hat{\sigma} \sin (\omega t - \delta) \tag{B.59}$$

Initially, since the starting conditions will not satisfy the above equations, there must be a transient response, before the steady pattern settles down. This time may be extended if the material properties change during the first few cycles. In the steady-state situation, we may put $\sigma = f(\epsilon, \omega)$, though we shall find that other forms of representation are more useful. It should be noted that some of the formalism of simple dynamic testing may be used when the strict conditions given above are not followed: other patterns of cyclic variation may be included in practice.

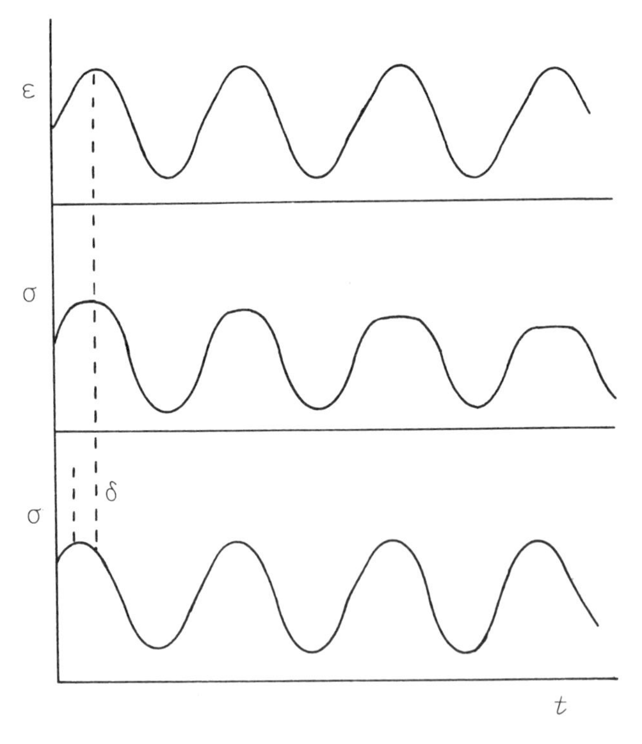

Fig. B.15 – Dynamic testing with (A) sinusoidal imposed strain, giving a general (B) or idealised (C) stress response.

B.5.2 Mathematical characteristics

We now turn to the definition of parameters which are used to characterise the results of experiments of the form described qualitatively in the last section.

In that section, it was convenient because of the simpler experimental arrangements, to think mainly of the simple extension of specimens. In this section, because of the simpler theoretical form, it is more convenient to think mainly of shear deformation, which involves only a change of shape, instead of extension, which involves both shape and volume changes. But analagous relations exist for all modes of deformation and in an anisotropic material there will also be different relations for different directions.

Creep

By an extension of the terminology of linear elasticity, we could define a modulus as:

$$\text{creep modulus} = G_c(t) = \sigma/\epsilon(t_c) \tag{B.60}$$

where $\epsilon(t_c)$ is the creep in time t_c under stress σ.

However to avoid introducing the reciprocal of the dependent variable $\epsilon(t_c)$, it is usually preferred to use:

$$\text{creep compliance} = J_c(t) = (G_c(t))^{-1} = \epsilon(t_c)/\sigma \tag{B.61}$$

Both the creep compliance and the creep modulus will be functions of time. The introduction of these functions is a way of removing the direct influence of stress in increaing the rate of creep: in a linear visco-elastic material $J_c(t)$ and $G_c(t)$ will be independent of σ.

Stress relaxation

In a similar way and for a similar reason, we can define:

$$\text{relaxation modulus} = G_r(t) = \sigma(t_r)/\epsilon \tag{B.62}$$

This is a quantity which is very commonly used to characterise the changing response of a polymeric material with changing times available for internal deformation of the structure. In some special circumstances, $G_r(t) \cong G_c(t) = (J_c(t))^{-1}$, but this is not true in general, because of the differences in loading histories, although the general pattern of change will usually be similar.

Stress-strain relations

The stress-strain relations obtained at different rates can be used to give values for moduli or compliances. The best forms are $G_s(\mathrm{d}\epsilon/\mathrm{d}t)$ or $J_s(\mathrm{d}\sigma/\mathrm{d}t)$. For linear viscoelastic materials, it is possible to calculate the relation of these quantities to $G_r(t)$ or $J_c(t)$, but generally this is difficult. It will also be seen that $G_s(\mathrm{d}\epsilon/\mathrm{d}t)$ and $J_s(\mathrm{d}\sigma/\mathrm{d}t)$ are related to rates, that is to reciprocals of time. The roughly equivalent times would be $\epsilon(\mathrm{d}\epsilon/\mathrm{d}t)^{-1}$ and $\sigma(\mathrm{d}\sigma/\mathrm{d}t)^{-1}$ where ϵ and σ are the levels of strain or stress involved. The former would usually be of the order of the reciprocal of the rate of strain in per cent per second.

Dynamic properties

With dynamic tests there are a variety of parameters which come out of different experimental measurements or theoretical concepts. In relating these parameters to one another, it is always tacitly assumed, that at least within the range of conditions of any particular test, the response of the material is linear. Of course, a formal definition of any particular parameter can always be applied to any form of variation: thus in the formulation of (B.58) and (B.59), the quantities $\hat{\sigma}$ and $\hat{\epsilon}$ would be defined as the peak values; and a phase lag δ could be defined, although its value might vary according to which parts of the cycle (for example maximum, minimum, zero or points of inflection) were chosen as the reference points. It is the simple interrelation of quantities which breaks down when the response is non-linear.

It is convenient to start from the forms given earlier in terms of peak stress and strain and a loss angle δ:

$$\epsilon = \hat{\epsilon} \sin \omega t \tag{B.58}$$

$$\sigma = \hat{\sigma} \sin (\omega t + \delta) \tag{B.59}$$

where the frequency of oscillation is ν Hz or $\omega = 2\pi\nu$ radians/sec. The roughly equivalent time will be ω^{-1} secs. In order to characterise the response, it will be necessary to define two material parameters. One such pair would be a modulus derived from the peak values, $G_m = \hat{\sigma}/\hat{\epsilon}$, and the phase difference δ.

We can however rearrange (B.59):

$$\begin{aligned}\sigma &= \hat{\sigma} \sin (\omega t + \delta) \\ &= \hat{\sigma}(\sin \omega t \cos \delta + \cos \omega t \sin \delta)\end{aligned} \tag{B.63}$$

Here the first term in the brackets gives the component of stress in phase with the strain and the second term gives the out-of-phase component. It is useful to define a modulus as the ratio of the in-phase components:

$$G = \frac{\hat{\sigma} \sin \omega t \cos \delta}{\hat{\epsilon} \sin \omega t} = \hat{\sigma} \cos \delta / \hat{\epsilon} = G_m \cos \delta \tag{B.64}$$

The leads to:

$$\sigma = G\hat{\epsilon}(\sin \omega t + \cos \omega t \tan \delta) \tag{B.65}$$

The out-of-phase component of stress is equal to $\tan \delta$ times the in-phase component; and the quantity $\tan \delta$, which is often reported is termed the loss factor.† The reason for this name is that we have:

$$\begin{aligned}\text{energy loss/unit volume/radian} &= \frac{1}{2\pi}\int_t^{(t+2\pi/\omega)} \sigma \mathrm{d}\epsilon \\ &= \frac{1}{2\pi}\int_t^{(t+2\pi/\omega)} G\epsilon(\sin \omega t + \cos \omega t \tan \delta)\, \omega\epsilon \cos \omega t \,\mathrm{d}t \\ &= \frac{1}{2} G\epsilon^2 \tan \delta\end{aligned} \tag{B.66}$$

† In the analagous electrical problem, it is common to use the power factor $= \cos\phi = \sin\delta$, where $\phi = \pi/2 - \delta =$ phase angle.

In an elastic Hookean solid, $\frac{1}{2}E\epsilon^2$ would be the maximum stored energy, and $\tan\delta$ is thus a measure of the fraction of the energy which is lost.

As an alternative to the sine and cosine representation used above, a complex number notation can be used. Here we put:

$$\epsilon = |\boldsymbol{\epsilon} \exp(i\omega t)| = |(\epsilon' + i\epsilon'') \exp(i\omega t)| \tag{B.67}$$

$$\sigma = |\boldsymbol{\sigma} \exp(i\omega t)| = |(\sigma' + i\sigma'') \exp(i\omega t)| \tag{B.68}$$

Here $\boldsymbol{\epsilon}$ and $\boldsymbol{\sigma}$ are complex numbers (vectors) with the real part representing an in-phase component and the imaginary part an out-of-phase component in a perpendicular direction. We can define a complex modulus by the relation:

$$\boldsymbol{\sigma} = \boldsymbol{G\epsilon} \tag{B.69}$$

where

$$\boldsymbol{G} = G' + iG'' \tag{B.70}$$

G' is the real part of the modulus and G'' is the imaginary part. The advantage of this notation is that it is possible to drop $\exp(i\omega t)$ from all the working, and use the complex stress, strain, moduli and compliances. When it is required to get out to observable quantities, it is only necessary to multiply by $\exp(i\omega t)$ and take the real part.†

Combination of the equations with the relation $\exp(i\omega t) = (\cos\omega t + i\sin\omega t)$ leads to the equation:

$$\sigma = \epsilon(G' \sin\omega t + G'' \cos\omega t) \tag{B.71}$$

This shows the real part G' as the in-phase modulus and the imaginary part G'' as the out-of-phase modulus. We also have:

$$\tan\delta = G''/G' \tag{B.72}$$

The interrelation of the various quantities is summarised geometrically in Fig. B.16, which also includes the parameters E_p, η_p, E_s, η_s from parallel and series combinations of idealised springs and dashpots.

As an alternative to representation in terms of moduli, there are corresponding compliance relations, which will be more appropriate when stress is the independent variable and strain is dependent. The complex quantities are reciprocals:

$$\text{compliance} = \boldsymbol{J} = 1/\boldsymbol{G} = J' - iJ'' \tag{B.73}$$

† There is a minor complication in comparison with the previous notation. Since $\exp(i\omega t) = \cos\omega t + i\sin\omega t$, it is the imaginary part which corresponds to the previous notation. However by an immaterial shift of time origin, (B.58) could have been put:

$$\epsilon = \epsilon \cos\omega t\ .$$

But the relations between the separate components are not so simple:

$$J' = \frac{(G')^{-1}}{1 + (G''/G')^2} \tag{B.74}$$

$$J'' = \frac{(G'')^{-1}}{1 + (G'/G'')^2} \tag{B.75}$$

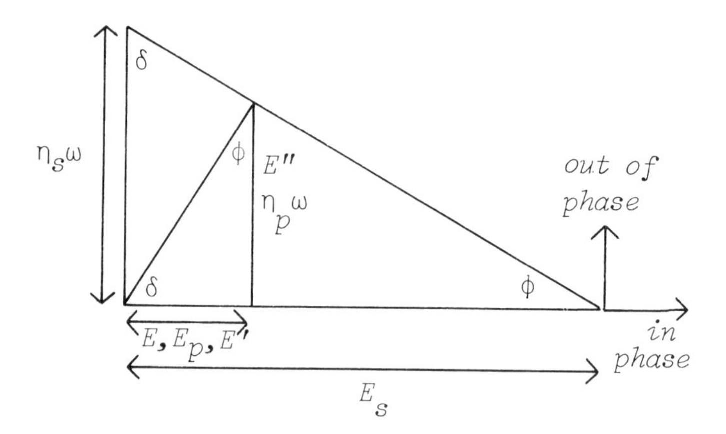

Fig. B.16 – Summary of relations between parameters.

B.5.3 Simple visco-elastic models

Simple visco-elastic models are useful for several reasons. Firstly, they can be used to give a qualitative demonstration of different types of rheological behaviour. Secondly, they are useful in a more limited sense as a way of characterising the behaviour in a particular test, and of working out the interrelation between observed quantities and the parameters discussed in the last section. Thirdly, they are the base from which more complex models can be built up.

The one-element models – the Hookean spring and the Newtonian dashpot – have already been mentioned as describing idealised elastic and viscous behaviour. There are two forms of two-element model. The Voigt or Kelvin model where the two units are in parallel, and Maxwell model where they are in series. The general responses of these models are illustrated in Fig. B.17. The only mode of deformation in the Voigt model is primary creep, and corresponding stress-strain and dynamic relations. On the other hand, the Maxwell model shows instantaneous deformation, secondary creep and stress relaxation.

The simplest model which gives all the types of visco-elastic behaviour contains four elements. Several equivalent arrangements are possible: two of

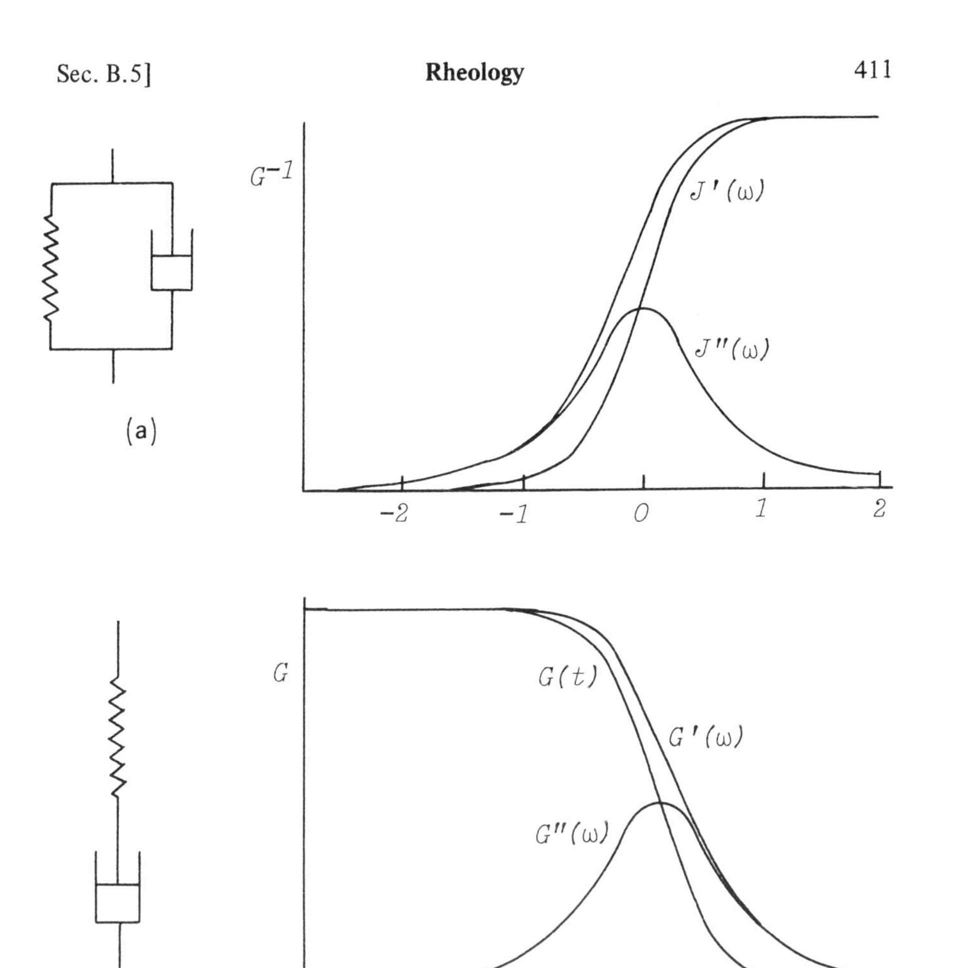

Fig. B.17 – (a) Voigt (Kelvin) model and responses. (b) Maxwell model and responses.

these are shown in Fig. B.18. We see that one can be regarded as a Maxwell model in series with a Voigt model. The other form is notable chiefly because the upper three-element model represents the complete recoverable, but time-dependent, behaviour while the lower dashpot gives the permanent deformation.

We must now look in more detail at the behaviour of the Voigt model. The differential equation which describes the model's behaviour (or which the model represents) is:

$$\sigma = G_{\mathrm{p}}\epsilon + \eta_{\mathrm{p}}\mathrm{d}\epsilon/\mathrm{d}t \tag{B.76}$$

where G_p is modulus of the spring constant and η_p is the viscous constant of the dashpot.

If we impose a cyclic strain, $\epsilon = \epsilon \sin \omega t$, we get:

$$\sigma = G_p \epsilon \sin \omega t + \eta_p \omega \epsilon \cos \omega t \tag{B.77}$$

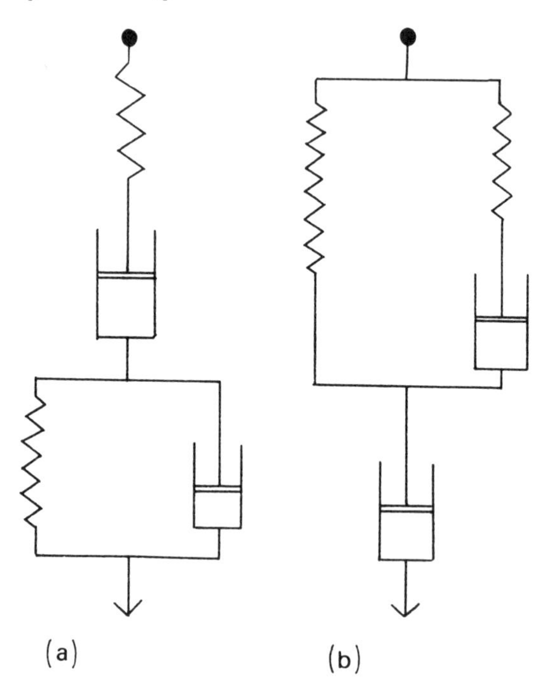

Fig. B.18 – Four-element models.

Comparison with the earlier equations shows that:

$$G_p = G = G' \tag{B.78}$$

$$\eta_p \omega = G \tan \delta = G'' \quad \text{or} \quad \tan \delta = \eta_p \omega / G_p \tag{B.79}$$

The Voigt model is thus a specially useful one to have in mind when considering the results of dynamic tests. It must however be firmly borne in mind that whereas in the model G_p and η_p are constants, in a real material the parameters vary. The identity of the parameters applies only for the particular test conditions.

Under a constant stress σ_0, (B.76) can be arranged:

$$\frac{G_p}{\eta_p} \, dt = \frac{d\epsilon}{\sigma_0 / G_p - \epsilon} \tag{B.80}$$

Integration, with a starting condition $\epsilon = 0$ at $t = 0$,

$$\frac{G_p}{\eta_p} t = -\log_e \left(1 - \frac{\epsilon G_p}{\sigma_0}\right) \tag{B.81}$$

or

$$\epsilon = (\sigma_0/G_p)(1 - \exp(-G_p t/\eta_p)) = (\sigma_0/G_p)(1 - \exp(-t/\tau)) \tag{B.82}$$

This is an equation for an exponential creep towards an ultimate strain σ_0/G_p. Thus the creep compliance at infinite time $J_c(\infty) = G_p^{-1}$. The relaxation time $\tau_p = \eta_p/G_p$. The use of a Voigt model thus suggests forms of relation between creep tests and dynamic tests.

We can also examine the behaviour of a Maxwell model in a dynamic test. The differential equation of the model is:

$$\frac{d\epsilon}{dt} = \frac{1}{G_s}\frac{d\sigma}{dt} + \frac{\sigma}{\eta_s} \tag{B.83}$$

Application of this equation leads to the results:

$$E_s = E' \sec^2 \delta \tag{B.84}$$

$$\tan \delta = E_s/\eta_s \omega \tag{B.85}$$

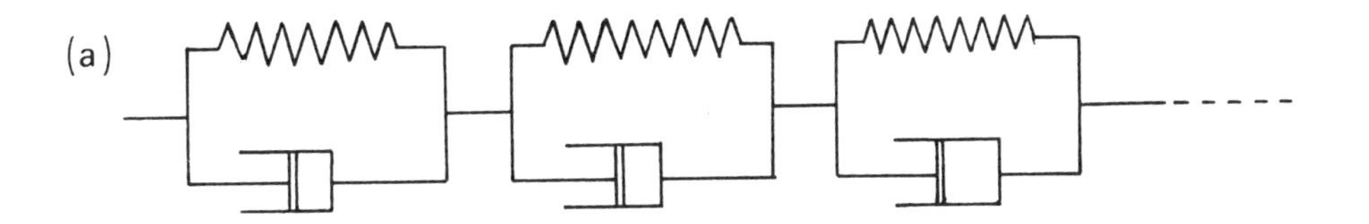

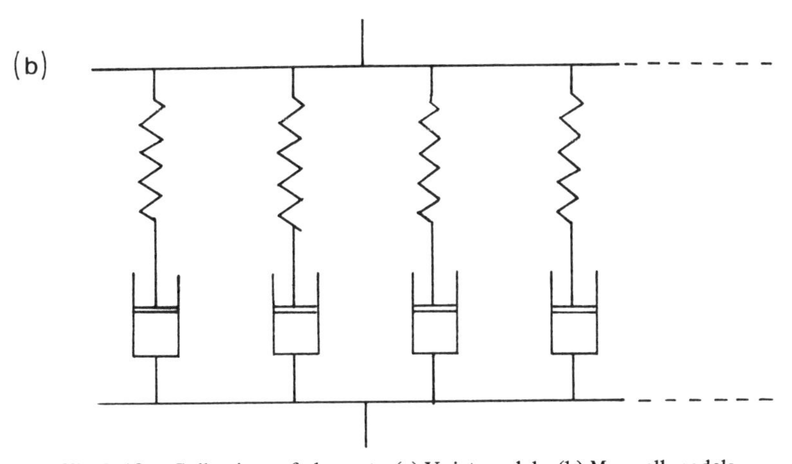

Fig. B.19 – Collections of elements. (a) Voigt models. (b) Maxwell models.

These quantities, although less useful in themselves, do give a link to stress relaxation. With constant strain ϵ_0:

$$\begin{aligned}\sigma &= E_s\epsilon_0 \exp(-E_s t/\eta_s) \\ &= E_s\epsilon_0 \exp(-t/\tau_s)\end{aligned} \tag{B.86}$$

where $\tau_s = \eta_s/E_s$ is the relaxation time constant.

The behaviour of more complicated collections of springs and dashpots can be worked out analytically or by computation. We note that combinations of Voigt or Maxwell models as in Fig. B.19 will give responses with a spectrum of time constants, and thus will represent a more general time-dependent response.

B.5.4 Linear viscoelasticity

The response of real materials is very different from the response of the simple models as shown in Fig. B.17 or B.18. More complicated models must be examined. The theory of linear viscoelasticity is one way of doing this. It brings in a great deal more complexity (and non-linearity) in time dependence, but retains the linearity of the relation between stress and strain. In other words if any sequence of applied stresses results in a given sequence of strains, then a sequence which differs only in that the stress is twice as large at each instant will cause the strain at each instant to be twice as large: conversely doubling the imposed strains in a sequence will double the resulting stresses.

While retaining the linearity between stress and strain, the theory of linear viscoelasticity allows for the most general non-linearity of time dependence. The clue to the situation is given in Fig. B.19 at the end of the last section. By a combination of elements, responses of different magnitudes with different time constants can be combined together. Thus a combination of two widely separated time constants would give the sort of response shown in Fig. B.20(a), with a spectrum marked by sharp transitions, while a collection of elements with time constants over a range would behave as in Fig. B.20(b).

The mathematical development of the subject goes beyond what is needed in this book. In summary, a collection of Maxwell elements in parallel gives a relaxation spectrum in discrete or continuous form:

$$G(t) = \Sigma G_r \exp(-t/\tau_{s,r}) \tag{B.87}$$

$$G(t) = G_e + \int_{-\infty}^{\infty} H \exp(-t/\tau)\,\mathrm{d}\ln\tau \tag{B.88}$$

A collection of Voigt elements in series gives the retardation spectrum:

$$J(t) = \Sigma J_r[1 - \exp(-t/\tau_{p,r})] \tag{B.89}$$

$$J(t) = J_g + \int_{-\infty}^{\infty} L[1 - \exp(-t/\tau)]\,\mathrm{d}\ln\tau + t/\eta_p \tag{B.90}$$

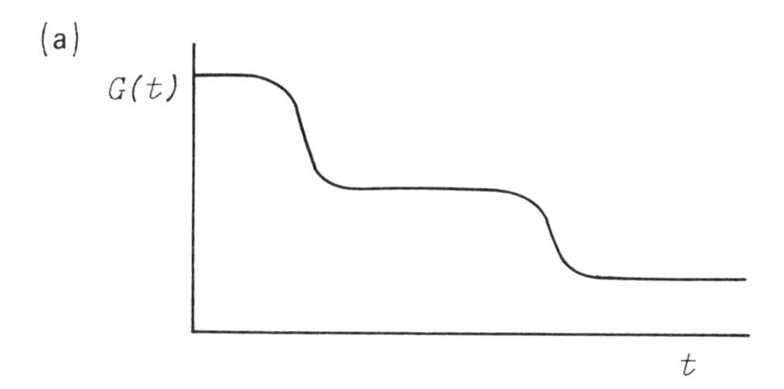

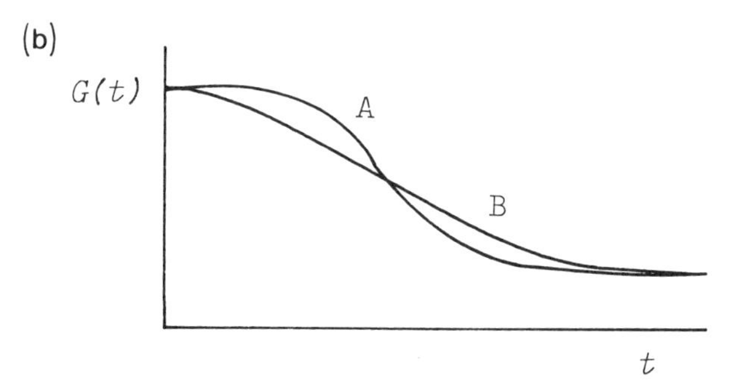

Fig. B.20 – (a) Behaviour of combination of two elements with widely spaced time constants. (b) Behaviour of a continuum of elements (A) over narrow band of time constants; (B) over a wide band.

An important consequence of linear viscoelasticity is the validity of Boltzmann's superposition principle. This states that the consequences of separate changes of load are independent and additive. This makes it much easier to handle the consequences of a sequence of changes since each step can be treated separately, and the overall effect obtained by summation.

B.5.5 Non-linear rheology

The theory of linear viscoelasticity is well developed with many useful forms for analysing the results of experiments, defining material behaviour and predicting response. The non-linear theory is still difficult and ripe for further development.

There are analytical treatments involving more complicated functions, but, if these are general, they are difficult to handle, and, if they are restricted, they are liable to be poorly representative.

Where there are theoretical reasons for having a different functional form, then these can be incorporated into simple combinations of elements. Thus non-linear rubber elasticity relations can be used for springs and the hyperbolic sine law for dashpots. Because this approach has a physical basis it offers more promise than arbitrary mathematical generalisation.

From an operational viewpoint, with powerful computing available, there is an alternative approach based on the four element model of Fig. B.18 as the basic form, capable of representing change over short time intervals, but with variable parameters, which alter with the situation. The problem then is to find suitable means of storing the parametric data in numerical form.

B.5.6 Experimental methods

Some general points on experimental methods may be useful. Creep and stress relaxation experiments cannot easily be carried out rapidly, but are available for times from a few seconds up to as long as it is practical to wait for results – a day (10^5 sec) is usually possible, and some experiments have lasted for months (10^7 secs). Dynamic testing (free oscillation, forced oscillation and wave propagation) is better for shorter times, since oscillation periods of more than a few seconds are inconvenient. The upper limit of frequency is given by the frequency of sound waves and the method can be used up to about 100 kc/s, equivalent to times of 10^{-5} secs. A range of about 10^{12} times is thus possible. Stress-strain curves can be measured at various rates ranging from tests at 1000% per sec (10^{-3} secs) to tests lasting many minutes (10^3 secs).

B.6 OTHER TOPICS

B.6.1 Yield criteria

Yielding, namely the onset of easy, usually irrecoverable, deformation above a critical level of stress, is an important feature of the behaviour of many materials. For a uniaxial applied stress, the only problem in defining a yield stress is to specify some particular point in the region of rapidly decreasing slope of the stress-strain curve.

With combined stresses, some more complicted yield criterion is required. In general on a multi-dimensional plot of stress there will be a locus surface defining the yield point.

It goes beyond the scope of this book to go into detail of this subject, which is treated in other texts such as J. G. Williams' *Stress Analysis of Polymers*. One of the simplest forms is the Tresca yield criterion which states that yield will occur when the maximum shear stress reaches some value, but there are more complicated forms which have greater applicability.

In anisotropic materials, the yield criteria will need to be related to direction.

B.6.2 Macroscopic features of yielding

Yielding can lead to macroscopic forms of deformation which are a consequence of the form of the stress-strain relations and thus only depend indirectly on the material structure.

If the force on the specimen reduces with increase of deformation, there will be instability. This must lead to the formation of a neck with a region of high elongation. If the force continues to decrease, the thinning can continue indefinitely. However, if there is a stiffening as a result of the drawing, then the yield may cease and the neck propagates along the specimen.

The choice between drawing at neck and drawing uniformly depends on the shape of the stress-strain curve.

B.6.3 Competition between modes of failure

Many materials show a brittle-ductile transition. Under some conditions they show crack propagation to catastrophic failure while at other conditions they show a ductile yielding. It is important to remember that these are competitive processes: if one occurs, it prevents the occurrence of the other. Consequently if the stresses needed to cause the alternative events vary differently with some other parameter, such as temperture or rate, then a transition can be expected for the reasons illustrated in Fig. B.21.

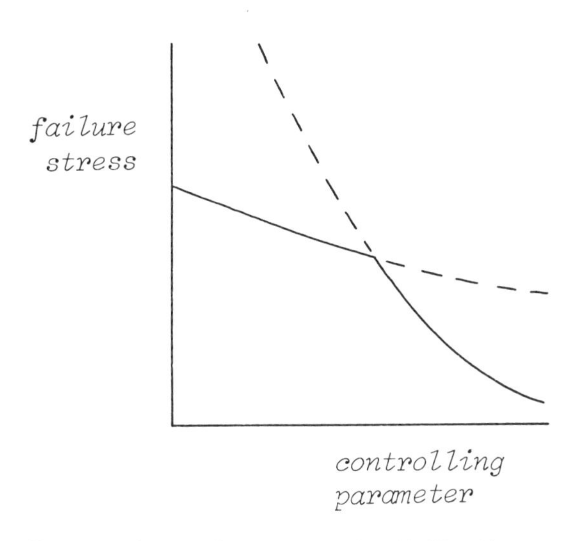

Fig. B.21 – Variation of brittle fracture (F) and yield (Y) with some controlling parameter.

This behaviour is but one example of competition between various modes of failure: thus there may be different forms of yield, different types of crack propagation, especially in oriented materials, and other effects in fatigue situations. The first to occur defines the form of failure in a particular region within boundaries defined on a failure map.

B.6.4 The 'geometric' consequences of drawing

Many polymer materials show yield followed by a large irrecoverable elongation. With the simple idealisation of the four-parameter model, this gives the results shown in Fig. B.22(a). However the materials at A, B, C . . . etc. which have been subject to prior extension, are stable and can be regarded as fresh specimens of drawn oriented material. If the stress and strain values are recalculated with the new initial dimensions, we obtain the results in Fig. B.22(b). If the stress-strain curves have a different non-linear shape, there will be differences in detail, but the general form of the transformation will be similar. The locus of breaking

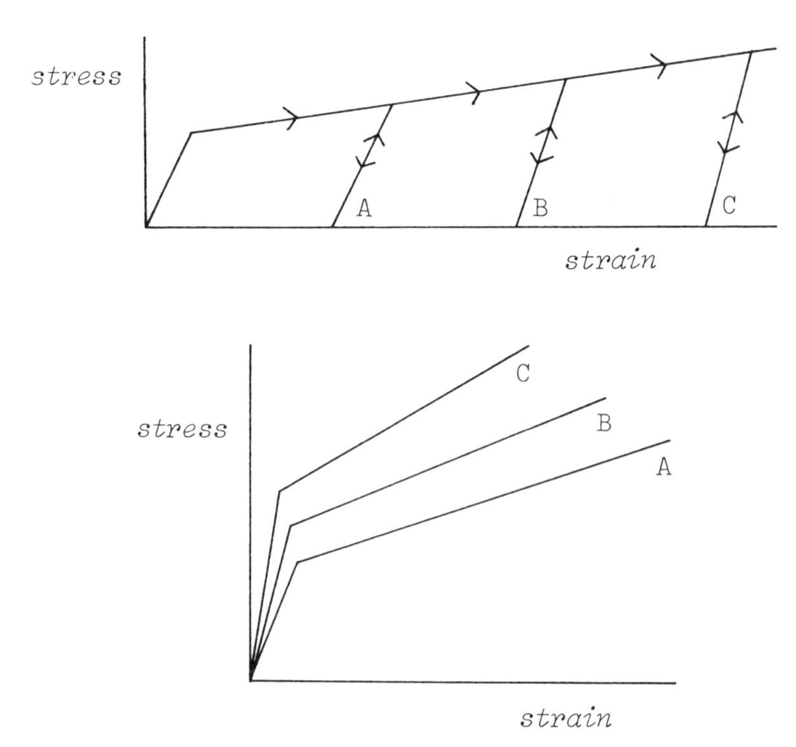

Fig. B.22 – Consequences of drawing. (a) Effect of varying amounts of draw. (b) Stress-strain curves after drawing.

points will, of course, be independent of the route, and is given by the following analysis:

Unoriented specimen

Initial length $= L_0$, initial area $= A_0$, breaking load $= F_b$, length at break $= L_b$.

Nominal tensile stress at break $= F_b/A_0 = S_0$

Breaking strain $= (L_b - L_0)/L_0 = b_0$

Drawn, oriented specimen, with draw strain ϵ_d

Drawn length $= L = (1 + \epsilon_d)L_0$

Area (for constant volume draw) $= A = A_0L_0/L = A_0(1 + \epsilon_d)^{-1}$

Nominal tensile stress at break $= S = F_b/A$

$= S_0(1 + \epsilon_d)$

Breaking strain $= b = (L_b - L)/L$

$= (b_0 - \epsilon_d)/(1 + \epsilon_d)$

Specimen at break

Area $= A_0(1 + b_0)^{-1}$

True stress at break $= S_t = S_0(1 + b_0)$

The locus of breaking points turns out to be given by:

$$S(1 + b) = S_0(1 + b_0) = \text{constant} = S_t$$

It must be emphasised that these predictions are the purely geometric consequences. Any more subtle effects due to changes of structure will be superimposed on them. However, conversely, the changes due to the geometric factors should not be attributed to any unusual structural effects.

B.6.5 The mechanical response of composite systems

Polymer materials are often composite. At a level which is macroscopic in relation to a definition of the components as mechanical continua, though truly microscopic in subdivision, there are systems of fibre-reinforced plastics or various filled systems. At a much finer level, where molecular effects may need to be considered although composite models are a useful approximation, there are bulk crystallised polymers and some copolymer systems.

The prediction of the properties of a composite material, even in the absence of any 'chemical' difficulties, is not easy except when the geometry is very simple. However even though a variety of mixture laws are applicable, the two simple extreme bounds give a useful framework.

In a series (lamellar) system, Fig. B.23(a), the stress is the same in both components, but the strain is different; and in a parallel (fibrillar) system, Fig. B.23(c), the reverse is true. This leads to the simple mixture laws for two components with moduli E_1 and E_2, compliances C_1 and C_2, and a volume fraction f for the first component:

parallel

$$E_p = fE_1 + (1 - f)E_2$$

$$C_p = \left(\frac{f}{C_1} + \frac{(1 - f)}{C_2}\right)^{-1}$$

series

$$E_s = \left(\frac{f}{E_1} + \frac{(1 - f)}{E_2}\right)^{-1}$$

$$C_s = fC_1 + (1 - f)C_2$$

As appears in Fig. B.23(d), the fibrillar material is dominated by the stiff component, while the lamellar is closer to the soft material.

With non-linear responses, the parallel response is given by averaging stresses at given strains, and the series response by averaging strains at given stresses, with consequences illustrated in Fig. B.23(e).

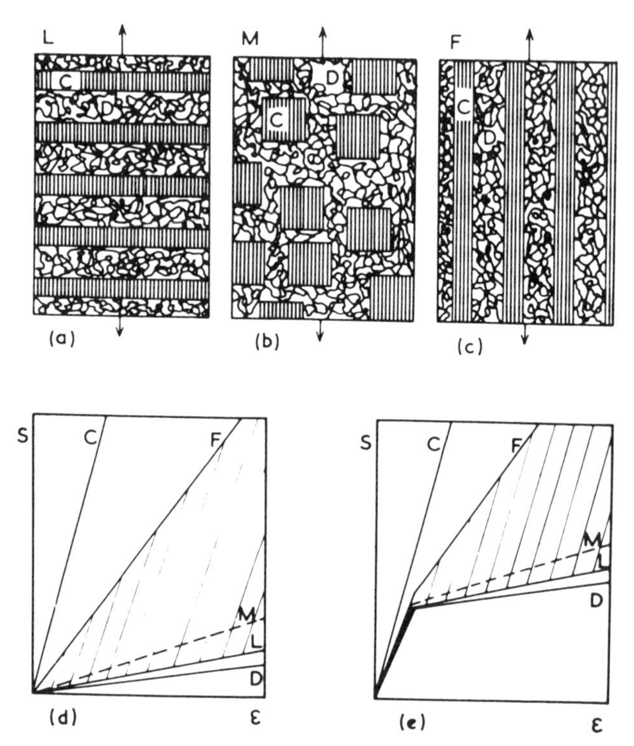

Fig. B.23 – Composite systems of stiff component (C) and soft component (D). (a) Lamellar (L). (b) Micellar (M). (c) Fibrillar (F). (d) Linear responses. (e) Non-linear responses.

A micellar structure, such as Fig. B.23(b), will lie somewhere between the two extremes, as suggested by the dotted lines in Fig. B.23(d) and (e), but exact prediction will require both detailed information on the composite geometry and elaborate analysis or computation.

APPENDIX C

The Chemical Constitution of some Polymers, and other matters of reference

> We may note then that polymerizations do not involve a single unique type of reactivity: they are for the most part merely ordinary reactions made manifold by polyfunctionality, and thus made capable of indefinite continuation in one, two, or three dimensions.
>
> W. H. Carothers, *Collected Papers*

C.1 A SAMPLE OF THE CHEMICAL DIVERSITY

C.1.1 Linear polymers

Even if one omits the macromolecules which may yet be synthesised, the inevitable differences in end-groups of long chains, and the variants arising in the different sorts of copolymer, there is still an enormous variety of polymers in the three categories of existing materials: known and unknown macromolecules in natural organisms; commercial synthetic polymers; and polymers synthesised experimentally in the laboratory. In this appendix, the chemistry of a few of these polymers, including those mentioned at various places in the book, will be summarised for reference.

The simplest family of addition polymers comes from the opening of double bonds in small molecules to give a long chain as typefied by the formation of polyethylene:

$$n\,CH_2{=}CH_2 \rightarrow (\text{-}CH_2\text{-}CH_2\text{-})_n$$

The family may be divided into various classes, and some examples are as follows:

polyolefin *polyethylene* (polythene, PE) $[\text{-}CH_2\text{-}CH_2\text{-}]_n$ or $[\text{-}CH_2\text{-}]_{2n}$

polypropylene (PP) $[\text{-}CH_2\text{-}\underset{\displaystyle CH_3}{\underset{|}{C}}H\text{-}]_n$

vinyl

polyvinyl chloride (PVC) $[-CH_2-CH(Cl)-]_n$

polystyrene $[-CH_2-CH(C_6H_5)-]_n$

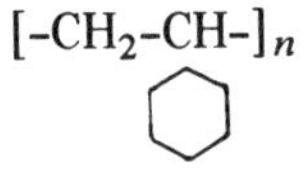

polyacrylonitrile (polyvinyl cyanide, PAN) $[-CH_2-CH(C{\equiv}N)-]_n$

1,2 polybutadiene $[-CH_2-CH(CH{=}CH_2)-]_n$

generally: $[-CH_2-CH(R)-]_n$

where R is -OH, *polyvinyl alcohol*
-COOH, *polyacrylic acid*
-COOX, *polyacrylates*
-COO.O.CH_3, *polyvinyl acetate*
etc.

polyvinyl formal $[-CH_2-CH-CH_2-CH-]_n$ with the two CH groups joined by $O-CH_2-O$

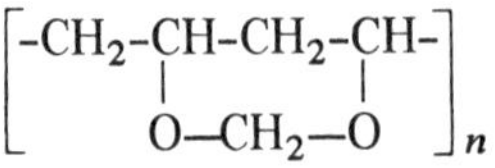

vinylidene

polyisobutylene $[-CH_2-C(CH_3)_2-]_n$

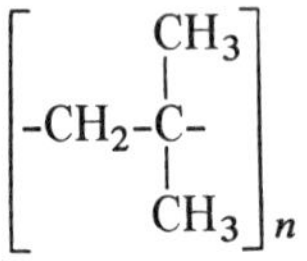

polyvinylidene chloride $[-CH_2-CCl_2-]_n$

others

polymethyl methacrylate (Perspex) $[-CH_2-C(CH_3)(COO.CH_3)-]_n$

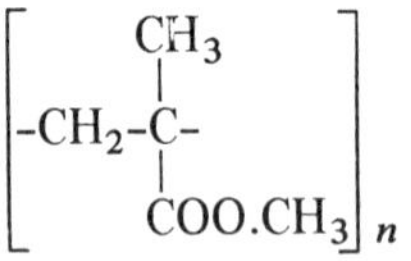

polytetrafluorethylene (Teflon) $[-CF_2-CF_2-]_n$ or $[-CF_2-]_{2n}$

Another group of polymers has double bonds in the main chain.

1,4 cis-polybutadiene $[-CH_2-CH{=}CH-CH_2-]_n$

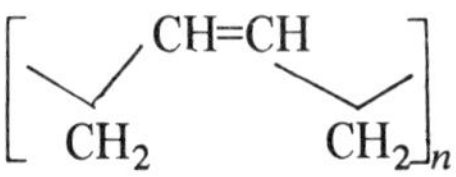

1,4 trans-polybutadiene		$[-CH_2-CH{=}CH-CH_2-]_n$ (trans)
cis-polyisoprene, (natural rubber, hevea, India rubber)		$[-CH_2-C(CH_3){=}CH-CH_2-]_n$ (cis)
trans-polyisoprene (gutta-percha, balata)		$[-CH_2-C(CH_3){=}CH-CH_2-]_n$ (trans)
polychloroprene (neoprene)		$[-CH_2 \cdot CCl{=}CH \cdot CH_2-]_n$

Condensation polymers are formed by reactions with the elimination of some other substance such as water. Common examples include the following:

polyamide (aliphatic)	*nylon 6*	$[-CO(CH_2)_5NH-]_n$
	nylon 66	$[-NH(CH_2)_6NH \cdot CO(CH_2)_4CO-]_n$
	generally: *nylon X*	$[-CO(CH_2)_{X-1}NH-]_n$
	nylon YZ	$[-NH(CH_2)_YNH \cdot CO(CH_2)_{Z-2}CO-]_n$
aramid (aromatic polyamide)	*poly-m-phenylene terephthalamide* (Nomex)	$[-NH-C_6H_4-NH \cdot CO-C_6H_4-CO-]_n$
	poly-p-phenylene terephthalamide (Kevlar)	$[-NH-C_6H_4-NH \cdot CO-C_6H_4-CO-]_n$
linear polyester	*polyethylene terephthalate* (PET, 2GT)	$[-CH_2 \cdot CH_2 \cdot CO-C_6H_4-CO-]_n$
polyurethane	generally:	$[-NH \cdot R_1 \cdot NH \cdot O \cdot CO \cdot R_2 \cdot CO-]_n$
polycarbonate	generally:	$[-O \cdot R \cdot O \cdot CO-]_n$

e.g. with R = bisphenol A

$$[-O-C_6H_4-C(CH_3)_2-C_6H_4-O \cdot CO-]_n$$

polypeptide generally: $\left[\text{-NH}\cdot\underset{\substack{|\\ \text{R}}}{\text{CH}}\cdot\text{CO-}\right]_n$

Some other synthetic linear polymers should also be mentioned.

polyoxymethylene (polyacetal) $[\text{-CH}_2\text{-O-}]_n$

polysulphide $[\text{-R-S}_X\text{-}]_n$

polysulphone e.g. $[\text{-O-C}_6\text{H}_4\text{-C(CH}_3)_2\text{-C}_6\text{H}_4\text{-O-C}_6\text{H}_4\text{-SO}_2\text{-C}_6\text{H}_4\text{-}]_n$

The natural macromolecules are generally more complicated. Two important families are the polysaccharides from the polymerisation of sugars and the proteins from the polymerisation of amino-acids.

cellulose *poly(1,4)D glucose*

$$\left[\text{-O}\cdot\text{CH}\begin{matrix}\text{CH(OH)}\cdot\text{CH(OH)}\\ \text{O}\text{———}\underset{\substack{|\\ \text{CH}_2\text{(OH)}}}{\text{CH}}\end{matrix}\text{CH}\cdot\text{O}\cdot\text{CH}\begin{matrix}\text{O}\text{———}\overset{\substack{\text{CH}_2\text{OH}\\ |}}{\text{CH}}\\ \text{CH(OH)}\cdot\text{CH(OH)}\end{matrix}\text{CH-}\right]_n$$

cellulose derivatives -OH groups substituted by other groups

e.g. *cellulose triacetate*, $\text{-O}\cdot\text{CO}\cdot\text{CH}_3$
(secondary acetate retains about 20% of -OH groups)

proteins generally $\left[\text{-NH}\cdot\underset{\substack{|\\ \text{R}_1}}{\text{CH}}\cdot\text{CO}\cdot\text{NH}\cdot\underset{\substack{|\\ \text{R}_2}}{\text{CH}}\cdot\text{CO}\cdot\text{NH}\cdot\underset{\substack{|\\ \text{R}_3}}{\text{CH}}\cdot\text{CO}\ldots\text{-}\right]_n$

where R_1, R_2, R_3 ... consistent of particular sequences of about 20 different groups.

In concluding this summary of the chemical composition of polymers, inorganic polymers should also be mentioned by example.

plastic sulphur $[\text{-S-}]_n$

silicone generally $\left[\begin{matrix}\text{R}_1\\ |\\ \text{Si-O-}\\ |\\ \text{R}_2\end{matrix}\right]_n$

C.1.2 Thermoset resins

It will be clear to the reader that, with the exception of some proteins which are cross-linked by cystine when R = $-CH_2.S.S.CH_2-$, all the polymers listed above are linear macromolecules. These are the most important in current scientific knowledge and practical importance. Furthermore the chemistry of the three-dimensional polymers, the thermo-setting resins, is rather more complicated and cannot be briefly summarised.

It is only possible here to give a rough identification of some common component groups in the thermosetting resins.

phenol-formaldehyde

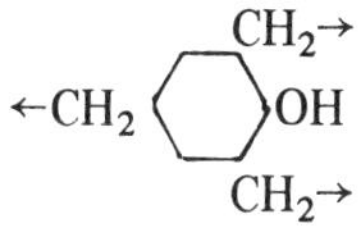

$$\leftarrow CH_2 - C_6H_2(OH)(CH_2\rightarrow)(CH_2\rightarrow)$$

urea formaldehyde

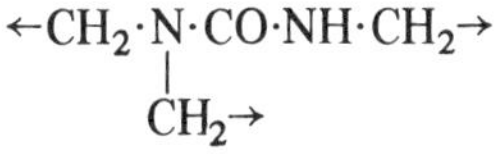

$$\leftarrow CH_2 \cdot \underset{\underset{CH_2\rightarrow}{|}}{N} \cdot CO \cdot NH \cdot CH_2 \rightarrow$$

melamine

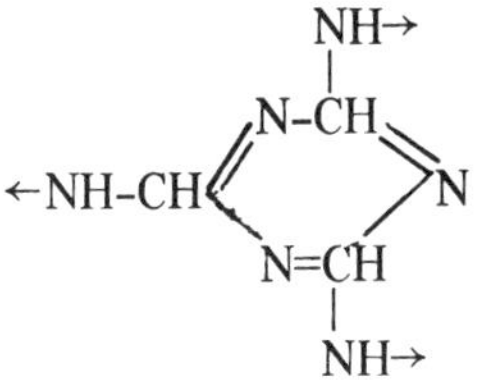

NH→ | N-CH ←NH-CH N N=CH | NH→

polyester

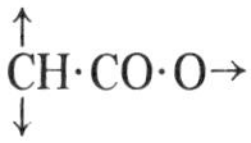

$$\underset{\downarrow}{\overset{\uparrow}{CH}} \cdot CO \cdot O \rightarrow$$

silicone

$$\leftarrow \underset{\underset{R_2\rightarrow}{|}}{\overset{\overset{R_1}{|}}{Si}} - O \rightarrow$$

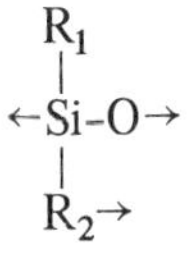

C.2 ATOMIC PARAMETERS

C.2.1 Geometric dimensions

The behaviour of polymers is very largely determined by the geometry of the packing together of atoms and molecules and the strength of the linking forces. It is therefore useful to have on reference some important numerical values.

The sizes of atoms are indicated by the following figures:

atom	*effective radius*, Å	*bond*	*length*, Å
H	0.28	C-H	1.09
F	0.64	C-C	1.54
O	0.66	C-N	1.47
N	0.70	C-O	1.42
C	0.77	C-F	1.38
Cl	0.99	C-Cl	1.72
S	1.04	N-H	1.02
P	1.10	O-H	0.96
Si	1.17	C=C	1.33
		C≡C	1.20
		C=O	1.24

The angular geometry of some simple single bonds, typefied by CH_4, NH_3 and H_2O, is summarised in Fig. C.1. Double and triple bonds of carbon lead to planar

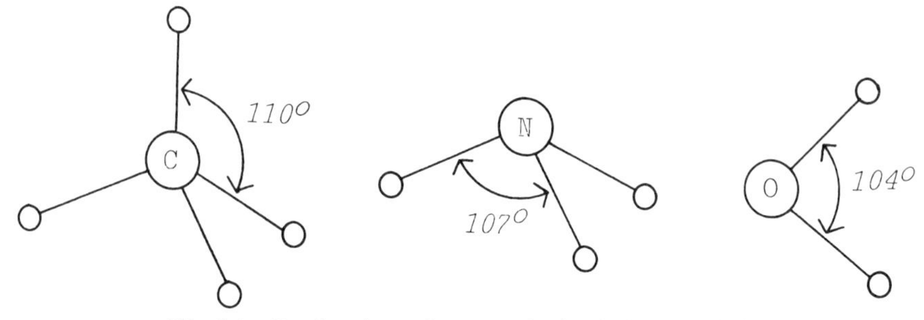

Fig. C.1 – Bond angles as they occur in simple single-bonded compounds of carbon, nitrogen and oxygen.

and linear forms as shown in Fig. C.2. However these simple shapes may be distorted to a minor degree to allow for more favourable packing, and in a major way when resonance occurs. An example of the latter effect is the fact that the six atoms in a peptide group,

-C.CO.NH.C-

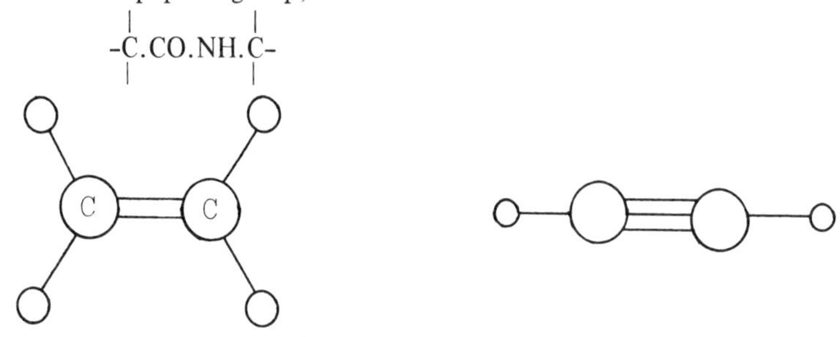

Fig. C.2 – The planar form with C=C, and the linear form with C≡C.

all lie in a plane, instead of the three-dimensional structure which would be expected from Fig. C.1 and C.2: the resonance is between -(C=O)-(N-H)- and -(C-O$^-$)=(N-H$^+$)-.

C.2.2 Energy values

The order of magnitude of energy levels is indicated by the following examples, with the figures given in kJ/mole.

covalent bond breakage	
C-C	20
C-H	25
C-O	20
C-N	17
C-F	25
bond rotation	
CH_3-CH_3	0.7
$(CH_3)_3$-C-$C(CH_3)_3$	1.5
CH_3-OH	0.25
CF_3-CF_3	1.0
C Cl_3-C Cl_3	3.0
around single bond in	
CH_3-CH=CH-CH_3, trans	0.5
CH_3-CH=CH-CH_3, cis	0.12
CH_3-C≡C-CH_3	<0.1
around double bond	
CH_2=CH_2	10
intermolecular attractions	
hydrogen bonds and strongly polar interactions	0.5 to 2.5
van der Waals bonds between non-polar groups	<0.2

C.3 SUGGESTIONS FOR FURTHER READING

The profusion of books on polymer science and technology is less than the number of polymer types, but is nevertheless extensive, and, once again, only a sample of the literature can be mentioned. The following is a list of books which have been found to be useful sources of more detailed information.

P. J. Flory, (1953), *Principles of Polymer Chemistry*
– the classic text on polymer physical chemistry.

H. G. Elias, (1977), *Macromolecules*, **1**, *Structure and Properties*, **2**, *Synthesis and Materials*
– a comprehensive modern text on the physical and organic chemistry.

H. G. Elias, (1977), *New Commercial Polymers*
– updating the list of polymers.
C. Tanford, (1961), *Physical Chemistry of Macromolecules*
– a bias towards natural polymers.
A. D. Jenkins, (ed.), (1972), *Polymer Science*, **1** and 2
– a useful compendium for reference.
J. Brandrup and E. Immergut, (eds.), (1966), *Polymer Handbook*
– a collection of numerical data.
M. V. Volkenstein, (1963), *Configurational statistics of polymeric chains.*
T. M. Birshtein and O. B. Ptitsyn, (1966), *Conformations of macromolecules*
– two Russian texts, emphasising the concept of rotation isomerism.
L. R. G. Treloar, (1975), *The physics of rubber elasticity*, 3rd edn.
J. D. Ferry, (1970), *Viscoelastic properties of polymers*, 2nd edn.
– two classic texts.
F. Bueche, (1962), *Physical properties of polymers.*
A. V. Tobolsky, (1960), *Properties and structure of polymers.*
L. E. Nielsen, (1962), *Mechanical properties of polymers*
– three interesting books by pioneers of the subject.
H. Morawetz, (1965), *Macromolecules in solution.*
E. G. Richards, (1980), *An introduction to the physical properties of large molecules in solution.*
P. H. Geil, (1963), *Polymer single crystals.*
D. C. Bassett, (1981), *Principles of polymer morphology.*
N. G. McCrum, B. E. Read, and G. Williams, (1970), *Anelastic and dielectric effects in polymeric solids.*
T. Murayama, (1978), *Dynamic mechanical analysis of polymer material.*
I. M. Ward, (1975), *Mechanical properties of solid polymers.*
W. E. Morton and J. W. S. Hearle, (1975), *Physical properties of textile fibres*, 2nd edn.
– good accounts of special topics.
R. A. Fava, (ed.), (1980), *Methods of experimental physics*, **16**, Polymers, Part A. Molecular structure and dynamics. Part B. Crystal structure and morphology. Part C. Physical properties
– good on techniques.
J. G. Williams, (1980), *Stress analysis of polymers*
– good on analysis and application.

C.4 SOURCES OF ILLUSTRATIVE DATA

This book has concentrated on general principles, but, in places, specific data has been used for illustration. The sources of these experimental results are as follows:

Figs. 3.12, 3.13(b), 3.14, 3.20, 3.30 – T. G. Fox, S. Gratch, and S. Loshalk, (1965), in *Rheology*, **1**, (ed. F. R. Eirich), Academic Press.

Figs. 3.13(a), 3.22, 5.11, 5.12 – J. D. Ferry, (1970), *Viscoelastic properties of polmers*, Wiley.

Fig. 3.15 – A Kaye quoted by A. S. Lodge, (1969), *Elastic Liquids*, Academic Press.

Figs. 4.2, 4.3, 4.4, 4.9, 4.12, 4.13, 4.15, 4.16 – L. R. G. Treloar, (1975), *Physics of Rubber Elasticity*, Oxford.

Figs. 4.17, 5.1(b) – A. V. Tobolsky, (1960), *Properties and Structure of polymers*, Wiley.

Fig. 5.1(a) – A. V. Tobolsky, (1958), in *Rheology*, 2, (ed. by F. R. Eirich), Academic Press.

Fig. 5.2 – T. Alfrey, (1948), *Mechanical behaviour of high polymers*, Interscience.

Fig. 5.3 – T. S. Carswell and H. K. Nason, (1944), *Modern Plastics*, **21**, 121.

Fig. 5.4 – W. Sommer, (1959), *Kolloid-Z*, **167**, 97.

Fig. 5.5 – M. Gordon and B. M. Grieveson, (1958), *J. Polymer Sci.*, **29**, 9.

Fig. 5.6 – K. Schmeider and K. Wolf, (1953), *Kolloid-Z*, **134**, 149.

Fig. 5.7(a) – B. Maxwell, (1956), *J. Polymer Sci.*, **20**, 551.

Fig. 5.7(b) – W. Dannhauser, W. C. Child and J. D. Ferry, (1958), *J. Colloid Sci.*, **13**, 103.

Fig. 5.8 – Y. Ishida, M. Matsuo and K. Yamafuji, (1962), *Kolloid-Z.*, **180**, 108.

Fig. 5.9 – W. Heijboer, (1956), *Kolloid-Z*, **148**, 36.

Fig. 5.10 – A. W. Nolle, (1950), *J. Polymer Sci.*, **5**, 1.

Fig. 5.14, 5.16 – P. Mears, (1957), *Trans. Faraday Soc.*, **53**, 31.

Fig. 5.17 – J. M. O'Reilly and F. E. Karasz, (1966), *J. Polymer Sci.*, **C.14**, 49.

Fig. 5.18(a) – R. J. Angelo, R. M. Ikeeda and M. L. Wallach, (1965), *Polymer*, **6**, 141.

Fig. 5.18(b) – K. H. Illers and B. E. Jenckel, (1958), *Rheol. Acta*, **1**, 322.

Fig. 5.18(c) – M. Takayanagi, (1963), *Mem. Fac. Eng. Kyushu University*, **1**, 23.

Fig. 6.5(a) – F. K. Knowles and A. G. H. Dietz, (1955), *Trans. ASME*, **77**, 177.

Fig. 5.6(b) – S. Rabinowitz and P. Beardmore, (1972), *CRC Crit. Rev. in Macromol. Sci.*, **1**, 1.

Fig. 6.6, 6.9 – P. B. Bowden and S. Raha, (1970), *Phil. Mag.*, **22**, 463.

Fig. 6.9 – R. A. Duckett, S. Rabinowitz and I. M. Ward, (1970), *J. Mat. Sci.*, **5**, 909.

Fig. 6.9 – C. Crowet and G. A. Homes, (1964), *App. Mat. Research*, **3**, 1.

Figs. 6.10, 8.30(d) – H. Schnell, (1964), *Chemistry and Physics of Polycarbonates*, Interscience.

Table 7.2 – I. Sakurada, T. Ito and K. Nakama, (1966), *J. Polymer Sci.*, **C15**, 75.

Fig. 7.40, 7.41 – A Nakajima and F. Hamada, (1972) – *J. Pure Appl. Chem.*, **31**, 1.

Fig. 7.44 – W. O. Statton and P. H. Geil, (1960), *J. Appl. Pol. Sci.*, **3**, 357.

Fig. 7.45 – E. W. Fischer and G. F. Schmidt, (1962), *Angew. Chem.*, **74**, 551.

Fig. 7.48, 7.49, 7.50 – P. H. Lindenmeyer, (1966), *J. Pol. Sci.*, **C20**, 145.
Fig. 7.64 – A. Keller, G. R. Lester and L. B. Morgan, (1954), *Phil. Trans.*, **A.247**, 1.
Fig. 7.66 – J. H. Magill, (1962), *Polymer*, **3**, 655.
Fig. 7.68 – J. D. Hoffman, *et al.*, (1969), *Kolloid-Z. Z. Polymer*, **251**, 564.
Table 8.1 – G. Wobser and S. Blasenbrey, (1970), *Kolloid-Z. Z. Polym.*, **241**, 1985.
Fig. 8.2 – R. Hoseman, (1962), *Polymer*, **3**, 349.
Fig. 8.10 – P. Predecki and W. O. Statton, (1960), *J. Appl. Phys.*, **37**, 4053.
Fig. 8.13 – E. W. Fischer, H. Goddar and G. F. Schmidt, (1968), *Makromol. Chem.*, **119**, 1970.
Fig. 8.24 – J. P. Bell, P. E. Slade and J. H. Dumbleton, (1968), *J. Polymer Sci.*, A-2, **6**, 1773.
Fig. 8.27 – A. Peterlin, (1965), *J. Polymer Sci.*, C, No. 9, 61.
Fig. 8.28 – B. C. Jariwala, Ph.D. thesis, Manchester.
Fig. 8.29 – R. Work, (1949), *Textile Res. J.*, **19**, 381.
Figs. 8.30(a), 8.31(a) – H. V. Boenig, (1966), *Polyolefins: structure and properties*, Elsevier.
Fig. 8.30(b) – I. Marshall and J. R. Whinfield, (1953), in *Fibres from Synthetic Polymer*, (ed. R. Hill), Elsevier.
Fig. 8.30(c), J. W. S. Hearle, P. K. Sen. Gupta and A. Matthews, (1971), *Fibre Sci. & Tech.*, **3**, 167.
Fig. 8.30(e) – R. Meredith, (1958), in *Rheology*, (ed. F. R. Eirich), 2, Academic Press.
Fig. 8.30(f), (i) – L. Konopasek, (1975), M. Sc. thesis, University of Manchester.
Fig. 8.30(g) – N. Wilson, (1967), *J. Textile Inst.*, **58**, 611.
Fig. 8.30(h) – J. B. Speakman, (1927), *J. Textile Inst.*, **18**, T.431.
Fig. 8.31(b) – R. Meredith, (1945), *J. Textile Inst.*, **36**, T.147.
Fig. 8.31(c) – A. J. Hughes and J. E. McIntyre, (1976), *Textile Progress*, **8**, No. 1.
Fig. 8.31(d), W. E. Morton and J. W. S. Hearle, (1975), *Physical properties of textile fibres*, Textile Institute.
Fig. 8.31(e), – B. S. Sprague, (1974), *The Solid State of Polymers*, Report of the U.S.-Japan Joint Seminar, Dekker.
Fig. 8.32 – I. Marshall and A. B. Thompson, (1954), *Proc. Roy. Soc.*, **A.221**, 541.

Index

T

U

V

W

X

Y

Z